D1731592

Edited by
Andrzej Cybulski, Jacob A. Moulijn,
and Andrzej Stankiewicz

Novel Concepts in Catalysis and Chemical Reactors

Related Titles

Sanchez Marcano, J. G., Tsotsis, T. T.

Catalytic Membranes and Membrane Reactors

Second Edition

2011

ISBN: 978-3-527-32362-3

Buchholz, K., Collins, J.

Concepts in Biotechnology

History, Science and Business

2010

ISBN: 978-3-527-31766-0

Hessel, V., Renken, A., Schouten, J. C., Yoshida, J.-I. (eds.)

Micro Process Engineering

A Comprehensive Handbook

3 volumes

2009

ISBN: 978-3-527-31550-5

Li, N. N., Fane, A.G., Winston Ho, W.S., Matsuura, T. (eds.)

Advanced Membrane Technology and Applications

2008

ISBN: 978-0-471-73167-2

Roberts, G. W.

Chemical Reactions and Chemical Reactors

2008

ISBN: 978-0-471-74220-3

Wirth, T. (ed.)

Microreactors in Organic Synthesis and Catalysis

2008

ISBN: 978-3-527-31869-8

Ertl, G., Knözinger, H., Schüth, F., Weitkamp, J. (eds.)

Handbook of Heterogeneous Catalysis

Second, Completely Revised and Enlarged Edition

8 Volumes

2008

ISBN: 978-3-527-31241-2

Niemantsverdriet, J. W.

Spectroscopy in Catalysis

An Introduction

Third, Completely Revised and Enlarged Edition

2007

ISBN: 978-3-527-31651-9

Edited by Andrzej Cybulski, Jacob A. Moulijn,
and Andrzej Stankiewicz

Novel Concepts in Catalysis and Chemical Reactors

Improving the Efficiency for the Future

WILEY-VCH Verlag GmbH & Co. KGaA

The Editors

Dr. Andrzej Cybulski†
CHEMIPAN
Polish Academy of Sciences
Institute of Physical Chemistry
Kasprzaka 44/52
01-224 Warszawa
Poland

Prof. Dr. Jacob A. Moulijn
University of Technology
Reactor/Cat. Engineering Group
Julianalaan 136
2628 BL Delft
The Netherlands

Prof. Dr. Andrzej Stankiewicz
Delft University of Technology
Process & Energy Department
Leeghwaterstraat 44
2628 CA Delft
The Netherlands

Library of Congress Card No.: applied for

British Library Cataloguing-in-Publication Data
A catalogue record for this book is available from the British Library.

Bibliographic information published by the Deutsche Nationalbibliothek
The Deutsche Nationalbibliothek lists this publication in the Deutsche Nationalbibliografie; detailed bibliographic data are available on the Internet at <http://dnb.d-nb.de>.

Composition Laserwords Private Ltd., Chennai, India
Printing and Bookbinding Fabulous Printers Pte Ltd
Cover Design Formgeber, Eppelheim

Printed in Singapore
Printed on acid-free paper

ISBN: 978-3-527-32469-9

"In memory of our too early deceased friend Andrzej Cybulski, the inspiring initiator of this book."

Jacob Moulijn
Andrzej Stankiewicz

Novel Concepts in Catalysis and Chemical Reactors: Improving the Efficiency for the Future.
Edited by Andrzej Cybulski, Jacob A. Moulijn, and Andrzej Stankiewicz

ISBN: 978-3-527-32469-9

Contents

Novel Concepts in Catalysis and Chemical Reactors: Improving the Efficiency for the Future.
Edited by Andrzej Cybulski, Jacob A. Moulijn, and Andrzej Stankiewicz

ISBN: 978-3-527-32469-9

Preface

The chemical process industry faces a tremendous challenge of supplying a growing and ever more demanding global population with the products it needs. The consumption of nonrenewable resources by the chemical and related industries is rapidly increasing, as on the one hand new markets open in different parts of the world for already existing products and on the other hand new types of chemical and biochemical products are brought each year to the market. However, the efficiency at which resources are converted into the final products is still dramatically low. According to the World Resources Institute, the average mass efficiency, at which the Earth's resources are currently converted into the final products, does not exceed 25%! The remaining 75% (or more) is a major contributor to pollution, waste, and environmental disturbances. We feel that in order to achieve a "steady-state" and retain the Earth's ecosphere in its present shape the resource efficiency of production processes has to increase drastically, say by a factor of three or even more. This is a figure that is both worrying and challenging to the chemists and chemical engineers.

The most obvious solution to this problem is to carry out chemical conversions at much higher yields and selectivity than is presently the case. Hence, more active and selective catalysts and more efficient ways to design better catalysts are needed. Moreover, efficient reactors with improved energy and mass transfer between reaction zone and surroundings, and eliminating the randomness and chaos (often harmful) that are characteristic of conventional reactors, are needed. Improved methods for scale-up, modeling, and design are vehicles in search for the optimal reactors. This is exactly what this book is all about. The aim of the book is to provide the reader with an up-to-date insight into the most important developments in the field of industrial catalysis and chemical reactor engineering contributing to the creation of a sustainable environment.

The search for better catalysts has been facilitated in recent years by molecular modeling. We are seeing here a step change. This is the subject of Chapter 1 (Molecular Catalytic Kinetics Concepts). New types of catalysts appeared to be more selective and active than conventional ones. Tuned mesoporous catalysts, gold catalysts, and metal organic frameworks (MOFs) that are discussed in Chapter 2 (Hierarchical Porous Zeolites by Demetallation, 3 (Preparation of Nanosized Gold Catalysts and Oxidation at Room Temperature), and 4 (The Fascinating Structure

Novel Concepts in Catalysis and Chemical Reactors: Improving the Efficiency for the Future.
Edited by Andrzej Cybulski, Jacob A. Moulijn, and Andrzej Stankiewicz

ISBN: 978-3-527-32469-9

and the Potential of Metal Organic Frameworks (MOFs)), respectively, belong to this class of catalysts. Enzymes as catalysts have been known for a long time. Recent developments in this field are remarkable, particularly in the fast-growing sector of fine chemicals and pharmaceuticals, and the future is bright. New achievements and the future in this area are dealt with in Chapter 5 (Enzymatic Catalysis Today and Tomorrow). The Fenton chemistry principle is applied to produce new solid catalysts that can be efficiently used for oxidation of organics in waste waters based upon the same Fenton chemistry. Synthesis of catalysts and their applications are presented in Chapter 6 (Oxidation Tools in Synthesis of Catalysts and Related Functional Materials). Fenton chemistry probably will find many more applications in the future because it represents a much milder step than the not very sophisticated generally applied calcination step.

In the above chapters, catalysis is the tool for improvements of chemical processes, thereby reducing consumption of raw materials and energy, and emissions to environment. Progress in the direct use of catalysis in processes for protection of the environment is presented in Chapter 7 (Challenges in Catalysis for Sustainability). Progress in catalytic processes aimed at the use of renewables and wastes for production of chemicals is presented in Chapter 8 (Catalytic Engineering in Processing of Biomass into Chemicals) with particular attention paid to processing of wood into chemicals.

A random character in a reaction zone can cause flow maldistribution and nonuniform residence time of reactants in the reaction zone, resulting in decreasing conversion and selectivity. Structured catalysts (monoliths and other structures) eliminate randomness in the reaction zone. New developments are presented in Chapter 9 (Structured Reactors, a Wealth of Opportunities). Selectivity and yield can also be improved by intensification and new ways for mass and energy exchange between the reaction zone and the surroundings. This enables a better control of temperature and concentration versus time profiles, thereby approaching the optimal reaction conditions. Membranes allow for cross-flow operation; that is, a controlled supply and withdrawal of reactants from the reaction zone. Progress and industrial prospects for zeolitic membranes is presented in Chapter 10 (Zeolitic Membranes in Catalysis; What is New and How Bright is the Future?). New developments in intensification of heat transfer between the reaction zone and the surroundings are discussed in Chapters 11 (Microstructures on Macroscale: Microchannel Reactors for Medium- and Large-size Processes and 12 (Intensification of Heat Transfer in Chemical Reactors: Heat Exchanger Reactors). Specific energy sources can provide much more energy than the conventional ones or provide energy in a particular form, increasing process rates and reactor capacities significantly. Such methods are presented in Chapter 13 (Reactors Using Alternative Energy Forms for Green Synthetic Routes, and New Functional Products). Fine chemicals and pharmaceuticals are predominantly manufactured in (semi)batch-operated reactors of low efficiency. Continuous reactors provide an inviting prospect. An evaluation based upon both technical and economic considerations is presented in Chapter 14 (Switching from Batch to Continuous Processing for Fine and Intermediate-scale Chemicals Manufacture). Progress in the search for the best reactor shape and operation conditions is illustrated with achievements in studies on micro- and macroscale phemomena in the reaction zone, see Chapter 15 (Progress in Methods for Identification of Micro- and Macroscale Physical Phenomena in Chemical Reactors: Improvements in Scale-up of Chemical Reactors).

Are the developments in industrial catalysis and chemical reactor engineering presented in this book sufficient to reach the ultimate goal of fully sustainable processing? Obviously not: although numerous developments in catalysis and chemical reactors have been reported in recent years, a number of big steps are still needed. Here is the editors' selection of topics that present significant steps, which bring us closer to that goal. We hope the topics chosen will inspire the reader to take further steps toward meeting the most important challenge to the mankind.

Andrzej Cybulski
Jacob Moulijn
Andrzej Stankiewicz

Note:

We sketched the above Preface with Andrzej together in early 2008, at the start of the present project. We have decided to leave it essentially unchanged. This book is above all *his book*. We miss him a lot.

Jacob Moulijn
Andrzej Stankiewicz

List of Contributors

Luc Alaerts
Katholieke Universiteit Leuven
Centre for Surface Chemistry
and Catalysis
Kasteelpark Arenberg 23
3001 Leuven
Belgium

Bengt Andersson
Chalmers University of
Technology
Department of Chemistry
and Biological Engineering
Kemigården 4
Göteborg
412 96, Sweden

Johan van den Bergh
Delft University of Technology
Chemical Engineering
Department
Faculty of Applied Sciences
Julianalaan 136
2682 BL Delft
The Netherlands

Michael Cabassud
Université de Toulouse
Laboratoire de Génie Chimique
UMR 5503 CNRS/INPT/UPS
Allee Emile Monso
31482 Toulouse
France

Derek Creaser
Chalmers University of
Technology
Department of Chemistry
and Biological Engineering
Kemigården 4
Göteborg
412 96, Sweden

Andrzej Cybulski†
Polish Academy of Sciences
CHEMIPAN
Institute of Physical Chemistry
Kasprzaka 44/52
01-224 Warszawa
Poland

Novel Concepts in Catalysis and Chemical Reactors: Improving the Efficiency for the Future.
Edited by Andrzej Cybulski, Jacob A. Moulijn, and Andrzej Stankiewicz

ISBN: 978-3-527-32469-9

Dirk E. De Vos
Katholieke Universiteit Leuven
Centre for Surface Chemistry
and Catalysis
Kasteelpark Arenberg 23
3001 Leuven
Belgium

Kari Eränen
Åbo Akademi
Process Chemistry Centre
Industrial Chemistry and
Reaction Engineering
20500 Turku/Åbo
Finland

Tom Van Gerven
Katholieke Universiteit Leuven
Faculty of Engineering
Department of Chemical
Engineering
Willem de Croylaan 46
3001 Leuven
Belgium

Christophe Gourdon
Université de Toulouse
Laboratoire de Génie Chimique
UMR 5503 CNRS/INPT/UPS
Allee Emile Monso
31482 Toulouse
France

Johan C. Groen
Delft Solids Solutions B.V
Rotterdamseweg 183c
2629 HD Delft
The Netherlands

Masatake Haruta
Tokyo Metropolitan University
Graduate School of Urban
Environmental Sciences
1-1 Minami-osawa
Hachioji
Tokyo 192-0397
Japan

and

CREST
Japan Science and
Technology Agency
4-1-8 Hon-cho
Kawaguchi
Saitama 332-0012
Japan

Mika Huuhtanen
University of Oulu
Department of Process and
Environmental Engineering
P.O. Box 4300
90014 Oulu
Finland

Tamao Ishida
Tokyo Metropolitan University
Graduate School of Urban
Environmental Sciences
1-1 Minami-osawa
Hachioji
Tokyo 192-0397
Japan

and

CREST
Japan Science and
Technology Agency
4-1-8 Hon-cho
Kawaguchi
Saitama 332-0012
Japan

Freek Kapteijn
Delft University of Technology
Chemical Engineering
Department
Faculty of Applied Sciences
Julianalaan 136
2682 BL Delft
The Netherlands

Riitta L. Keiski
University of Oulu
Department of Process and
Environmental Engineering
P.O. Box 4300
90014 Oulu
Finland

Piotr Kiełbasiński
Polish Academy of Sciences
Centre of Molecular and
Macromolecular Studies
Sienkiewicza 112
90-363 Łódź
Poland

Tanja Kolli
University of Oulu
Department of Process and
Environmental Engineering
P.O. Box 4300
90014 Oulu
Finland

Narendra Kumar
Åbo Akademi
Process Chemistry Centre
Industrial Chemistry and
Reaction Engineering
20500 Turku/Åbo
Finland

Jan J. Lerou
Velocys Inc
7950 Corporate Blvd.
Plain City
Ohio 43064
USA

Päivi Mäki-Arvela
Åbo Akademi
Process Chemistry Centre
Industrial Chemistry and
Reaction Engineering
20500 Turku/Åbo
Finland

Ignacio Melián-Cabrera
University of Groningen
Institute of Technology and
Management
Department of Chemical
Engineering
Nijenborgh 4
9747 AG Groningen
The Netherlands

Jyri-Pekka Mikkola
Åbo Akademi
Process Chemistry Centre
Industrial Chemistry and
Reaction Engineering
20500 Turku/Åbo
Finland

and

University of Umeå
Department of Chemistry
Technical Chemistry
Chemical Biological Centre
90871 Umeå
Sweden

Jacob A. Moulijn
Delft University of Technology
DelftChemTech
Julianalaan 136
2628 BL Delft
The Netherlands

Guido Mul
Delft University of Technology
DelftChemTech Department
Julianalaan 136
2628 BL Delft
The Netherlands

Dmitry Murzin
Åbo Akademi
Process Chemistry Centre
Industrial Chemistry and
Reaction Engineering
20500 Turku/Åbo
Finland

Norikazu Nishiyama
Osaka University
Division of Chemical Engineering
Graduate School of Engineering
Science
1-3 Machikaneyama Toyonaka
Osaka 560-8531
Japan

Satu Ojala
University of Oulu
Department of Process and
Environmental Engineering
P.O. Box 4300
90014 Oulu
Finland

Ryszard Ostaszewski
Polish Academy of Sciences
Institute of Organic Chemistry
Kasprzaka 44-52
01-224 Warsaw
Poland

Javier Pérez-Ramírez
ETH Zurich
Institute for Chemical and
Bioengineering
Department of Chemistry and
Applied Biosciences
HCI E125
Wolfgang–Pauli–Strasse 10
8093 Zurich
Switzerland

Eva Pongrácz
University of Oulu
Thule Institute Centre of
Northern Environmental
Technology
(NorTech Oulu)
P.O. Box 4300
90014 Oulu
Finland

David W. Rooney
Queen's University Belfast
School of Chemistry and
Chemical Engineering
Belfast
BT9 5AG
Northern Ireland
UK

Tapio Salmi
Åbo Akademi
Process Chemistry Centre
Industrial Chemistry and
Reaction Engineering
20500 Turku/Åbo
Finland

Rutger A. van Santen
Eindhoven University of Technology
Department of Chemical Engineering and Chemistry
Schuit Institute of Catalysis
Laboratory of Inorganic Chemistry and Catalysis
Den Dolech 2
5612 AZ Eindhoven
The Netherlands

Andrzej Stankiewicz
Delft University of Technology
Process & Energy Department
Leeghwaterstraat 44
2628 CA Delft
The Netherlands

E. Hugh Stitt
Johnson Matthey Technology Centre
PO Box 1
Billingham
Teeside TS 23 1LB
UK

Wiktor Szymański
Warsaw University of Technology
Faculty of Chemistry
Noakowskiego 3
00-664 Warsaw
Poland

Takashi Takei
Tokyo Metropolitan University
Graduate School of Urban Environmental Sciences
1-1 Minami-osawa
Hachioji
Tokyo 192-0397
Japan

and

CREST
Japan Science and Technology Agency
4-1-8 Hon-cho
Kawaguchi
Saitama 332-0012
Japan

Anna Lee Y. Tonkovich
Velocys Inc
7950 Corporate Blvd.
Plain City
Ohio 43064
USA

Johan Wärnå
Åbo Akademi
Process Chemistry Centre
Industrial Chemistry and Reaction Engineering
20500 Turku/Åbo
Finland

1
Molecular Catalytic Kinetics Concepts

Rutger A. van Santen

1.1
Key Principles of Heterogeneous Catalysis

We discuss the following topics in the subsequent sections:

- Sabatier principle and volcano curve;
- Brønsted–Evans–Polanyi (BEP) linear activation energy–reaction energy relationships;
- compensation effect in catalytic kinetics;
- micropore size dependence in zeolite catalysis;
- structure sensitivity and insensitivity in transition-metal catalysis;
- transition-state stabilization rules.

The molecular interpretation of major topics in catalytic kinetics will be highlighted based on insights on the properties of transition-state intermediates as deduced from computational chemical density functional theory (DFT) calculations.

1.2
Elementary Rate Constants and Catalytic Cycle

A catalytic reaction is composed of several reaction steps. Molecules have to adsorb to the catalyst and become activated, and product molecules have to desorb. The catalytic reaction is a reaction cycle of elementary reaction steps. The catalytic center is regenerated after reaction. This is the basis of the key molecular principle of catalysis: the Sabatier principle. According to this principle, the rate of a catalytic reaction has a maximum when the rate of activation and the rate of product desorption balance.

The time constant of a heterogeneous catalytic reaction is typically a second. This implies that the catalytic event is much slower than diffusion (10^{-6} s) or

Novel Concepts in Catalysis and Chemical Reactors: Improving the Efficiency for the Future.
Edited by Andrzej Cybulski, Jacob A. Moulijn, and Andrzej Stankiewicz

ISBN: 978-3-527-32469-9

elementary reaction steps ($10^{-4} - 10^{-2}$ s). Activation energies of elementary reaction steps are typically in the order of 100 kJ mol^{-1}. The overall catalytic reaction cycle is slower than elementary reaction steps because usually several reaction steps compete and surfaces tend to be covered with an overlayer of reaction intermediates.

Clearly, catalytic rate constants are much slower than vibrational and rotational processes that take care of energy transfer between the reacting molecules (10^{-12} s). For this reason, transition reaction rate expressions can be used to compute the reaction rate constants of the elementary reaction steps.

Eyring's transition-state reaction rate expression is

$$r_{\mathrm{TST}} = \frac{kT}{h}\Gamma\frac{Q^{\#}}{Q_0}e^{-\frac{E_{\mathrm{bar}}}{kT}} \tag{1.1a}$$

$$Q = \prod_i \frac{e^{-\frac{1}{2}\frac{h\nu_i}{kT}}}{1 - e^{-\frac{h\nu_i}{kT}}} \tag{1.1b}$$

$Q^{\#}$ is the partition function of transition state and Q_0 is the partition function of ground state, k is Boltzmann's constant, and h is Planck's constant.

The *transition-state energy* is defined as the saddle point of the energy of the system when plotted as a function of the reaction coordinates illustrated in Figure 1.1.

Γ is the probability that reaction coordinate passes the transition-state barrier when the system is in activated state. It is the product of a dynamical correction and the tunneling probability. Whereas statistical mechanics can be used to evaluate the pre-exponent and activation energy, Γ has to be evaluated by molecular dynamics techniques because of the very short timescale of the system in the activated state. For surface reactions not involving hydrogen, Γ is usually close to 1.

Most of the currently used computational chemistry programs provide energies and vibrational frequencies for ground as well as transition states.

A very useful analysis of catalytic reactions is provided for by the construction of so-called *volcano plots* (Figure 1.2). In a volcano plot, the catalytic rate of a reaction normalized per unit reactive surface area is plotted as a function of the adsorption energy of the reactant, product molecule, or reaction intermediates.

A volcano plot correlates a kinetic parameter, such as the activation energy, with a thermodynamic parameter, such as the adsorption energy. The maximum in the volcano plot corresponds to the Sabatier principle maximum, where the rate of activation of reactant molecules and the desorption of product molecules balance.

1.3
Linear Activation Energy–Reaction Energy Relationships

The Sabatier principle deals with the relation between catalytic reaction rate and adsorption energies of surface reaction intermediates. A very useful relation often

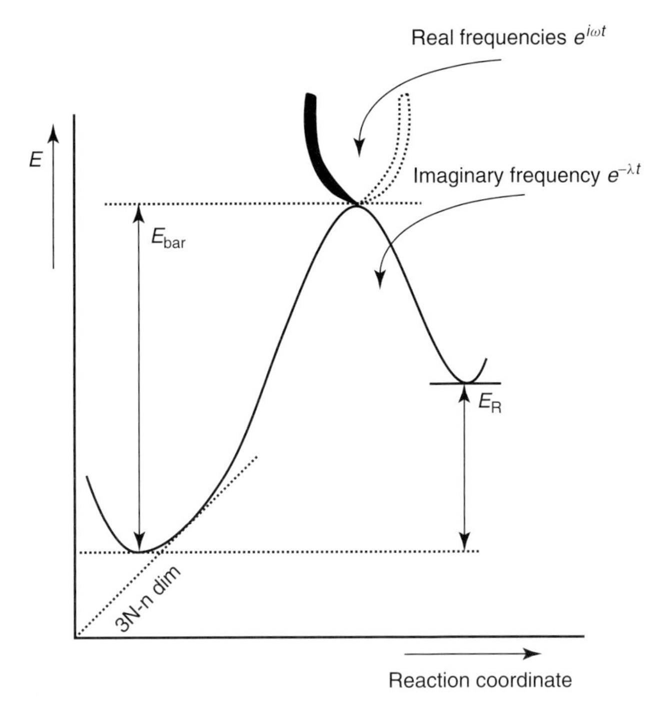

Figure 1.1 Transition-state saddle point diagram. Schematic representation of potential energy as a function of reaction coordinate.

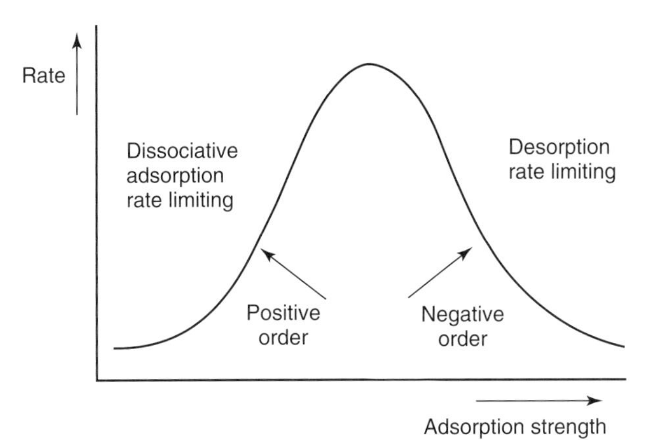

Figure 1.2 Volcano plot illustrating the Sabatier principle. Catalytic rate is maximum at optimum adsorption strength. On the left of the Sabatier maximum, rate has a positive order in reactant concentration, and on the right of Sabatier maximum the rate has a negative order.

exists between the activation energy of elementary surface reaction steps, such as adsorbate bond dissociation or adsorbed fragment recombination and corresponding reaction energies. These give the Brønsted–Evans–Polanyi relations.

For the forward dissociation reaction, the BEP relation is

$$\delta E^{\#}_{\text{diss}} = \alpha \delta E_{\text{react}} \tag{1.2a}$$

Then for the backward recombination reaction Eq. (1.2b) has to hold:

$$\delta E^{\#}_{\text{rec}} = -(1-\alpha)\, \delta E_{\text{react}} \tag{1.2b}$$

Owing to microscopic reversibility, the proportionality constants of the forward and backward reactions are related. These relations are illustrated in Figure 1.3.

The original ideas of Evans and Polanyi [1] to explain such a linear relation between activation energy and reaction energy can be illustrated through a two-dimensional analysis of two crossing potential energy curves.

The two curves in Figure 1.4 represent the energy of a chemical bond that is activated before and after a reaction. The difference between the locations of the potential energy minima is the reaction coordinate x_0.

If one assumes the potential energy curves to have a similar parabolic dependence on the displacement of the atoms, a simple relation can be deduced between activation energy, the crossing point energy of the two curves, and the reaction energy. One then finds for α:

$$E_{\text{act}} = E^{\circ}_{\text{act}} \left(\frac{\Delta E_r}{4E^{\circ}_{\text{act}}} + 1 \right)^2 \tag{1.3a}$$

$$\alpha = \frac{\delta E_{\text{act}}}{\delta E_r} = \frac{1}{2}\left(1 + \frac{\Delta E_r}{4E^{\circ}_{\text{act}}}\right) \tag{1.3b}$$

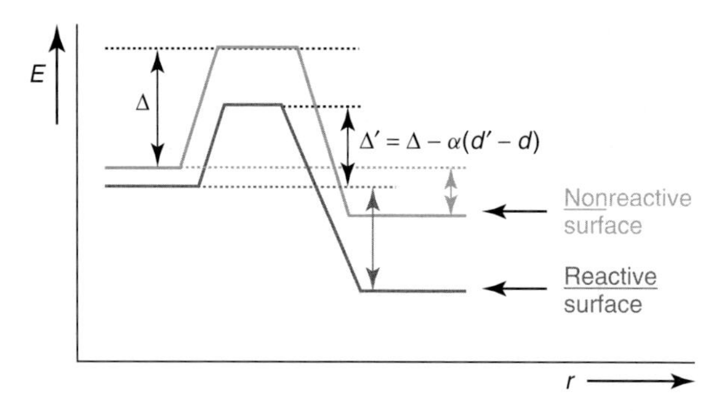

Figure 1.3 Illustration of the BEP relation $\Delta' = \Delta - \alpha\,(d' - d)$.

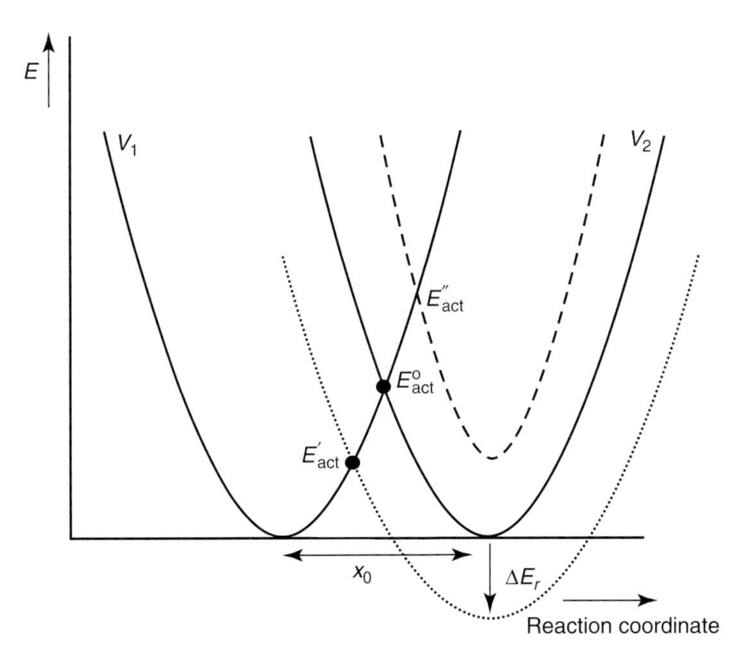

Figure 1.4 Two-dimensional curve-crossing potential energy diagram of reacting system with similar potential energies before and after reaction (schematic).

One notes that the proportionality constant, α, depends on the reaction energy, ΔE_r. Therefore, Eq. (1.3) is not strictly a linear relation between activation energy change and reaction energy. In the extreme limit of high exothermicity of the reaction energy $\alpha = 0$, and the crossing point of the two curves is at the minimum of curve V_1. In this case the transition state is called *early*. Its structure is close to that of the reactant state.

In the limit of high endothermicity $\alpha = 1$ and now the crossing point is close to the minimum of curve V_2. The transition-state structure is now close to that of the final state. The transition state can now be considered to be late. This analysis is important, since it illustrates why α varies between 0 and 1. Often α is simply assumed to be equal to 0.5.

The BEP relation is only expected to hold as long as one compares systems in which the reaction path of the reacting molecules is similar. An illustration is provided in Figure 1.5 [2].

In this figure, the activation energies of N_2 dissociation are compared for the different reaction centers: the (111) surface structure of an fcc crystal and a stepped surface. Activation energies with respect to the energy of the gas-phase molecule are related to the adsorption energies of the N atoms. As often found for bond activating surface reactions, a value of α close to 1 is obtained. It implies that the electronic interactions between the surface and the reactant in the transition state and product state are similar. The bond strength of the chemical bond

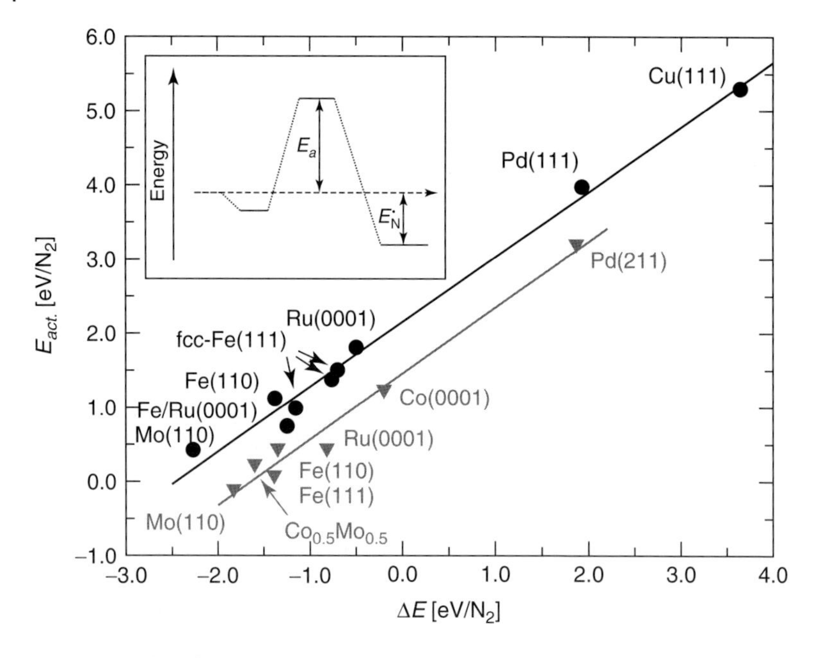

Figure 1.5 Plot of computed reaction barriers for dissociation at $E_{act.}$ for N_2 dissociation as a function of nitrogen atom adsorption energy on surface terrace and stepped surface [2]. The upper curve is for surface terrace of (111) type of fcc crystals, and the lower curve presents data on the stepped surfaces.

that is activated is substantially weakened due to the strong interaction with the metal surface. The structure of the transition state is close to that of the product state.

This is illustrated in Figure 1.6 for the dissociation of CO [3]. As a consequence of the high value of α, the proportionality constant of recombination is usually approximately 0.2, reflecting a weakening of the adatom surface bonds in transition state by this small amount. It implies that typically one of the six surface bonds is broken in the transition state compared to the adsorption state of the two atoms before recombination.

1.4
Microkinetic Expressions; Derivation of Volcano Curve

In microkinetics, overall rate expressions are deduced from the rates of elementary rate constants within a molecular mechanistic scheme of the reaction. We will use the methanation reaction as an example to illustrate the

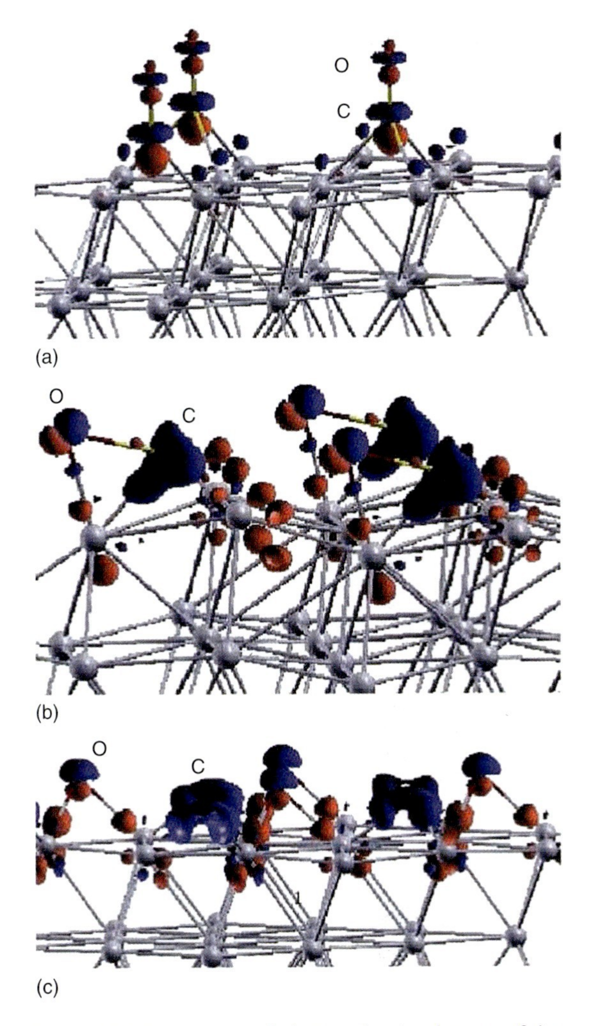

Figure 1.6 Structures and electron density changes of dissociating CO on Ru(0001) surface: (a) adsorbed CO, (b) transition state for dissociation, and (c) dissociated state.

physicochemical basis of the Sabatier volcano curve. The corresponding elementary reactions are

$$CO_g \underset{k_{diss}}{\overset{k_{ads}}{\rightleftarrows}} CO_{ads} \tag{1.4a}$$

$$CO_{ads} \xrightarrow{k_{diss}} C_{ads} + O_{ads} \tag{1.4b}$$

$$C_{ads} + 2H_2 \underset{r_H}{\longrightarrow} CH_4 \nearrow \tag{1.4c}$$

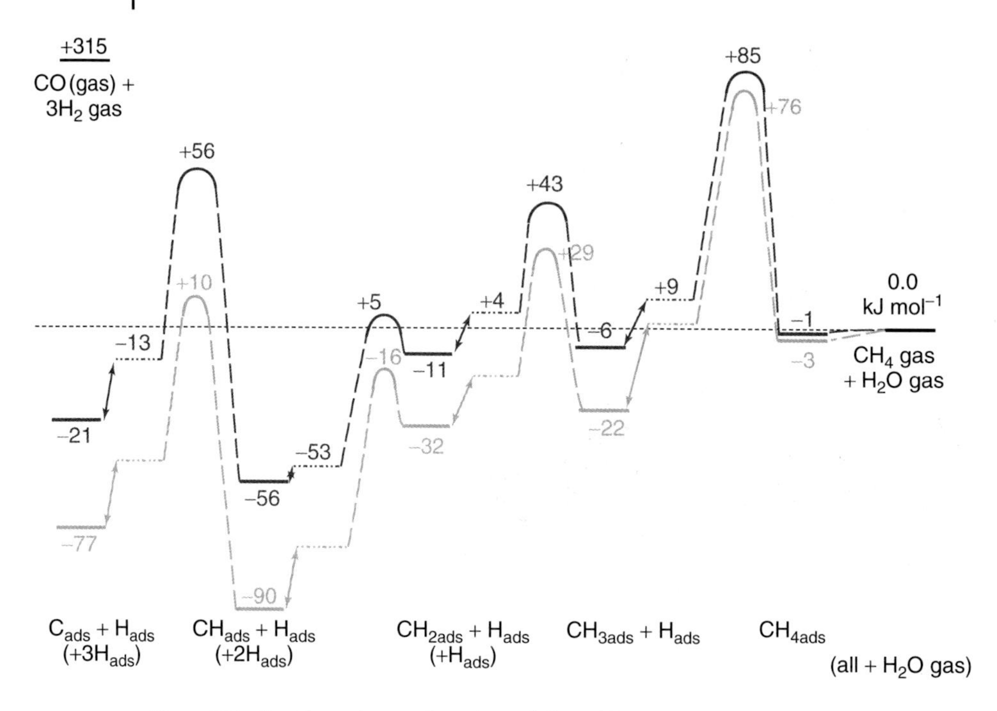

Figure 1.7 Reaction scheme of reaction of C_{ads} with H_{ads} to produce CH_4 on Ru(0001) surface. Coverages of 25% and 11% are compared. Energies with respect to gas phase [4].

Surface carbon hydrogenation occurs through a sequence of hydrogenation steps in which $CH_{x,ads}$ species are formed with increasing hydrogenation. r_H, the rate of C_{ads} hydrogenation, depends implicitly on hydrogen pressure.

For Ru(0001) the corresponding reaction energy scheme is shown in Figure 1.7 [4]. The relative energies of the different reaction intermediates, C_{ads} or CH_{ads}, may strongly depend on the type of surface and metal. When for different surfaces or metals the relative interaction with H_{ads} increases C_{ads} may for instance become more stable than CH. This is found for more coordinative unsaturated surfaces or more reactive metals.

We present expressions for reaction rates and steady-state concentrations using the simplified assumption that C_{ads} hydrogenation to CH_4 occurs in one reaction step. We also assume that O_{ads} removal is fast and that hydrogen adsorption is not influenced by the other adsorbates.

Then the activation energy for methane production from C_{ads} is the overall activation energy for the hydrogenation of C_{ads} to CH_4, and Eq. (1.5) gives the rate of methane production:

$$R_{CH_4} = r_H \cdot \theta_c \tag{1.5}$$

A closed expression for θ_C can be deduced:

$$\theta_C = 1 + \frac{1}{2}\lambda - \frac{1}{2}\sqrt{\lambda^2 + 4\lambda} \tag{1.6a}$$

$$\approx \frac{1}{1+\lambda} \tag{1.6b}$$

with

$$\lambda = \frac{r_H}{k_{diss}} \frac{\left(K_{ads}^{CO} \cdot [CO] + 1\right)^2}{\left(K_{ads}^{CO}[CO]\right)} \tag{1.7a}$$

$$= A\frac{r_H}{k_{diss}} \tag{1.7b}$$

One notes that the coverage of C_{ads} depends on two important parameters: the ratio ρ of the rate of hydrogenation of C_{ads} to give methane and the rate constant of CO dissociation:

$$\rho = \frac{r_H}{k_{diss}} \tag{1.8}$$

and the equilibrium constant of CO adsorption, K_{eq}^{CO}. The coverage with C_{ads} increases with decreasing value of ρ. This implies a high rate of k_{diss} and slow rate of C_{ads} hydrogenation. The strong pressure dependence of CO relates to the need of neighboring vacant sites for CO dissociation.

Beyond a particular value of K_{eq}^{CO} the surface coverage with C_{ads} decreases because CO dissociation becomes inhibited.

In order to proceed, one needs to know the relation between the rate constants and reaction energies. This determines the functional behavior of ρ.

We use the linear activation energy–reaction energy relationships as deduced from the BEP relation and write expressions for k_{diss}, r_H, and λ:

$$k_{diss} = \nu_0 e^{-\frac{E_{diss}^0}{kT}} \cdot e^{-\alpha\frac{E'_{ads}}{kT}} \tag{1.9a}$$

$$= \nu'_0 e^{-\alpha'\frac{E_{ads}}{kT}} \tag{1.9b}$$

$$r_H = r'_H e^{x\frac{E_{ads}}{kT}} \tag{1.9c}$$

$$\lambda = A\frac{r'_H}{\nu'_0} e^{(x+\alpha')\frac{E_{ads}}{kT}} \tag{1.10}$$

The dissociation rate of CO_{ads} will increase with increasing exothermicity of the reaction energy. One can use the adsorption energy of the carbon atom as the standard measure.

From chemisorption theory we know that adatom adsorption energies will decrease in a row of the periodic system of the group VIII metals when the position of the element moves to the right. The rate of hydrogenation of C_{ads} will decrease with increasing adsorption energy of C_{ads} and hence will decrease in the same order with element position in the periodic system.

We now study the consequences of these BEP choices to the dependence of predicted rate of methane production on E_{ads}. Making the additional simplifying assumption that the adsorption energy parameters in Eqs. (1.9b) and (1.9c) are the same, one finds for the rate of methane production an expression

$$R_{CH_4} = C\frac{\lambda^{\frac{x}{x+\alpha'}}}{1+\lambda} \tag{1.11a}$$

with

$$C = r_H'\left(\frac{r_H'}{r_0'}A\right)^{-\frac{x}{x+\alpha'}} \tag{1.11b}$$

In Eq. (1.11b) the constant A depends on the equilibrium constant K_{eq}^{CO}. This will vary also with the adsorption energy of C or O, but will be much less sensitive to these variations than the activation energies of CO dissociation and hydrogenation [5].

The dependence of R_{CH_4} on λ is sketched in Figure 1.8.

Equation (1.11a) will have only a maximum as long as

$$\frac{x}{x+\alpha'} < 1 \tag{1.12}$$

Within our model this condition is always satisfied. We find an interesting result that the Sabatier volcano maximum is found when

$$\lambda_{max} = \frac{x}{\alpha'} \tag{1.13}$$

The controlling parameters that determine the volcano curve are the BEP constants: k_{diss} and r_H. It is exclusively determined by the value of ρ. It expresses the compromise of the opposing elementary rate events: dissociation versus product

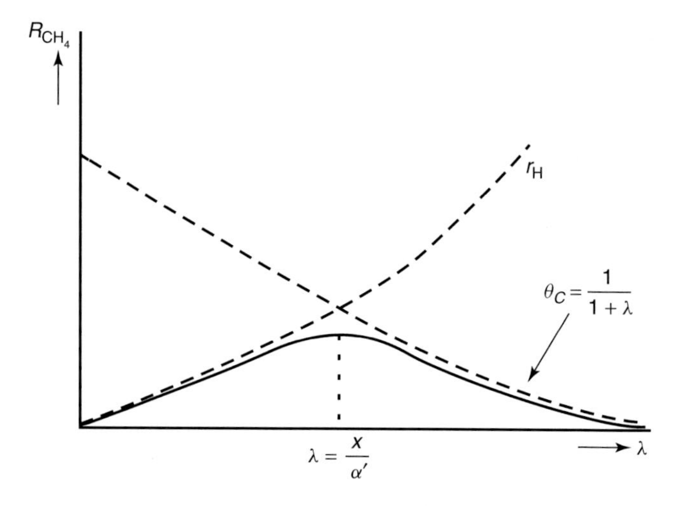

Figure 1.8 Dependence of R_{CH_4} on λ (schematic).

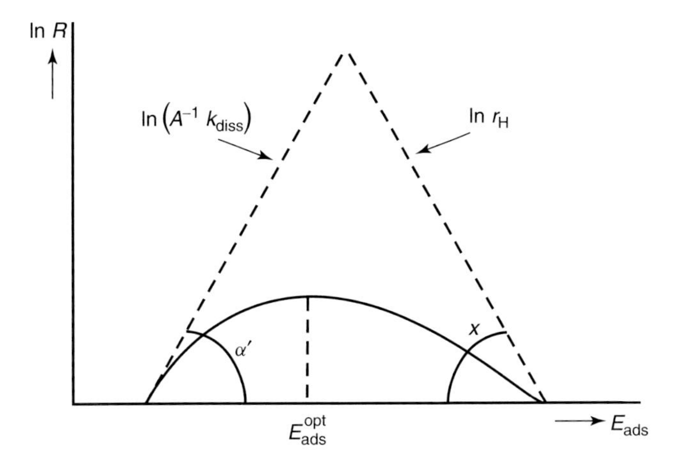

Figure 1.9 Volcano curve dependence of rate of methanation, R, on E_{ads} (schematic).

formation. The CO partial pressure determines the dependence of the rate on gas-phase pressure. It controls λ through changes of parameter A. Note that variation of A would also give volcano-type behavior. This is obviously found when the rate is plotted as a function of gas-phase pressure A. It relates to the blocking of surface sites by increasing adsorption of CO molecules. Volcano-type behavior is illustrated in Figure 1.9 for constant CO pressure.

Equation (1.5) then allows us to deduce the optimum value of E_{ads} of the Sabatier maximum rate. It can be deduced from Eq. (1.12a). The latter depends on the CO pressure through A.

The volcano curve is bounded by the rate of dissociative adsorption of CO and hydrogenation of adsorbed carbon. This is illustrated in Figure 1.9.

For relevant related treatments see [6]. Whereas the above discussion limits itself to the conversion of CO to a single product, the treatment can be easily extended to a selectivity problem.

Interestingly, one can easily deduce an expression for the relative rate of coke formation as compared to that of methanation. The rate of initial coke formation depends on the combination probability of carbon atoms and hence is given by

$$R_{C-C} = r_{CC} \cdot \theta_C^2 \tag{1.14}$$

The relative rate of coke versus methane formation then follows from

$$\frac{R_{CC}}{R_{CH_4}} = \frac{r_{CC} \cdot \theta_C^2}{r_H \cdot \theta_C} \tag{1.15a}$$

$$= \frac{r_{CC}}{r_H} \cdot \theta_C \tag{1.15b}$$

Interestingly as we have seen this may have a maximum as a function of the metal–carbon bond energy.

The occurrence of a maximum depends on the BEP parameter α'' of the C–C bond formation rate. Volcano-type behavior for the selectivity is found as long as

Eq. (1.16a) is satisfied.

$$2x + \alpha > \alpha'' > x \tag{1.16a}$$

Then,

$$\lambda_{\max}^{\text{sel}} = \frac{\alpha'' - x}{2x + \alpha - \alpha''} \tag{1.16b}$$

r_{CC} and r_H decrease when the carbon adsorption energy increases. Volcano-type behavior of the selectivity to coke formation is found when the activation energy of C–C bond formation decreases faster with increasing metal–carbon bond energy than with the rate of methane formation. Equation (1.16b) indicates that the rate of the nonselective C–C bond forming reaction is slow when θ_C is high and when the metal–carbon bond is so strong that methane formation exceeds the carbon–carbon bond formation. The other extreme is the case of very slow CO dissociation, where θ_C is so small that the rate of C–C bond formation is minimized.

This analysis indicates the importance of a proper understanding of BEP relations for surface reactions. It enables a prediction not only of conversion rates but also of selectivity trends.

1.5 Compensation Effect

For catalytic reactions and systems that are related through Sabatier-type relations based on kinetic relationships as expressed by Eqs. (1.5) and (1.6), one can also deduce that a so-called *compensation effect* exists. According to the compensation effect there is a linear relation between the change in the apparent activation energy of a reaction and the logarithm of its corresponding pre-exponent in the Arrhenius reaction rate expression.

The occurrence of a compensation effect can be readily deduced from Eqs. (1.6) and (1.7). The physical basis of the compensation effect is similar to that of the Sabatier volcano curve. When reaction conditions or catalytic reactivity of a surface changes, the surface coverage of the catalyst is modified. This change in surface coverage changes the rate through change in the reaction order of a reaction.

In Eq. (1.5) the surface coverage is given by θ_C, and θ_C is related to parameter λ of Eq. (1.7). Equation (1.5) can be rewritten to show explicitly its dependence on gas-phase concentration. Equation (1.17a) gives the result. This expression can be related to practical kinetic expressions by writing it as a power law as is done in Eq. (1.18b). Power-law-type rate expressions present the rate of a reaction as a function of the reaction order. In Eq. (1.17b) the reaction order is m in H_2 and $-n$ in CO.

$$R_H = \frac{r_H k_{\text{diss}} K_{\text{ads}}^{\text{CO}} [\text{CO}]}{k_{\text{diss}} K_{\text{ads}}^{\text{CO}} [\text{CO}] + r_H \left(K_{\text{ads}}^{\text{CO}} [\text{CO}] + 1\right)^2} \tag{1.17a}$$

$$\approx k_H^l k_{\text{diss}}^{-n} \left(K_{\text{ads}}^{\text{CO}}\right)^{-n} [\text{CO}]^{-n} [\text{H}_2]^m$$

$$l \le 1 \tag{1.17b}$$

Power law expressions are useful as long as the approximate orders of reactant concentration are constant over a particular concentration course. A change in the order of the reaction corresponds to a change in the surface concentration of a particular reactant. A low reaction order usually implies a high surface concentration, a low reaction order, and a low surface reaction of the corresponding adsorbed intermediates. In order to deduce (Eq. (1.17b)) the rate of surface carbon hydrogenation, the power law of Eq. (1.18) has been used.

$$r_{\mathrm{H}} = k_{\mathrm{H}} \cdot [\mathrm{H_2}]^t\,;\; m = t \cdot l \tag{1.18}$$

From Eq. (1.17b) the apparent activation energy as well as the pre-exponent can be readily deduced. They are given in Eq. (1.19).

$$\begin{aligned} E_{\mathrm{app}} &= lE_{\mathrm{act}}^{\mathrm{H}} - n\left\{E_{\mathrm{act}}(\mathrm{diss}) + E_{\mathrm{ads}}\,(\mathrm{CO})\right\} \\ &= lE_{\mathrm{act}}^{\mathrm{H}} - n\Delta E_{\mathrm{app}} \end{aligned} \tag{1.19a}$$

$$\ln A_{\mathrm{app}} = \ln \Gamma \cdot \frac{ekT}{h} + \frac{l}{k}\Delta S_{\mathrm{act}}^{\mathrm{H}} - \frac{n}{k}\Delta S_{\mathrm{app}} \tag{1.19b}$$

$$\Delta S_{\mathrm{app}} = \Delta S_{\mathrm{act}}\,(\mathrm{diss}) + \Delta S_{\mathrm{ads}}(\mathrm{CO}) \tag{1.19c}$$

The orders of the reaction appear as coefficients of activation energies and adsorption energies and their corresponding entropies. For more detailed discussions see [7].

A consequence of the compensation effect is the presence of an isokinetic temperature. For a particular reaction, the logarithm of the rate of a reaction measured at different conditions versus $1/T$ should cross at the same (isokinetic) temperature. For conditions with varying n, this isokinetic temperature easily follows from Eq. (1.19) and is given by

$$T_{\mathrm{iso}} = \frac{\Delta E_{\mathrm{app}}}{k\Delta S_{\mathrm{app}}} \tag{1.20}$$

It is important to realize that the compensation effect in catalysis refers to the overall catalytic reactions.

The activation energies of elementary reaction steps may sometimes show a relationship between activation energy changes and activation entropies.

A reaction with a high activation energy tends to have a weaker interaction with the surface and hence will have enhanced mobility that is reflected in a larger activation entropy. For this reason, the pre-exponents of surface desorption rate constants are 10^4-10^6 larger than the pre-exponents of surface reaction rates.

In classical reaction rate theory expressions, this directly follows from the frequency–pre-exponent relationship:

$$k_{\mathrm{class}} = \nu \frac{r_t^2}{r_i^2} e^{-\frac{E_{\mathrm{act}}}{kT}} \tag{1.21}$$

A high frequency of vibration between surface and reactant implies a strong bond, which will give a high activation energy. Hence, increase in pre-exponent and corresponding activation energies counteract. Equation (1.21) is the rate expression for a weakly bonded complex. The bond frequency is ν, and r_i and r_t are the initial and transition-state radii. Equation (1.21) is valid for a freely rotating diatomic complex.

Compensation-type behavior is quite general and has been extensively studied, especially in transition-metal catalysis [8a], sulfide catalysis [8b], and zeolite catalysis [7].

In the next section, we present a short discussion of compensation-type behavior in zeolite catalysis.

1.6 Hydrocarbon Conversion Catalyzed by Zeolites

As a further illustration of the compensation effect, we use solid-acid-catalyzed hydrocarbon activation by microporous zeolites. A classical issue in zeolite catalysis is the relationship between overall rate of a catalytic reaction and the match of shape and size between adsorbate and zeolite micropore.

For a monomolecular reaction, such as the cracking of hydrocarbons by protonic zeolites, the rate expression is very similar to the one in Eq. (1.5). The rate of the reaction is now proportional to the concentration of molecules at the reaction center, the proton of the zeolite, Eq. (1.22a).

$$r = k_{act} \cdot \theta \tag{1.22a}$$

$$= k_{act} \cdot \frac{K_{eq}[C_H]}{1 + K_{eq}[C_H]} \tag{1.22b}$$

Assuming adsorption to behave according to the Langmuir adsorption isotherm, we get Eq. (1.22b). Both the rate constant of proton activation k_{act} and the equilibrium constant of adsorption K_{eq} depend on cavity details.

Quantum-chemical studies have indeed shown that the presence of a surrounding cavity lowers the barriers of charge separation that occur when a molecule is activated by zeolitic protons [9] as shown in Figure 1.10.

The presence of the zeolite cavity dramatically lowers the activation energy for the protonation of toluene. It is mainly due to screening of the charges in the transition state due to the polarizable lattice oxygen atoms. In the transition state, a positive charge develops on protonated toluene.

This reduction in activation energy will occur only when the structure of the transition state complex fits well in the zeolite cavity. This is the case for the protonated toluene example in the zeolite mordenite channel. The structure of the transition state complex in the cluster simulation and zeolite can be observed to be very similar to the one in Figure 1.10.

The activation energy will be strongly increased when there is a mismatch between transition-state-complex shape and cavity. The rate constant then typically behaves as indicated in the following equation:

$$k_{act} = k_{act}^{n,st} e^{\frac{\Delta G_{st}}{kT}} \tag{1.23}$$

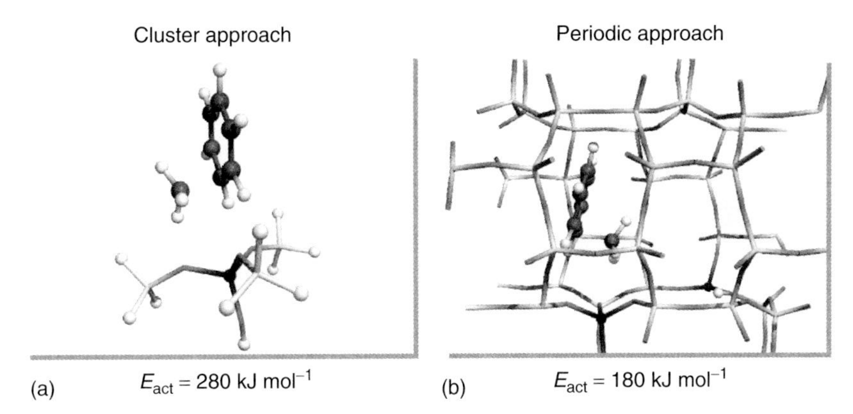

Figure 1.10 Comparison of the activation energy of protonation of toluene by (a) an acidic cluster versus that of activation of the same molecule by (b) Mordenite (MOR) [9a].

ΔG_{st} is the difference in free energy due to steric constants in reactant and transition state. $k^{n,st}$ is the rate constant of the nonsterically constrained reaction. The contribution of the steric component to the transition-state energy cannot be deduced accurately from DFT calculations because van der Waals energies are poorly computed. Force field methods have to be used to properly account for such interactions.

For adsorption in zeolites, the biased Monte Carlo method as developed by Smit is an excellent method to determine the free energies of molecules adsorbed on zeolites [9b]. This method can be used to compute the concentration of molecules adsorbed on zeolites, as we discuss below.

We will use this method to deduce ΔG_{st} for hydrogen transfer reaction. The free energies of adsorption of reacting molecules such as propylene and butane are compared with the free energies of reaction intermediate molecules that are analogous to the intermediates formed in the transition state. A C–C bond replaces the C–H–C bond. An example of such a transition state and analog intermediate is given in Figure 1.11.

In Table 1.1 a comparison is made of the differences in free energies for two different zeolites. Note the large repulsive energies computed for the intermediates and their sensitivity to zeolite structure.

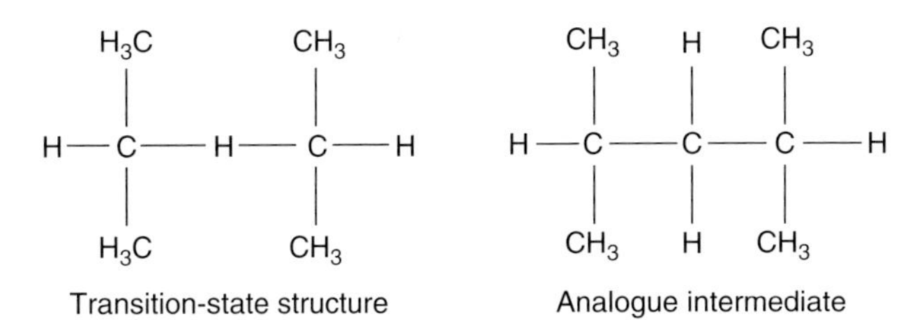

Figure 1.11 Transition state for hydrogen transfer and its analog intermediate (schematic).

Table 1.1 Configurationally biased Monte Carlo simulations of the adsorption enthalpies of hydrocarbons for two zeolites.

	ΔH_{ads} (kJ mol^{-1})	
	MFI	**Chabasite**
Propane	−41.0	−34.6
n-Butane	−44.1	−47.0
C_6	−50.7	−65.1
C_8	−49.9	−43.1
$\Delta E(C_6 - 2C_3)$	+31.3	+4.1
$\Delta E(C_8 - 2C_4)$	+38.3	+50.9

An alternative view to interpreting ΔG_{st} is the realization that reactants before getting activated to a particular reaction have to be present in a conformation such that a particular reaction can occur. The actual activation of reacting molecule or molecules is not strongly affected by this state. Calculations on the activation of the different isomers of xylene have indeed demonstrated that differences in the energies of the pretransition-state configurations dominate the activation energy differences [10], and hence the Maxwell–Boltzmann term, Eq. (1.23), has to be interpreted as the relative probability that a particular intermediate pretransition-state structure is realized in zeolite [11].

This is the reason that for complex cracking reactions in zeolites the product pattern can be predicted from a simulation of the free energies of the corresponding intermediate molecules in the zeolite [11].

As long as there are no important steric contributions to the transition-state energies, the elementary rate constant of Eq. (1.22) does not sensitively depend on the detailed shape of the zeolite cavity. Then the dominant contribution is due to the coverage dependent term θ.

This has been demonstrated by a comparison of the cracking rates of small linear hydrocarbons in ZSM-5 [12] and also for reactions in different zeolites for the hydroisomerization of hexane [13]. Differences in catalytic conversion appear to be mainly due to differences in θ.

The apparent activation energies can be deduced from Eq. (1.22b). The corresponding expression is given by Eq. (1.24a):

$$E_{app} = E_{act} + E_{ads}(1 - \theta) \tag{1.24a}$$

$$r \approx k_{act}\, K_{eq}^{(1-\theta)} [C_H]^{1-\theta} \qquad 0 < \theta < 1 \tag{1.24b}$$

In the absence of steric constraints in Eq. (1.24a) E_{act} will not vary. E_{ads} and θ are the parameters that significantly change with hydrocarbon chain length or zeolite.

Table 1.2 Calculated heats of adsorption and adsorption constants for various hydrocarbons in zeolites with different channel dimensions.

	ΔH_{ads}(kJ mol^{-1})	K_{ads}(T = 513K) (mmol (g Pa)$^{-1}$)
	Simulation	**Simulation**
n-Pentane/TON	−63.6	4.8×10^{-6}
n-Pentane/MOR	−61.5	4.8×10^{-5}
n-Hexane/TON	−76.3	1.25×10^{-5}
n-Hexane/MOR	−69.5	1.25×10^{-4}

Since the interaction of linear hydrocarbons is dominated by the van der Waals interaction with the zeolite, the apparent activation energies for cracking decrease linearly with chain length. In some cases, differences in the overall rate are not dominated by differences in the heat of adsorption but instead are dominated by differences in the enthrones of adsorbed molecules.

One notes in Table 1.2 a uniform increase in the adsorption energies of the alkanes when the microspore size decreases (compare 12-ring-channel zeolite MOR with 10-ring-channel TON). However, at the temperature of hydroisomerization the equilibrium constant for adsorption is less in the narrow-pore zeolite than in the wide-pore system. This difference is due to the more limited mobility of the hydrocarbon in the narrow-pore material. This can be used to compute Eq. (1.22b) with the result that the overall hydroisomerization rate in the narrow-pore material is lower than that in the wide-pore material. This entropy-difference-dominated effect is reflected in a substantially decreased hydrocarbon concentration in the narrow-pore material.

1.7 Structure Sensitive and Insensitive Reactions

A classical issue in transition-metal catalysis is the dependence of catalytic activity on changes in the particle size of the metal clusters in the nanosize region [14].

As illustrated in Figure 1.12, three types of behavior can be observed. The most significant surface feature that changes with metal particle size is the ratio of corner, edge, and terrace surface atoms.

The increase in the rate of case II is related with an increase in the relative ratio of the edge and corner atoms over the decreasing number of terrace atoms. This increase in reactivity relates to the increased degree of coordinative unsaturation of the edge and corner atoms.

Important changes in the electronic structure occur. Electron delocalization decreases, which is reflected in a narrowing, especially, of the d-valence electron

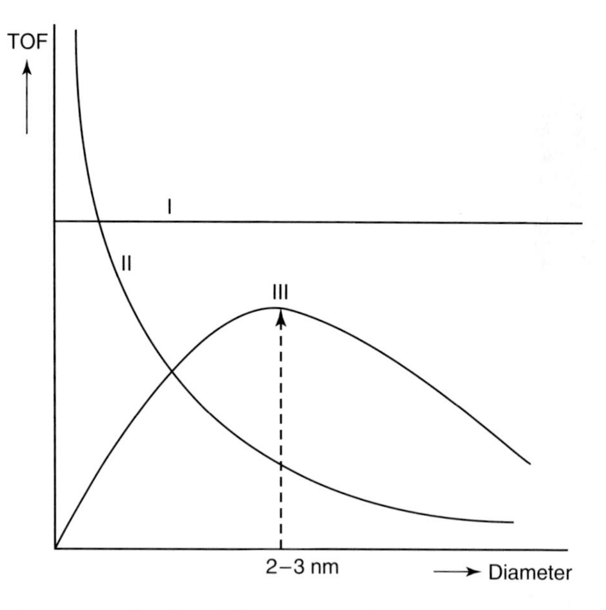

Figure 1.12 The three different types of cluster size dependence of catalytic conversion. Rates are considered normalized per exposed surface atom (schematic).

bands and a corresponding upward shift of the d-valence electron band center due to increased electron–electron interactions [5].

The decrease in bandwidth is proportional to

$$\sqrt{N_s} \tag{1.25}$$

Within the tight binding approximation, it implies a decrease in electron localization energy:

$$\Delta E_{\text{loc}} \approx \left(\sqrt{N_S'} - \sqrt{N_S}\right)\beta \tag{1.26}$$

for the surface atom with the lowest number of nearest neighbor surface atoms N_S' compared to that of a surface atom with N_s neighbors. β is a measure of the interatomic overlap energy.

The dependence on electron localization energy can also be illustrated by the use of the bond order conservation principle. This principle gives an approximate recipe to estimate changes in bond strength when coordination of a surface atom or adsorbate attachment changes [5, 15].

According to this principle, the valence of an atom is considered a constant. When more atoms coordinate to the same atom, the valence has to be distributed over more bonds, and hence the strength per bond decreases. When the chemical bonds are equivalent, the bond strength of an individual bond $\varepsilon(n)$ depends in the following way on the corresponding bond strength of a complex with a single

Figure 1.13 Transition-state configuration of methane activation on Ru(1120) surface.

bond (ε_0):

$$\varepsilon(n) = \varepsilon_0 \left(\frac{2n-1}{n^2} \right) \tag{1.27}$$

n is the total number of bonds with the metal atom. The surface atom metal coordination number is given by

$$N = n - 1 \tag{1.28}$$

Class II dependence for the activation of a chemical bond as a function of surface metal atom coordinative unsaturation is typically found for chemical bonds of σ character, such as the CH or C–C bond in an alkane. Activation of such bonds usually occurs atop of a metal atom. The transition-state configuration for methane on a Ru surface illustrates this (Figure 1.13).

The data presented in Table 1.3 illustrate the dependence of the activation energy of methane on the edge or corner (kink) atom position of some transition-metal surfaces.

The BEP α value for methane activation is close to 1. As a consequence of the BEP value for hydrogenation of adsorbed methyl, the reverse reaction should be nearly zero.

The dependence on decreasing particle size that results for this recombination reaction is the same as Class I in Figure 1.12. The differences between the activation

Table 1.3 Methane activation on edge and corner atoms (kilojoules per mol).

Ru(0001)[a]	76
Ru(1120)[b]	56
Rh(111)[c]	67
Rh step[c]	32
Rh kink[c]	20
Pd(111)[c]	66
Pd step[c]	38
Pd kink[c]	41

[a](0001) Ciobica *et al.* [4].
[b](1120) Ciobica and van Santen [16].
[c]Liu and Hu [17].

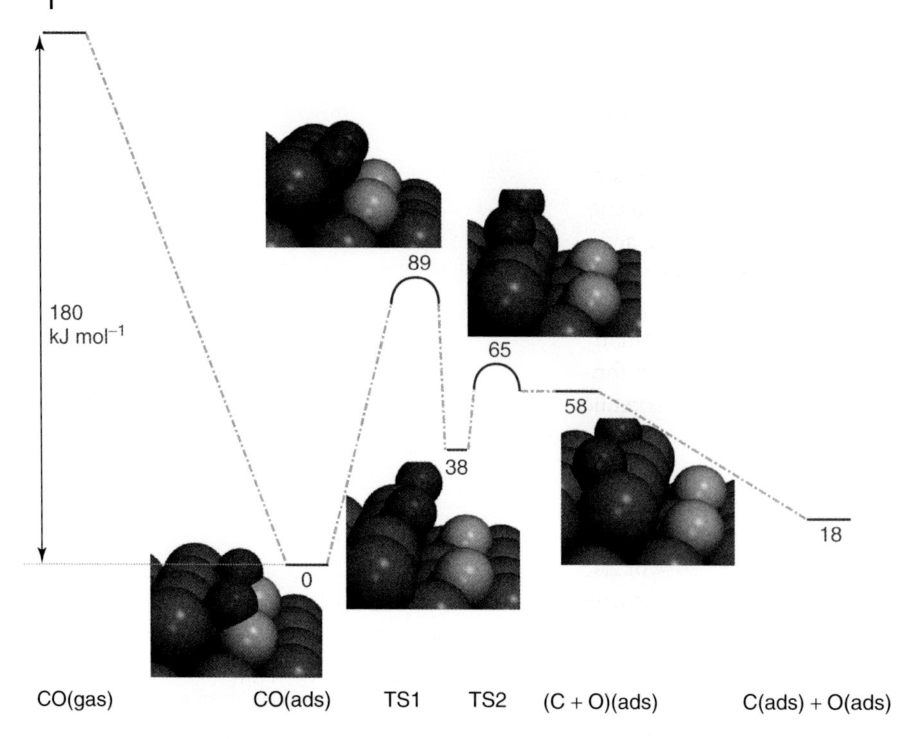

Figure 1.14 Energetics (kilojoules per mole) and structure of CO dissociating from Ru step-edge site [16].

energies for dissociation hence closely relate to the differences in energy of the adsorbed product fragments methyl$_{ads}$ and H_{ads}. The rate of this recombination reaction has become independent of surface atom coordination number.

Class II and Class I behavior are found to be closely related and are each other's complement. In practical catalysis, Class I behavior is typically found for hydrogenation reactions.

Class III-type behavior is representative of reactions in which π bonds have to be broken. It is the typical behavior of reactions in which CO or N_2 bond activation is rate limiting.

The activation energy of such molecules depends strongly on the structure of the catalytically active center. The structures of reactant, transition state as well as product state at a step-edge site are shown for CO dissociation in Figure 1.14.

Surface step-edge sites have substantially lowered activation energies compared to the activation energies of the same dissociation reactions on surface terraces (compare 91 kJ mol^{-1} in Figure 1.14 with 215 kJ mol^{-1} on Ru(0001) surface). This lowered barrier is due to several factors that affect chemical bonding. Because of multiple contacts of the CO molecule at the step-edge site, there is substantially more back donation into the CO bond weakening $2\pi^*$ orbitals. As a consequence of stretching the CO bond to its transition-state distance, only a small extension is required. Thirdly, in the transition state the oxygen and carbon atoms do not share bonding with the same surface metal atom. Such sharing is an important reason for enhanced barrier energies at terraces (due to the bond order conservation principle).

This is discussed more extensively in the next section.

Whereas the adsorption energies of the adsorbed molecules and fragment atoms only slightly change, the activation barriers at step sites are substantially reduced compared to those at the terrace. Different from activation of σ-type bonds, activation of π bonds at different sites proceeds through elementary reaction steps for which there is no relation between reaction energy and activation barrier. The activation barrier for the forward dissociation barrier as well as for the reverse recombination barrier is reduced for step-edge sites.

Interestingly, when the particle size of metal nanoparticles becomes less than 2 nm, terraces become so small that they cannot anymore support the presence of step-edge site metal atom configurations. This can be observed from Figure 1.15, which shows a cubo-octahedron just large enough to support a step-edge site.

Class III-type behavior is the consequence of this impossibility to create step-edge-type sites on smaller particles. Larger particles will also support the step-edge sites. Details may vary. Surface step directions can have a different orientation and so does the coordinative unsaturation of the atoms that participate in the ensemble of atoms that form the reactive center. This will enhance the activation barrier compared to that on the smaller clusters. Recombination as well as dissociation reactions of π molecular bonds will show Class III-type behavior.

The different BEP behavior for the activation of σ versus π bonds, basic to the very different Class I and Class II particle size dependence compared to Class III particle size dependence, is summarized in Figure 1.16 [14].

Whereas Class I and Class II behavior are intrinsically related through microscopic reversibility, Class III-type behavior implies that there is no BEP relation between the changes in activation energy and structure.

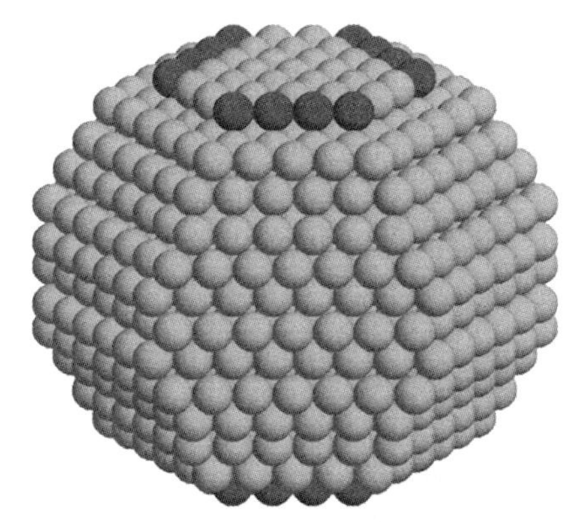

Figure 1.15 Cubo-octahedron with step-edge sites [18].

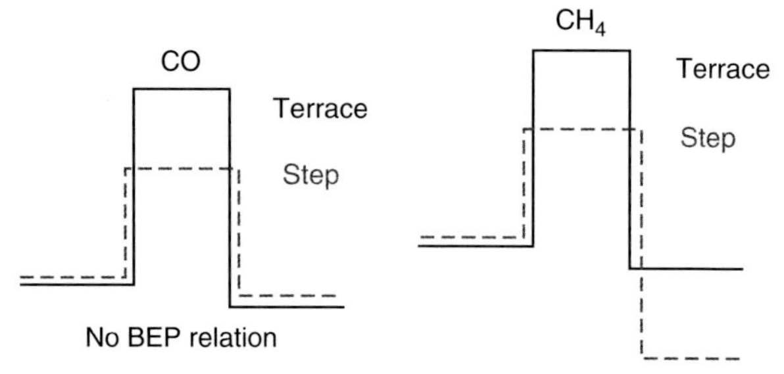

Figure 1.16 Changes in methane versus CO activation barriers as a function of surface structure (schematic).

When the selectivity of a reaction is controlled by differences in the way molecules are activated on different sites, the probability of the presence of different sites becomes important. An example again can be taken from the activation of CO. For methanation, activation of the CO bond is essential. This will proceed with low barriers at step-edge-type sites. If one is interested in the production of methanol, catalytic surfaces are preferred, which do not allow for easy CO dissociation. This will typically be the case for terrace sites. The selectivity of the reaction to produce methanol will then be given by an expression as in Eq. (1.29a):

$$S = \frac{x_1 r_1}{x_1 r_1 + x_2 r_2} \tag{1.29a}$$

$$\frac{r_1}{r_2} = \frac{r_{CH_3OH} \cdot \theta_{CO}}{k_{diss}\theta_{CO}(1-\theta)} \approx \frac{r_{CH_3OH}}{k_{diss}}\left(1 + K_{ads}^{CO}[CO]\right) \tag{1.29b}$$

In this expression, x_1 and x_2 are the fractions of terrace versus step-edge sites, r_1 is net rate of conversion of adsorbed CO to methanol on a terrace site, and r_2 is the rate of CO dissociation at a step-edge-type site. Increased CO pressure will also enhance the selectivity, because it will block dissociation of CO.

1.8 The Nonmetal Atom Sharing Rule of Low-Barrier Transition States

1.8.1 Introduction

As we discussed in the previous section, the primary parameter that determines the interaction strength between an adsorbate and a (transition) metal surface is the coordinative unsaturation of the surface metal atoms. The lower the coordination number of a surface atom, the larger the interaction with interacting adsorbates.

We discussed that for methane activation this leads to lowering of the activation energy compared to the reactivity of terrace, edge, or corner atoms successively.

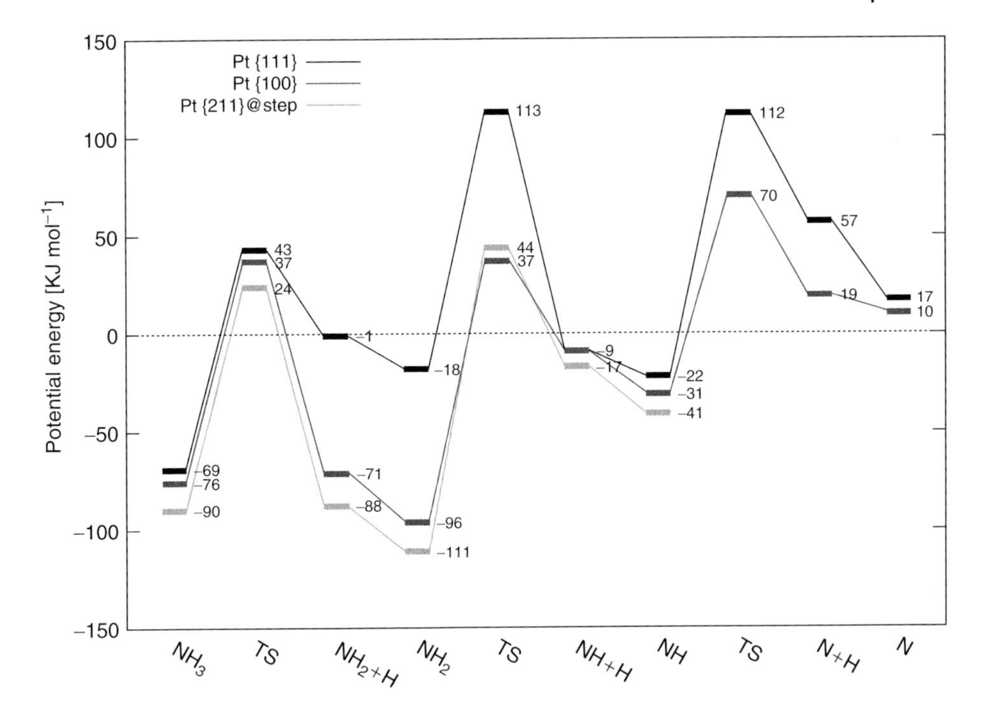

Figure 1.17 Reaction energy diagram of NH_3 activation compared on different surfaces (energies in kilojoules per mole).

Very important to this behavior is the cleavage of the CH bond over a single metal atom.

This trend in activation energy decrease with increasing degree of surface atom unsaturation is not general. It does not apply to the activation energies of NH_3 or H_2O activation, notwithstanding that again essentially only the interaction with a single surface metal atom is relevant and the bonds to be activated are also σ-type bonds.

The trend is illustrated for ammonia activation in Figure 1.17 [19]. In this figure, the activation energies of ammonia activation are compared for stepped and nonstepped surfaces of Pt. Similarly as also found for H_2O activation [20], the dissociation barrier is found to be invariant to surface structural changes. This is very different compared to the earlier discussed activation of methane that shows a very strong structural dependence.

It is due to the already significant surface atom–molecule interaction present in the case of ammonia and water before reaction, when the molecules adsorb strongly through their respective molecular lone-pair orbitals.

During the stretching of the XH bond, the molecule maintains its original coordination. Only when the XH bond is broken, the surface NH_2 or OH fragment may move to a higher coordination site.

Whereas now the bond cleavage reaction is nonsurface dependent, the reverse reaction clearly is. The stronger the NH_2 and NH fragments bind, the higher the barrier for the recombination reaction. In the case of methane activation we found the reverse situation. Both situations are consistent with microscopic reversibility.

Interestingly, this situation is very different when we consider activation of NH_3 or H_2O by coadsorbed O. This would typically occur in the Ostwald reaction that oxidizes ammonia to NO or the methane reforming reaction in which CH_4 reacts with O_2 or H_2O to give CO, CO_2, and H_2.

Ammonia activation by Pt, to be discussed in the next section, is an interesting example, because it illustrates the basic principle that provides chemical direction to the identification of surface topologies that give low reaction barriers in surface reactions. This holds specifically for elementary reactions that require a surface ensemble of atoms.

The next section introduces the topological concept of low-barrier transition states through the prevention of formation of shared bonds between reacting surface adsorbates and surface metal atoms.

1.8.2 Ammonia Oxidation

Figure 1.18 compares the activation energies of direct activation of NH_3 and its activation by coadsorbed O on Pt(111) [21]. As can be observed in this figure, reaction with coadsorbed O only lowers the barrier for NH_3 activation by coadsorbed O. The other NH_x intermediates have similar activation energies in the absence of coadsorbed oxygen.

The key difference between the activation energies of the NH_3–O interaction on the (111) surface and the interaction with the other NH_2 fragments is the different topology of the corresponding transition states. Since only NH_3 adsorbs atop, but oxygen requires higher coordination, only the transition state of NH_3–O is realized without binding to the same metal atom of the surface fragment nitrogen atom and coadsorbed O. Competitive adsorption to the same metal atom weakens the adsorbate bonds, and hence a repulsive interaction between reacting fragments arises. The essential chemical bonding feature on which this effect is based is bond order conservation, as discussed before in the context of the explanation of the increase in chemical reactivity with decreased surface atom coordination.

The difference in reactivity of adsorbed NH_x fragments with O as observed by a comparison of Figures 1.18 and 1.19 is striking. On the (100) surface, the activation of the NH_x fragments with x equal to 2 or 1 is also decreased when reacting with coadsorbed O.

As can be seen from Figure 1.20 [22], those transition states that do not share binding to the same surface metal atom have low barriers. The fcc(100) surface has the unique property that the reaction can occur through motion over the square hollow with bonds that remain directed toward the corner atoms of the square atom arrangement on the surface. This is a unique and important feature of reactions that require in their transition states interactions with several surface atoms.

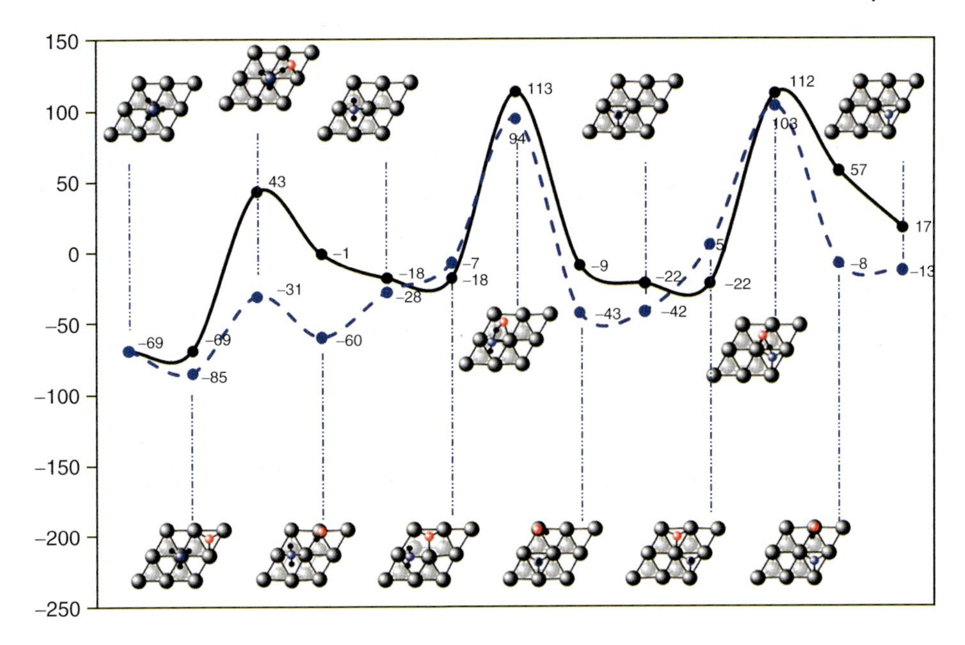

Figure 1.18 Reaction energy diagram that compares direct NH_3 activation and NH_3 (–) activation through reaction with adsorbed O (– – –) on Pt(111) and the corresponding structures for direct NH_3 activation (energies in kilojoules per mole).

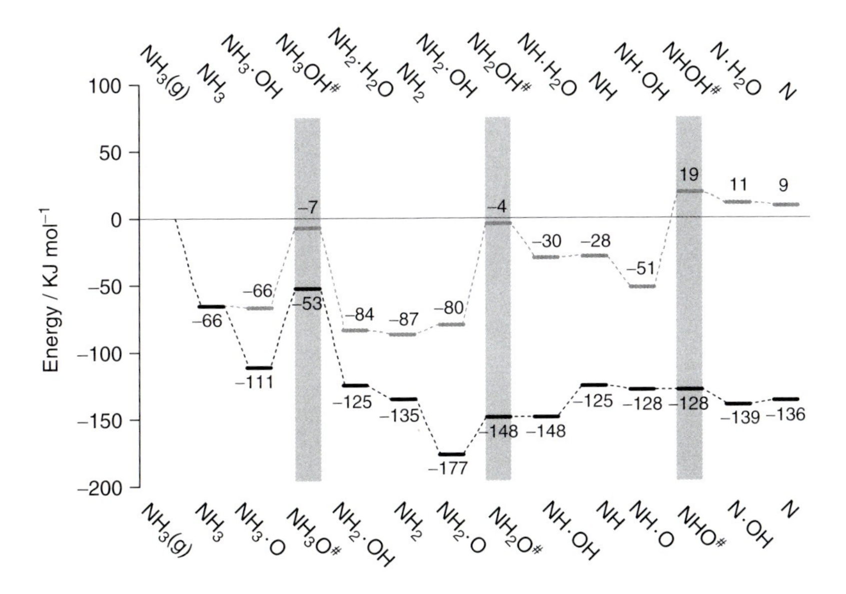

Figure 1.19 Reaction energy diagram of NH_3–O and NH_3–OH activation on Pt(100) [22].

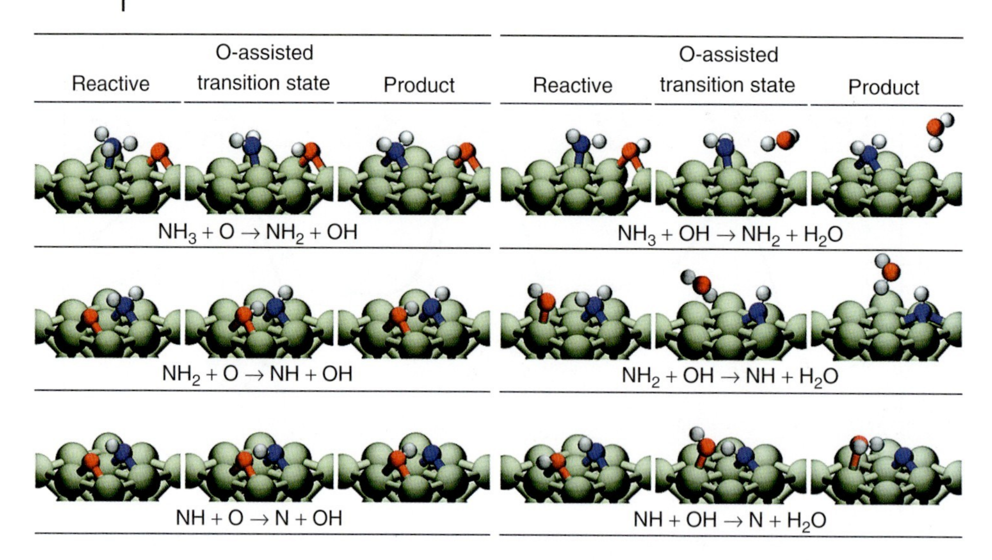

Figure 1.20 The structures of reaction intermediates that correspond to Figure 1.19 [22].

This is characteristic for the activation of molecular π bonds. The same principle was found to apply for stepped surfaces. It is also characteristic of other reactions with surface fragments that have strong repulsive interactions when they share binding with the same surface atom, for example, NH and O, which typically prefer bonding in high-coordination sites.

The relevance of the same nonsurface metal atom sharing principle in transition states is nicely illustrated by the similar lowering of the transition state for NH activation by O in a step site as for the (100) surface, as illustrated in Figure 1.21 [19]. Similarly, OH formation by recombination of oxygen and hydrogen is substantially lower at a step edge than on the (111) terrace.

The kinetics of H–O recombination is very important in the reforming reaction of methane to produce CO and H_2. When more weakly bonded O_{ads} recombines with H_{ads} (preferred on Pt), the main product next to CO will be H_2. On planar Rh with a stronger M–O bond interaction, this reaction is suppressed and therefore H_2 is the main product [23]. Clearly this selectivity will be dramatically affected by the presence of surface steps.

Figure 1.22 schematically summarizes the principle of the preferred transition states without sharing of a common metal atom. Whereas we have earlier discussed surface sensitivity as a function of the relative ratio of particle surface edge sites and surface terrace atoms, the discussion given above provides a principle for particle size shape differences.

Particles of face centered cubic (FCC) crystal would be exclusively terminated by (100) surfaces, whereas cubo-octahedron-type particles may have a dominance of the more stable (111) surfaces.

As first noted by Neurock *et al.* [24], the Pt(100) surface provides sites for extremely low barriers of NO and N_2 recombination. For NO, the energetics on

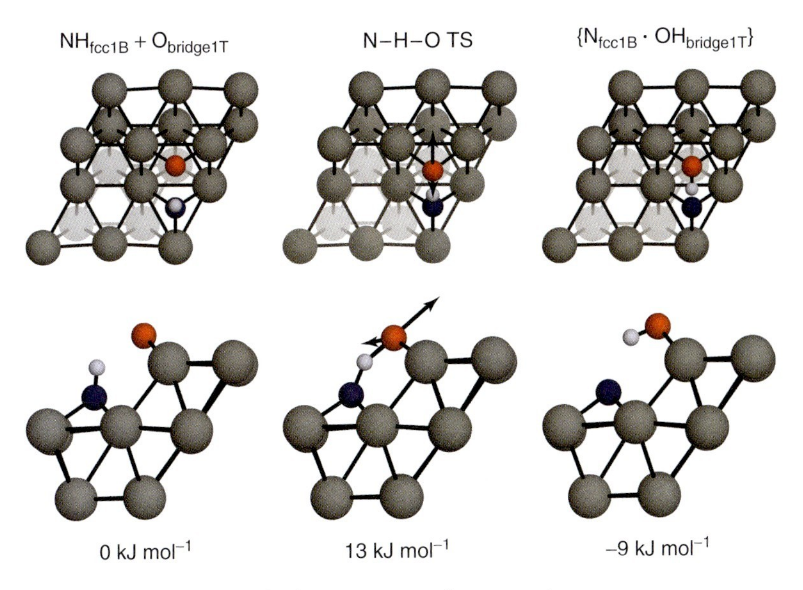

Figure 1.21 Structures and relative energies of NH_{ads} and O_{ads} reaction intermediates along {211}Pt step.

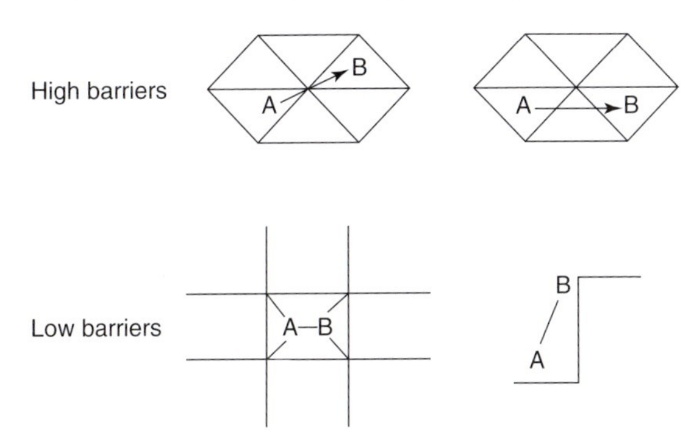

Figure 1.22 Structures of high-barrier and low-barrier transition states of surface bond cleavage reactions.

different surfaces is compared in Figure 1.23. The barriers for recombination to give NO are low on (100) surfaces but higher on the stepped as well as nonstepped (111) surfaces. The lower barrier on the (100) surface compared to the stepped (111) surface is due to destabilization of N_{ads} on the (111) compared to the (100) surface. N_{ads} prefers triangular coordination.

The adsorption energy of N_2 is also low, but that of NO on the (100) surface is substantial. Notwithstanding the very similar activation energies for N_2 and NO formation (see Tables 1.4 and 1.5), the strong interaction of NO with both surfaces implies that the selectivity of the reaction toward N_2 will be high at low temperatures. The NO once formed will not desorb and can only be removed as N_2O.

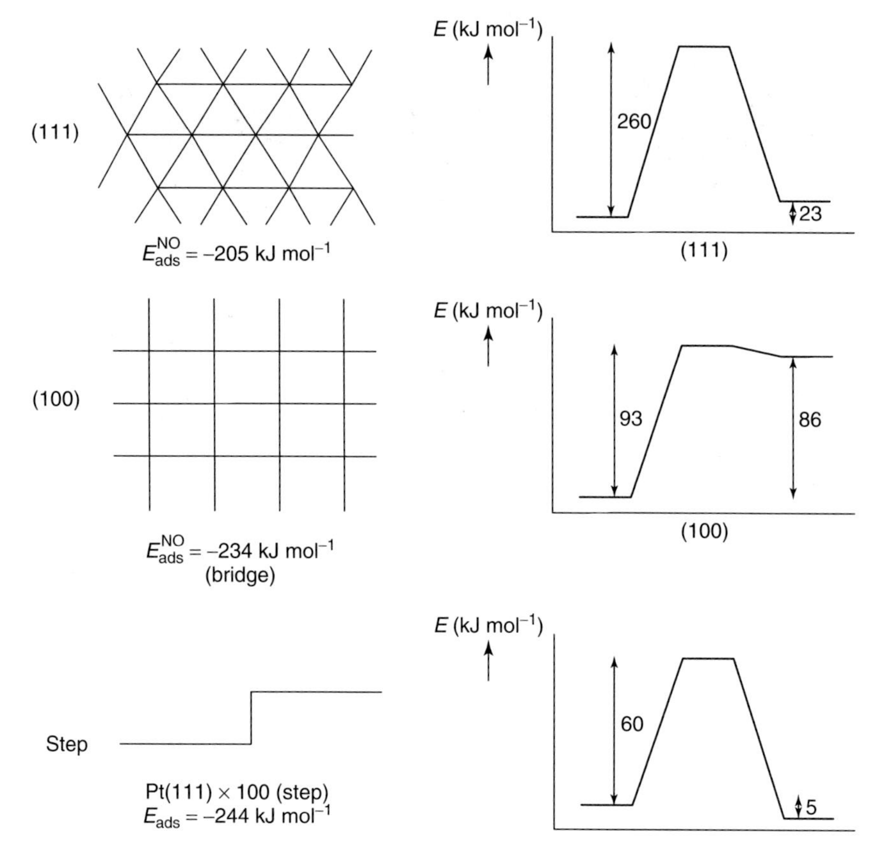

Figure 1.23 A comparison of NO activation on the (111), (100), and stepped (111) surface.

Table 1.4 The activation energies on the Pt(100) surface.

	E_{act} (kJ mol^{-1})
$N + N \rightarrow N_{2gas}$	36
$N + O \rightarrow NO_{ads}$	20
$NO_{ads} \rightarrow NO_{gas}$	204
$N_a + NO_a \rightarrow N_2O_{gas}$	67

At low temperatures, reaction towards N_2 and N_2O product formation preferentially occurs at the (100) surface, and hence a significant particle shape sensitivity is predicted. At higher temperatures when NO readily desorbs, overall activation barriers on the different surfaces tend to become similar and hence surface sensitivity becomes less. The high selectivity toward NO at higher temperatures relates

Table 1.5 The activation energies on the Pt (111) surface.

	E_{act} (kJ mol^{-1})
$N + N \rightarrow N_{2gas}$: Pt(111)	148
$N + O \rightarrow NO_{ads}$: Pt(111)	189
$NO_{ads} \rightarrow NO_{gas}$	187

to the rapid reaction of N_{ads} with coadsorbed O, whose coverage dominates and hence competitive N_2 formation has a slow rate.

1.9 Summary

Using microkinetic expressions, we have discussed the most important catalytic concepts that describe heterogeneous catalytic reactions. We have related these concepts with the energies, entropies, and transition-state features that are accessible through current state-of-the-art DFT techniques.

Whereas it is very useful to relate reaction mechanistic proposals with catalytic kinetics, one has to be aware that DFT-predicted energies typically have an error of at least 10 kJ mol^{-1}.

Predictive kinetics requires accuracies that are an order of magnitude more precise. There are many examples that predict overall kinetics quite accurately. This is then due to a fortuitous cancellation of errors that needs to be understood well for each case.

We did not extensively discuss the consequences of lateral interactions of surface species adsorbed in adsorption overlayers. They lead to changes in the effective activation energies mainly because of consequences to the interaction energies in coadsorbed pretransition states. At lower temperatures, it can also lead to surface overlayer pattern formation due to phase separation. Such effects cannot be captured by mean-field statistical methods such as the microkinetics approaches but require treatment by dynamic Monte Carlo techniques as discussed in [25].

References

1. Evans, M.G. and Polanyi, M. (1938) *Trans. Faraday Soc.*, **34**, 11.
2. Dahl, S., Logadottir, A., Jacobsen, C.J.H., and Nørskov, J.K. (2001) *Appl. Catal. A Gen.*, **222**, 19.
3. Shetty, S.G., Jansen, A.P.J., and van Santen, R.A. (2008) *J. Phys. Chem. C*, **112**, 14027.
4. Ciobica, I.M., Fréchard, F., van Santen, R.A., Kleijn, A.W., and Hafner, J. (2000) *J. Phys. Chem. B*, **104**, 3364.
5. van Santen, R.A. and Neurock, M. (2006) *Molecular Heterogeneous Catalysis*, Wiley-VCH Verlag GmbH, Weinheim.
6. (a) Bligaard, T., Nørskov, J.K., Dahl, S., Matthiesen, J., Christensen, C.M., and

Sehested, J. (2004) *J. Catal.*, **222**, 206; (b) Kaszetelan, S. (1992) *Appl. Catal. A Gen.*, **83**, L1.

7. Bond, G.C., Keane, M.A., Kral, H., and Lercher, J.A. (2000) *Catal. Rev.-Sci. Eng.*, **42** (3), 323.
8. (a) Bligaard, T., Honkala, K., Logadottir, A., Nørskov, J.R., Dahl, S., and Jacobson, C.J.H. (2007) *J. Phys. Chem. B*, **107**, 9325; (b) Toulhoat, H. and Raybaud, P. (2003) *J. Catal.*, **216**, 63.
9. (a) Rozanka, X. and van Santen, R.A. (2004) in *Computer Modelling of Microporous Materials*, Chapter 9 (eds C.R.A. Catlow, R.A. van Santen, and B. Smit), Elsevier; (b) Smit, B. (2004) in *Computer Modelling of Microporous Materials*, Chapter 2 (eds C.R.A.Catlow, R.A. van Santen, and B. Smit), Elsevier.
10. Vos, A.M., Rozanska, X., Schoonheydt, R.A., van Santen, R.A., Hutschka, F., and Hafner, J. (2001) *J. Am. Chem. Soc.*, **123**, 2799.
11. Smit, B. and Maesen, T.L.M. (2008) *Nature*, **451**, 671.
12. (a) Haag, W.O. (1994), in *Zeolites and Related Microporous Materials: State of the Art 1994 Chapter V* (eds J. Weitkamp, H.G. Karge, H. Pfeifer, and W. Hölderick), Elsevier, p. 1375; (b) Narbeshuber, Th.F., Vinety, H., and Lercher, J. (1995) *J. Catal.*, **157**, 338.
13. de Gauw, F.J.J.M.M., van Grondelle, J., and van Santen, R.A. (2002) *J. Catal.*, **206**, 295.
14. van Santen, R.A. (2008) *Acc. Chem. Res.*, **42**, 57.
15. Shustorovich, E. (1990) *Adv. Catal.*, **37**, 101.
16. Ciobica, I.M. and van Santen, R.A. (2002) *J. Phys. Chem. B*, **106**, 6200.
17. Liu, Z.-P. and Hu, P. (2003) *J. Am. Chem. Soc.*, **125**, 1958.
18. Honkala, K., Hellman, A., Remediakis, I.N., Logadottir, A., Carlsson, A., Dahl, S., Christensen, C.H., and Nørskov, J.K. (2005) *Science*, **307**, 555.
19. Offermans, W.K., Jansen, A.P.J., van Santen, R.A., Novell-Leruth, G., Ricart, J., and Pérez-Ramirez, J. (2007) *J. Phys. Chem. B*, **111**, 17551.
20. van Grootel, P.W., Hensen, E.J.M., and van Santen, R.A., submitted.
21. (a) Offermans, W.K., Jansen, A.P.J., and van Santen, R.A. (2006) *Surf. Sci.*, **600**, 1714; (b) Baerns, M., Imbihl, R., Kondratenko, V.A., Kraehert, R., Offermans, W.K., van Santen, R.A., and Scheibe, A. (2005) *J. Catal.*, **232**, 226; (c) van Santen, R.A., Offermans, W.K., Ricart, J.M., Novell-Leruth, G., and Pérez-Ramirez, J. (2008) *J. Phys.: Conf. Ser.*, **117**, 012028.
22. Novell-Leruth, G., Ricart, J.M., and Pérez-Ramírez, J. (2008) *J. Phys. Chem. C*, **112**, 13554.
23. Hickman, D.A. and Schmidt, L. (1993) *Science*, **249**, 343.
24. Ge, Q. and Neurock, M. (2004) *J. Am. Chem. Soc.*, **126**, 1551.
25. Gelten, R.J., van Santen, R.A., and Jansen, A.P.J. (1999) in *Molecular Dynamics, from Classical to Quantum Methods*, Chapter 18 (eds P.B. Balbuena and J.M. Seminario), Elsevier Science B.V.

2
Hierarchical Porous Zeolites by Demetallation

Johan C. Groen and Javier Pérez-Ramírez

2.1
Zeolites and Catalyst Effectiveness

The introduction of the first synthetic zeolites in the 1950s and the discovery of high-silica zeolites in the 1970s brought a paradigm shift in the field of porous materials. Zeolites are crystalline silicates and aluminosilicates with a unique combination of properties like high surface area, well-defined microporosity, high (hydro)thermal stability, intrinsic acidity, and the ability to confine active metal species. The multidimensional network of pores with molecular dimensions can serve as "microreactors" due to confinement effects in which the activity and selectivity can be enhanced in the presence of active species. As a consequence, zeolites are frequently applied in catalytic conversions where acidity and shape selectivity are required and, although these materials are applied in many catalytic and separation processes, future challenges exist to further tailor and customize the properties of zeolites [1]. An important aspect where progress can be expected concerns the purely microporous character of zeolites that frequently results in active sites that are difficult to access and adversely impacts on the molecular transport of reactants and products. A more efficient performance of the zeolitic catalyst can be envisioned upon enhanced accessibility to the active species and/or shortening of the diffusion path length in the micropores.

In classical chemical engineering, the degree of catalyst utilization is described by the effectiveness factor (Figure 2.1). Full utilization of the catalyst particle ($\eta \to 1$) represents a situation in which the observed reaction rate equals the intrinsic reaction rate due to operation in the chemical regime, that is, free of any diffusion constraints. In terms of intraparticle transport, this is attained at low values of the Thiele modulus ($\phi \to 0$). Contrarily, $\phi = 10$ renders $\eta = 0.1$, meaning that only 10% of the catalyst volume is effectively used in the reaction. Transport limitations negatively impact not only on activity but frequently also on selectivity and stability (lifetime), that is, the three distinctive features of any catalyst. Since the intrinsic rate coefficient k_v is fixed for a given reaction and zeolite, keeping the Thiele modulus small implies the practice of two basic strategies: shortening

Novel Concepts in Catalysis and Chemical Reactors: Improving the Efficiency for the Future.
Edited by Andrzej Cybulski, Jacob A. Moulijn, and Andrzej Stankiewicz

ISBN: 978-3-527-32469-9

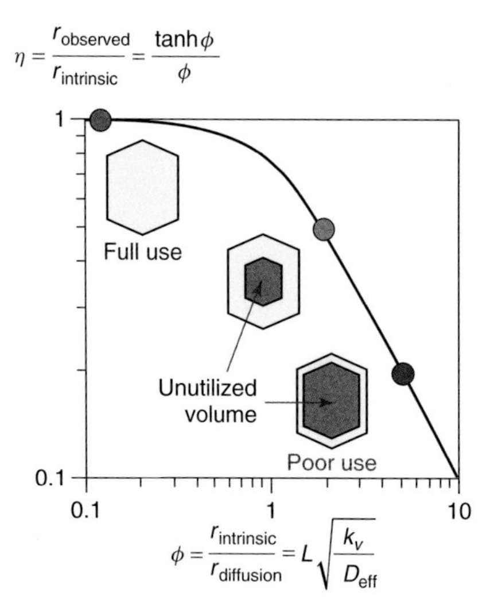

Figure 2.1 Dependence of the effectiveness factor on the Thiele modulus for a first-order irreversible reaction. Steady-state diffusion and reaction, slab model, and isothermal conditions are assumed.

the diffusion length L and/or enhancing the effective diffusivity D_{eff} in the zeolite pores. In the latter line of thinking, ordered mesoporous materials (OMMs) with regular pores in the typical range of 2–15 nm were intensively developed during the 1990s [2–6]. MCM-41 is prototypical in this category, displaying one-dimensional ordered arrays of nonintersecting hexagonal channels with controlled size in the range of 2–10 nm. The diffusion regime in mesoporous catalysts is typically bulk or Knudsen diffusion and this leads to diffusivities several orders of magnitude higher than in micropores, which often display an activated (configurational) diffusion mechanism. However, the a priori high optimism for moving the zeolite catalysis to the mesoscale by using OMMs has so far not crystallized in industrial applications due to the limited success in mimicking the unique functionalities of zeolites. Consequently, research has primarily aimed at effective chemical modification of the amorphous walls in OMMs by, for example, grafting [7] or crystallization [8, 9] to generate active sites equivalent to those in zeolites.

From this key learning, the scientific community started to look for alternative strategies that would lead to improved accessibility of the active sites confined in zeolites. Two fundamentally different approaches can be adopted: (i) increasing the width of the micropores or (ii) shortening the micropore diffusion path length. For several decades, researchers have pursued the preparation of new "large-cavity" and "wide-pore" zeolites (up to 1.25 nm), containing rings of 12 or more T-atoms. Most of these low-framework density structures, among many others VPI-5 [10], UTD-1 [11], and ECR-34 [12], suffer from similar problems as OMMs, that is, low thermal stability, low acidity, and unidirectional pore systems. Recently, wide-pore zeolites with multidirectional channels have indeed been obtained, for example, ITQ-15 [13], ITQ-21 [14], and ITQ-33 [15].

For a given zeolite framework, the basic strategy to change the diffusion path length is that of altering crystal size and morphology using particular crystallization conditions. With the aim of shortening the diffusion path lengths in micropores of existing zeolites, "hierarchical" systems have been developed and they are attracting rapidly growing attention. Broadly speaking, materials with structural hierarchy exhibit structure on more than one length scale [16]. Hierarchical porous materials integrate multiple levels of porosity. In zeolites, this can be attained by decreasing the crystal size or by introducing a secondary (meso)pore system within an individual zeolite crystal. Importantly, for a material to be entitled hierarchical, it is required that each level of porosity has a distinct function; the functionality is the differentiating feature with respect to a disordered porous material.

2.2 Hierarchical Zeolites

Generally speaking, hierarchical porous solids can be characterized by the number of porosity levels in the material and their individual geometry. The prime aim of hierarchical zeolites is that of coupling the catalytic features of micropores in a single material and the improved access and transport consequence of additional pores of larger size. However, the connectivity between the various levels of pores is vital to maximize the benefits of hierarchy in catalyzed reactions. Interconnected hierarchy refers to the network of voids generated in the intercrystalline space by fragmentation of the microporous crystal (Figure 2.2a) into nanocrystals (Figure 2.2b). Intraconnected makes reference to the occurrence of mesopores

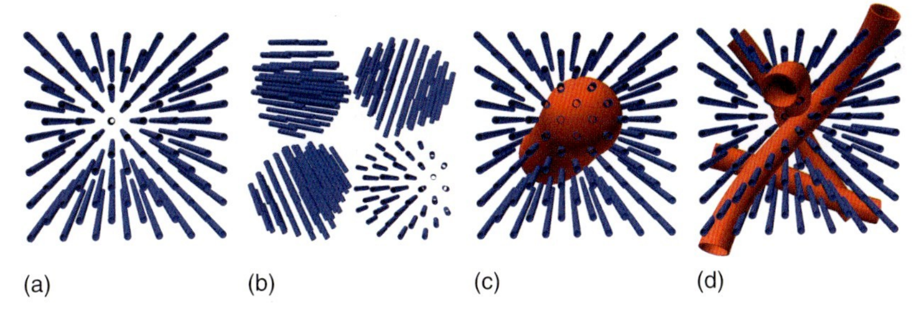

Figure 2.2 Classification of different types of porous materials. (a) A purely microporous zeolite is considered as a non-hierarchical system according to the single level of porosity. (b) Fragmentation of the zeolite into nanocrystals engenders a network of mesopores constituting the intercrystalline space, leading to an interconnected hierarchical system. Intraconnected hierarchical systems are shown in (c) and (d). In these schemes, micropores and larger pores coexist within the zeolite crystal. Two extreme cases can be devised in this category: (c) mesovoids that are occluded in the microporous matrix and (d) mesopores that can be entered from the external surface of the zeolite crystals.

in the microporous crystal (Figure 2.2c–d). The schematic representations in Figure 2.2b, c and d shorten in a similar way the length of the micropores with respect to Figure 2.2a as a result of smaller crystals or intracrystalline voids. A shorter diffusion length is necessary to increase the catalyst effectiveness, but it is not a sufficient condition. For example, the system in Figure 2.2d, which could well represent a hollow zeolite crystal, is transport-wise ineffective. This is due to the fact that the mesovoids are entrapped in the microporous matrix and thus only accessible via the micropores. On the other hand, the mesopores in Figure 2.2c are directly accessible from the outer surface of the zeolite crystal, similar to the intercrystalline space in nanocrystals (Figure 2.2b). In the latter two cases, the condition that mesopores enhance the molecular transport to/from the active sites in the micropores has been satisfied. Thus, introducing mesopores in zeolites could be infertile for the application if not properly located in the crystal. Consequently, engineering hierarchical materials in general and zeolites in particular requires a careful design aimed not only at extensively generating large pores but also principally at locating them in harmony with the micropores.

Different synthetic methodologies can be pursued to prepare hierarchical porous zeolites, which can be discriminated as bottom-up and top-down approaches. Whereas bottom-up approaches frequently make use of additional templates, top-down routes employ preformed zeolites that are modified by preferential extraction of one constituent via a postsynthesis treatment. For the sake of conciseness, we restrict ourselves here to the discussion of the latter route. Regarding bottom-up approaches, recently published reviews provide state-of-the-art information on these methodologies [8, 9, 17–19].

2.3
Mesoporous Zeolites by Demetallation

By demetallation, one constituent is preferentially extracted from a preformed zeolite material to form mesoporous zeolite crystals. Existing and emerging demetallation strategies basically comprise dealumination, detitanation, and desilication.

The traditional method for introducing intracrystalline pores in zeolites is by dealumination, which involves preferential extraction of aluminum from zeolite crystals by steaming, acid leaching, or chemical treatment [20, 21]. Almost all the work on dealumination has been performed on zeolite Y and mordenite, that is, low-silica zeolite, although ZSM-5, beta, ferrierite, omega, and mazzite have been investigated too. A crucial aspect of mesoporous zeolites prepared by dealumination stems from the type of porosity and their actual benefits in alleviating mass-transfer constraints due to the shortened diffusion pathlength. de Jong *et al.* [22] conducted seminal work on the characterization of mesopores in dealuminated Y and mordenite zeolites by electron tomography, a form of 3D-TEM (Figure 2.3). This technique provides valuable information on the three-dimensional shape and

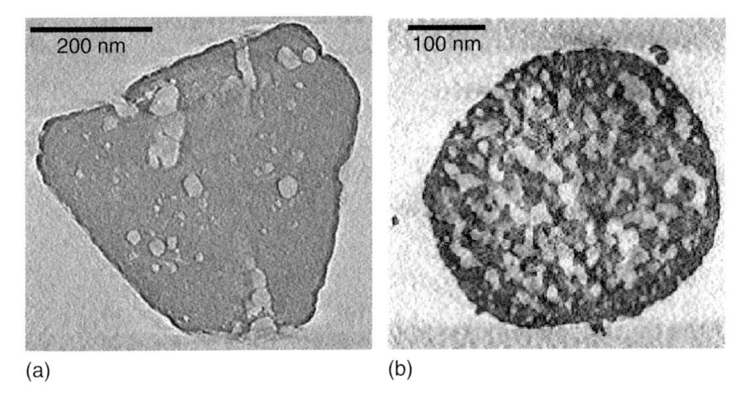

Figure 2.3 3D-TEM reconstruction of (a) severely steamed and subsequently acid-leached Y zeolite [22] and (b) desilicated ZSM-5 zeolite crystal [24]. The mesopores in the crystal are shown as lighter gray tones.

connectivity of the mesopores. Surprisingly, most of the mesopores were present as cavities rather than as cylindrical pores connecting the external surface with the interior of the crystallite, similar to the model in Figure 2.2d. In the specific case of Y zeolite, the authors closed the paper with the statement: "However, the shape of the mesopores also raises the question what extent the accessibility and diffusion are enhanced by the formation of these cavities" [22]. Four years later, Kortunov *et al.* [23] confirmed the concerns of de Jong *et al.* Diffusion studies by the pulsed field gradient (PFG) NMR technique concluded that the mesopores formed by the steaming of Y zeolite are essentially of no influence for the intracrystalline diffusion of probe molecules (*n*-octane and 1,3,5-triisopropylbenzene). This result can be understood by assuming that mesopores in USY crystals do not form a connected (i.e., percolation) network allowing diffusion of guest molecules through the crystals via only mesopores. Although not addressed explicitly in the literature, the generation of mesopores by removal of other framework trivalent cations such as Fe, Ga, and B is expected to follow parallels with aluminum extraction with respect to extent and type of mesoporosity.

Besides extraction of trivalent cations from the zeolite lattice, tetravalent cations can also be selectively dislodged. Recently, treatment of ETS-10, a titanosilicate material, in aqueous H_2O_2 solution induced structural defects due to partial removal of Ti atoms, resulting in the interruption of titania chains and the consequent formation of larger micropores (supermicropores) without substantial degradation of crystallinity [25]. Application of microwaves resulted in the formation of larger mesopores [26]. However, the number of materials exhibiting both silicon and titanium in the zeolite lattice are limited, which makes this treatment limitedly applicable. In contrast, selective extraction of silicon would enable extrapolation of such methodology to a variety of zeolite topologies, even of different chemical composition. This particular treatment is extensively discussed in the next section.

2.4
Desilication

2.4.1
Introduction

The potential of desilication, that is, dissolution of silicon by hydrolysis in aqueous medium (Figure 2.4), for controlled porosity development was unrecognized for a long time. Despite numerous works available on (partial) dissolution of silicon from amorphous materials, only in the late 1980s information was reported on the extraction of silicon from crystalline zeolites. Dissolution studies of silicalite-1, the purely siliceous form of zeolite ZSM-5, with concentrated HF and NaOH solutions have revealed a progressive dissolution of the zeolite crystals [27]. As aluminum is amphoteric and can be dissolved both at low and high pH, it can be anticipated that a similar reasoning would apply for aluminosilicate zeolites. However, Dessau *et al.* [28] reported, in the early 1990s, an excessive but anisotropic dissolution of ZSM-5 zeolite crystals upon treatment in hot alkaline Na_2CO_3 solution. That observation has been speculatively attributed to the presence of an aluminum gradient in the zeolite crystals, although no conclusive evidence was given for this hypothesis. In separate studies [29, 30], careful physicochemical characterization of large ZSM-5 crystals synthesized with TPA^+ ions as template has revealed that the outer surface is often rich in aluminum while the center is Al-poor. The presence of aluminum in the zeolite framework appears to be crucial in the desilication process in alkaline medium. A distinct role of framework aluminum on the rate of dissolution of MFI zeolites was also observed in 1997 by Čižmek *et al.* [31] They found a preferential silicon dissolution in ZSM-5 upon treatment in 5 M NaOH; the concentration of extracted silicon was at least 1000 times higher than the concentration of extracted aluminum, and the overall rate of dissolution increased dramatically as the framework aluminum content decreased. In all these earlier works, a peculiar role of aluminum was embedded; however, no distinct attention was paid to this phenomenon with regard to the structural, morphological, and textural changes of the treated materials. Can controlled silicon extraction be obtained and can this

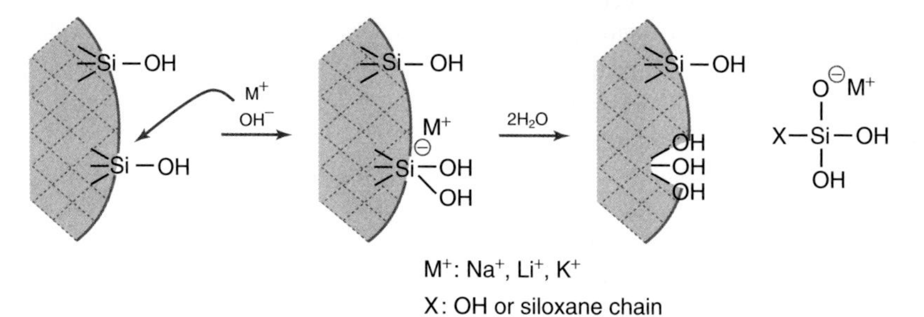

Figure 2.4 Schematic representation of silicon hydrolysis from the zeolite framework in aqueous alkali hydroxide solution.

approach be used to generate mesoporosity in the microporous matrix in order to prepare hierarchical zeolites?

The first paper highlighting the presence of mesopores in microporous ZSM-5 zeolites by framework silicon extraction in alkaline medium (0.2 M NaOH) appeared in the year 2000 by the group of Matsukata [32]. Although the newly obtained mesoporosity was initially attributed to intercrystalline pores by dissolution of crystal boundaries, subsequent systematic studies by Groen *et al.* [33, 34] over ZSM-5 confirmed that controlled desilication mainly induces intracrystalline mesoporosity while preserving most of the original microporosity. A combinatorial-type program was conducted to elucidate the role of both treatment variables such as time, temperature, and stirring speed and material-related parameters like framework Si/Al ratio, crystal size, and different framework types [35, 36]. The most striking result of the screening activities was the key role of the framework Si/Al ratio that highly determines the alkaline treatment's chance of success for controlled mesoporosity development. The alkaline-assisted hydrolysis of the Si–O–Si bonds from the zeolite framework can be directed toward mesoporosity formation when operating in an appropriate window of Si/Al ratios.

2.4.2
Role of the Trivalent Cation

2.4.2.1 Framework Aluminum

A remarkable evolution in mesopore surface area occurs upon systematic treatment of a series of commercial ZSM-5 zeolites of a wide range of framework Si/Al ratios (15–1000) in 0.2 M NaOH solutions. Upon alkaline treatment of zeolites with a low Si/Al ratio, hardly any mesoporosity ($S_{meso} < 20\ m^2\ g^{-1}$) is obtained and a limited extraction of silicon and aluminum is present in the filtrate. In the Si/Al range of 25–50, a greatly enhanced mesopore surface area, which exceeds $200\ m^2\ g^{-1}$, has been achieved, and the amount of silicon extracted is seven times higher than in for ZSM-5 zeolite with Si/Al = 15, whereas the concentration of aluminum in the filtrate still remains very low. Clearly, the extraction of framework silicon is highly favorable over that of aluminum. Surprisingly, at higher Si/Al ratios (Si/Al > 50), a less-pronounced increase in mesoporosity ($S_{meso} < 100\ m^2\ g^{-1}$) is obtained, despite the substantial silicon extraction that even goes beyond that of the samples with maximum mesopore surface area in the Si/Al range of 25–50. This has been mechanistically explained by the pore-directing role of framework aluminum in the desilication process [36].

The presence of tetrahedrally coordinated aluminum regulates the process of Si extraction and mechanism of mesopore formation as schematically presented in Figure 2.5. As a result of the negatively charged AlO_4^- tetrahedra, hydrolysis of the Si–O–Al bond in the presence of OH^- is hindered compared with the relatively easy cleavage of the Si–O–Si bond in the absence of neighboring Al [31, 37]. Materials with a relatively high density of framework Al sites (low Si/Al ratio) are relatively inert to Si extraction, as most of the Si atoms are stabilized by the nearby AlO_4^- tetrahedra. Consequently, these materials show a relatively low degree of Si

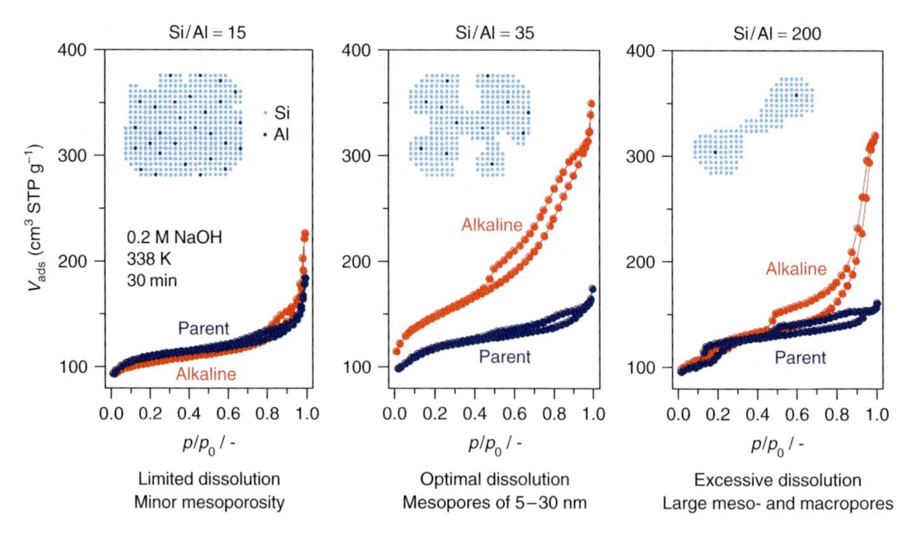

Figure 2.5 N_2 adsorption isotherms and schematized silicon dissolution (inset) upon alkaline treatment of ZSM-5 zeolites with different framework Si/Al ratios, highlighting the crucial role of framework aluminum.

dissolution and limited mesopore formation. On the contrary, the high density of Si atoms in zeolites with a high Si/Al ratio (low Al content) leads to substantial Si extraction and porosity development. The question is whether formation of these large pores is desirable, since pores in the lower nanometer size range will already provide adequate transport characteristics to and from the active sites. Moreover, formation of larger pores requires excessive extraction of valuable zeolite material. An intermediate framework Al content, equivalent to a molar Si/Al ratio in the range 25–50, is optimal and leads to a relatively high degree of selective Si dissolution from which well-controlled mesopores originate. Formation of substantial mesoporosity in the nanometer size range requires the dissolution of a significant volume of the zeolite framework and consequently should be accompanied by the removal of both framework Si and framework Al. However, only a small fraction of the expected Al is typically measured in the filtrate after alkaline treatment [32, 35]. This suggests that not all the Al removed from the framework during the alkaline treatment remains in the liquid phase, but is somehow reincorporated in the solid. Particularly when the pH decreases due to consumption of OH^- ions during the alkaline treatment, the solubility of Al decreases and deposition/reinsertion is promoted [38, 39]. In fact, various reports [40, 41] have shown that the majority of Al is in framework positions after alkaline treatment, implying that the Brønsted acidic properties are mostly maintained in the desilicated zeolites, which is highly favorable for subsequent catalytic applications requiring these types of active sites.

2.4.2.2 **Nonframework Aluminum**

Not only the concentration but also the nature of the aluminum and its location (next section) plays a role in desilication by alkaline treatment. The presence of a distinct concentration of extra-framework Al species in steamed samples has been shown to suppress silicon extraction and associated extra-porosity development upon alkaline treatment of such zeolites. In line with the reasoning of realumination in the previous section, the dislodged extra-framework aluminum species (partially) realuminate during the alkaline treatment, which leads to a locally high concentration of framework aluminum. This high concentration of realuminated species in turn suppresses extraction of neighboring silicon species. The concept of realumination of zeolites in alkaline medium has also been proposed by Lietz *et al.* [42], who observed the disappearance of the IR absorption band at 3660 cm^{-1}, which is characteristic of extra-framework aluminum species, in steamed zeolites after treatment in NaOH. In addition, the absorption band at 3610 cm^{-1} associated with Brønsted acid sites, that is, framework Al, was recovered. Selective removal of the nonframework species in steamed zeolites by mild treatment in oxalic acid recovers its proactivity for mesopore formation upon desilication [34]. The so-obtained zeolite presents larger mesopores due to its lower framework Si/Al ratio. These observations enable to conclude that both the concentration and the nature of the trivalent aluminum cations play an important role in the desilication process.

2.4.2.3 Distribution of Aluminum in the Framework

The synthetic protocol of zeolites determines the incorporation of the cations in the zeolite lattice during crystallization. It has been reported that, for example, the use of TPAOH as structure-directing agent for synthesis of MFI-type zeolites induces an anisotropic incorporation of aluminum in the zeolite framework, especially in large crystals [29, 43]. Taking into account the crucial role of framework aluminum, it can be expected that a heterogeneous framework Al incorporation will also induce a nonuniform extraction of silicon during alkaline treatment. Indeed, recently, SEM-EDX investigations on uniformly sized large ZSM-5 crystals have convincingly shown that particularly the regions rich in aluminum are mostly preserved upon treatment in NaOH solution, while the Al-poor regions are extensively dissolved [24]. Point analysis revealed an up to 30 times higher concentration of aluminum at the outer surface as compared to the center of the crystal, the latter being extensively dissolved upon treatment in NaOH solution. This is in line with previous observations by Dessau and coworkers [28] who reported a preferential destruction of the crystals interior upon treatment in alkaline Na_2CO_3 solution (viz. Figure 2.2d). On the contrary, if a different template, in this case 1,6-hexanediol, is used for the synthesis of the same zeolite topology (MFI), a uniform incorporation of aluminum can be achieved [30, 44]. The hierarchical system obtained upon alkaline treatment of such crystals resembles the one in Figure 2.2c with accessible mesopores that effectively improve the overall accessibility.

2.4.2.4 Different Zeolite Framework Topologies

Most of the papers available on desilication are devoted to the alkaline treatment of ZSM-5 zeolites. Since a large number of zeolites consist of aluminosilicate frameworks and framework Al plays a crucial role in the mesoporosity development during desilication, this methodology should be suitable for extrapolation to other zeolite structures. Besides MFI zeolites, successful application of this methodology has been reported for mordenite [45], beta [41], and ZSM-12 [46], where in all cases the efficiency for mesoporosity development was optimal in the Si/Al range of 20–50, analogously to ZSM zeolites. The stability of Al in the zeolite framework turns out to be of additional importance. The relative low stability of the BEA framework leads to a more difficult control of the porosity development and preservation of the intrinsic zeolite parameters [41].

2.4.2.5 Framework Trivalent Cations Different from Aluminum

So far, hierarchical zeolite crystals made by desilication and other routes such as supramolecular and solid templating are mostly restricted to pure silica and aluminosilicate frameworks [17, 19]. Preparation of hierarchical zeolites with trivalent heteroatoms other than aluminum, such as gallium, iron, or boron, will increase the impact of these materials in catalysis. The particular acidic properties associated with Ga, B, and Fe species in framework or extra-framework positions of the zeolite catalyze a number of reactions involving hydrocarbons, such as aromatization, cracking, dehydrogenation, and oxidation [47–50]. Besides, processing of bulkier molecules will then be no longer only restricted to Ga (or B or Fe)-containing amorphous mesostructured materials [51, 52] but can also be executed by these, typically more stable, crystalline mesoporous zeolites. From the previous section it is clear that lattice aluminum regulates the alkaline-assisted hydrolysis of silicon, which is essential to fabricate mesoporous aluminosilicate zeolites. Recently, the scope of desilication to synthesize highly mesoporous zeolite crystals has been expanded to metallosilicate zeolites with varying trivalent framework cations [53]. An important observation is that the pore-inducing role of the framework trivalent cation in the development of mesoporous zeolites by selective silicon extraction can be extrapolated to Ga, B, and Fe. Apart from the prerequisite that a trivalent metal should be present in the zeolite framework in order to generate extra-porosity development, its distribution in the crystal determines the spatial location of the newly generated porosity.

2.4.3
Accessibility and Transport

The aim of the newly introduced mesoporosity is to enhance the utilization of the microporous network by improved accessibility of the active sites that are mostly present in the micropores. Although numerous papers have reported on the improved catalytic performance of desilicated zeolites in catalysis (details in Section 2.4.5), only few works are available that really tackle the hierarchical nature of the desilicated zeolites and demonstrate that selective silicon removal leads to an enhanced physical transport in the zeolite crystals.

2.4.3.1 Characterization of Hierarchical Structures: from Argon Adsorption to MIP

As zeolites are known for their well-defined micropore size and the associated ability to host shape-selective transformations, it is of distinct importance that the micropores in the hierarchical systems have the same size as those in the parent material. To assess the micropore size in microporous metal oxide materials and the thereof derived hierarchical systems, high-resolution low-pressure argon adsorption is the technique that is the first choice. It has been shown for different zeolite framework types that the alkaline-post treatment introduces intracrystalline mesoporosity, but it does not alter the pristine micropore size [34, 45]. Although the micropore volume in the hierarchical zeolites is typically lower than in the parent zeolites, the micropore size is the same in both systems. In order to achieve enhanced access to the micropores, it is required that the newly introduced mesoporosity in the hierarchical system is accessible from the external surface of the zeolite crystal (Figure 2.2c). Application of mercury intrusion porosimetry has proven that most of the created mesoporosity, as evidenced by physical gas adsorption with nitrogen, can also be measured by intrusion of mercury. This is conclusive evidence that the mesoporosity is directly accessible and optimally facilitates access to the micropores [44, 54].

2.4.3.2 **Visualization by Tomography**

Whether the acquired mesoporosity is homogeneously distributed throughout the crystal volume or anisotropically introduced obviously depends on the applied synthetic protocol. If the aluminum is uniformly incorporated in the zeolite framework, the mesoporosity created by alkaline treatment is homogeneously distributed as has been visualized in alkaline-treated submicrometer ZSM-5 isolated crystals by means of 3D-TEM (Figure 2.3b) [24]. It was concluded that the so-obtained hierarchical systems contain a high degree of interconnected and accessible mesoporosity. This is in contrast to alkaline treatment of crystals with distinct Al gradient, where the porosity was preferentially located in the core of the crystals. As the spatial distribution of aluminum can be controlled in the synthesis [30], an optimal tailoring of the final porous properties upon desilication of zeolite crystals can be envisaged depending on one's requirements.

2.4.3.3 **Diffusion Studies**

Only rather recently, studies have become available that demonstrate that the newly introduced mesoporosity achieved by selective silicon removal leads to improved physical transport in the resulting zeolite crystals. Wei *et al.* [46] reported diffusion studies of long-chain hydrocarbons (*n*-heptane, 1,3-dimethylcyclohexane, and *n*-undecane) in mesoporous ZSM-12 and concluded that there was a 2–4 times accelerated mass transport in the hierarchical ZSM-5 systems compared to the parent samples. Groen *et al.* [44] communicated an impressive 3 orders of magnitude enhanced diffusion of the bulky neopentane in alkaline-treated mesoporous ZSM-5 crystals by transient uptake experiments in a microbalance. Similarly, diffusion and adsorption studies of cumene in mesopore-structured ZSM-5 by Zhao *et al.* [55] revealed a 2–3 orders of magnitude improved mass transport upon introduction of mesoporosity in ZSM-5 zeolite. Summarizing, various groups have independently shown that generation of mesoporosity in ZSM-5 and ZSM-12 zeolites by desilication greatly benefits the molecular transport inside the zeolite crystal. This should be attributed to improved accessibility and a distinct shortening of the micropores, while the original micropore size was preserved. It can be anticipated that other alkaline-treated zeolite topologies will also benefit from the hierarchical nature of these materials, particularly when it comes to originally one-dimensional systems.

2.4.4 Design of Hierarchical Zeolites

2.4.4.1 Combination of Post-synthesis Treatments

The previous sections have shown that desilication of ZSM-5 zeolites results in combined micro- and mesoporous materials with a high degree of tunable porosity and fully preserved Brønsted acidic properties. In contrast, dealumination hardly induces any mesoporosity in ZSM-5 zeolites, due to the relatively low concentration of framework aluminum that can be extracted, but obviously impacts on the acidic properties. Combination of both treatments enables an independent tailoring of the porous and acidic properties providing a refined flexibility in zeolite catalyst design. Indeed, desilication followed by a steam treatment to induce dealumination creates mesoporous zeolites with extra-framework aluminum species providing Lewis acidic functions [56].

However, if first silicon is leached from the framework by treatment in alkaline medium followed by a steam treatment, inducing supplementary extraction of framework aluminum, will the resulting zeolite still exhibit an acceptable (hydro)thermal stability? Subjecting a mesoporous ZSM-5 zeolite to extreme hydrothermal treatment conditions of 500 mbar steam at 1073 K for 50 h obviously impacts on the Brønsted acidic properties of the zeolite, which completely vanish (Figure 2.6a), although X-ray diffraction measurements have confirmed the preservation of the lattice properties [56]. Besides the acidic properties, the harsh steam treatment also alters the existing mesoporosity to somewhat larger mesopores as compared to the alkaline-treated zeolite (Figure 2.6b). Despite this modification, the sample still displays a substantial mesopore surface area coupled to a slightly decreased micropore volume, and a preserved characteristic micropore size. The hierarchically porous ZSM-5 zeolites can, in principle, function as (hydro)thermally stable materials with shape-selective properties similar to conventional ZSM-5, while having highly improved accessibility to the active sites and a shortened micropore diffusion path length.

2.4.4.2 Tunable Intracrystalline Mesoporosity by Partial Detemplation–Desilication

Hitherto, the opportunity to tailor the mesoporosity in zeolite frameworks by desilication has predominantly been guided by the Si/M^{3+} ratio. Increasing the tuning capabilities of the desilication treatment would be highly attractive for optimization purposes for zeolites with $Si/Al > 50$ that suffer excessive silicon dissolution leading to undesired macropores and for zeolites with a relatively low stability of the tetrahedral cations in the zeolite lattice (e.g., BEA topology) [41]. Partial detemplation of zeolites followed by desilication in alkaline medium has recently been demonstrated as a powerful and elegant approach to design hierarchical

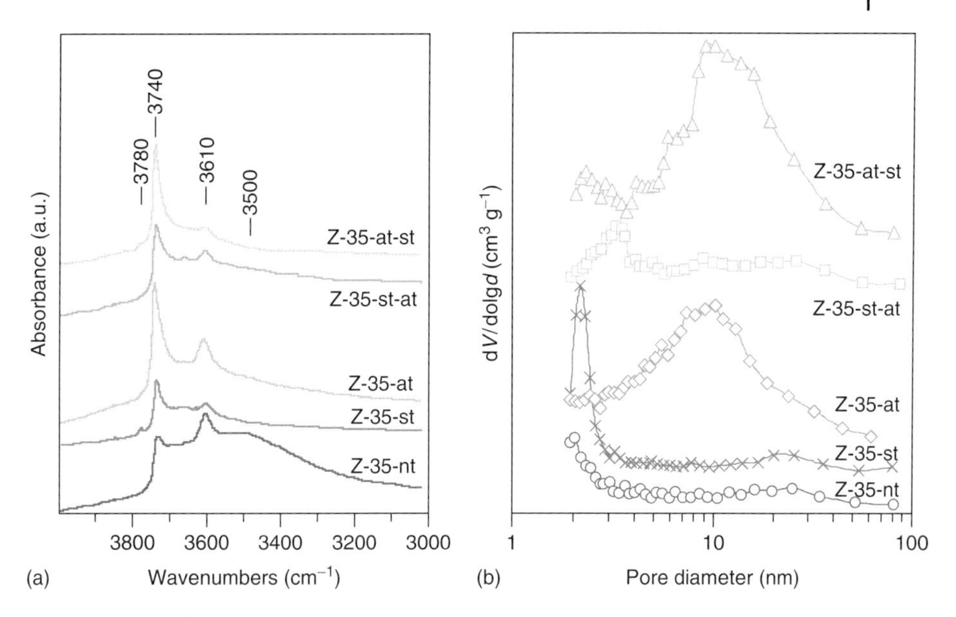

Figure 2.6 (a) Infrared spectra in the OH stretching region and (b) BJH pore size distributions of ZSM-5 with starting Si/Al ratio of 35. The acronyms nt, st, and at refer to the nontreated, steam-treated, and alkaline-treated samples, respectively.

zeolites with tailored degree of mesoporosity [57]. This achievement has been illustrated for large beta crystals and is based on the fact that the template-containing zeolite is virtually inert to Si leaching upon treatment in aqueous NaOH solutions. (Figure 2.7) Partial removal of the structure-directing agent creates regions in the crystal susceptible to mesopore formation by subsequent desilication, while template-containing regions are protected from silicon extraction. Consequently, the degree of mesoporosity in the hierarchical zeolite upon alkaline treatment is largely determined by the detemplation conditions, namely the calcination temperature. In the case of beta zeolite, variation of the calcination temperature in the range 230–550 °C enables mesopore formation control upon alkaline treatment in the range of 20–230 m^2 g^{-1}.

2.4.4.3 **Nanocrystals of Zero-Dimensional Zeolites**

The problem of accessibility in microporous solids is extreme in zero-dimensional zeolite structures such as clathrasils, that is, zeolite-related materials consisting of window-connected cages. The pore openings in these caged structures are restricted to six-membered rings of [SiO_4] units at most, which corresponds to pore diameters of approximately 0.2 nm [58]. These pores are too small for the removal of templates and, afterward, are impenetrable to typical sorptive molecules for characterization such as N_2 and Ar or reactants such as hydrocarbons. Therefore, the intrinsic

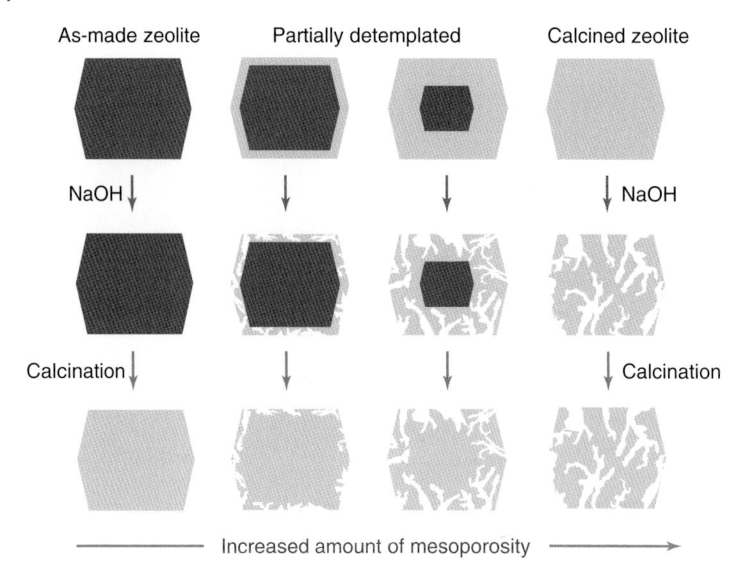

Figure 2.7 Schematic representation of the partial detemplation and desilication treatment to tailor mesoporosity development in zeolite crystals.

topology of clathrasils makes them "blinded" porous materials without current perspective for catalytic applications. Controlled desilication of aluminum-containing octadecasil by treatment in alkaline medium has led to the formation of uniform octadecasil nanocrystals (10–25 nm) with high external surface areas (up to $200\,m^2\,g^{-1}$) [59]. The high external surface area is the result of intercrystalline porosity among the octadecasil nanocrystals, and thus the hierarchical system obtained is similar to the one schematized in Figure 2.2a. This approach brings us closer to functional clathrasils, thus clearing the way for applications in catalysis.

2.4.4.4 **Hollow Zeolite Crystals**

Silicon extraction in alkaline medium has also proven to be a suitable methodology to create hollow zeolite crystals or zeolite crystals with a highly porous interior. To this end, Al-zoned crystals have been used. The high aluminum concentration at the outer surface suppresses silicon from being extracted in the outer rim and silicon dissolution merely occurs in the Al-poor center of the crystals [24]. A different approach has been successfully applied by Wang and Tuel [60] in which silicalite-1 nanocrystals have been partially dissolved in alkaline medium and in the presence of TPAOH, inducing recrystallization. In this way, hollow pure-silica nanoboxes or – in the presence of an additional aluminum source – hollow ZSM-5 crystals with tunable wall thicknesses were attained. A slightly different approach, applied by the same authors, comprised the post-treatment of TS-1, a titanosilicate, in highly alkaline TPAOH solution. This led to the formation of hollow TS-1 crystals with large intracrystalline voids due to combined dissolution and recrystallization [61].

2.4.5
Functionality

2.4.5.1 Catalyst Preparation: Improved Ion-Exchange

Zeolite-based catalysts have proven industrial possibilities for various redox reactions such as direct N_2O decomposition and N_2O-assisted oxidation of benzene to phenol. In industrial applications, aqueous or wet ion exchange is an attractive methodology, as it involves few catalyst preparation steps and the process parameters (e.g., pH, temperature, concentration) are easily accessible. However, during preparation of Fe-ZSM5 by ion-exchange, typically significantly less than half of the full exchange capacity can be realized. This problem stems from the zeolite crystal and/or agglomerate size. The rather low surface-diffusion coefficients for zeolites (typically 10^{-8} cm^2 s^{-1}) coupled to the diffusion path length, which is in the order of microns, makes the exchange rather slow. This can be overcome by shortening the diffusion path length by desilication, leading to deagglomeration and/or mesopore formation. Melián-Cabrera *et al.* [62] have shown that application of a desilication treatment prior to the ion-exchange step substantially enhances the degree of ion-exchange of Fe-MFI zeolites. The full exchange capacity of zeolites has been achieved and iron was fully incorporated by liquid-phase ion exchange on ZSM-5 without the formation of Fe-oxides. This in turn has increased the activity of these catalysts for direct N_2O decomposition.

2.4.5.2 Catalysis

As shown in Table 2.1, the improved catalytic performance of alkaline-treated zeolites compared to the parent purely microporous counterparts has been demonstrated decidedly by different groups active in academia and in industry. The positive effect is reflected in the enhanced activity, selectivity, and/or lifetime (coking resistance) of the hierarchical systems. The examples listed embrace not only a variety of zeolite topologies (MFI, MOR, MTW, BEA, and AST) but also reactions involving lighter hydrocarbons as well as bulky molecules. This illustrates the potential of the desilication treatment, although more work is to be done in optimizing the catalytic system for the wide variety of applications.

2.5
Conclusions and Outlook

Zeolites are crystalline metallosilicates featuring ordered micropores that enable shape-selective transformations. Owing to their frequent operation in the so-called transport limitation regime, the activity and occasionally also the selectivity and lifetime of zeolites in catalytic reactions are largely limited. It is therefore of great importance to increase the accessibility and molecular transport to/from the active sites in zeolites to their full catalytic potential. During the last decade, hierarchical zeolites have emerged as an important class of materials leading to improved catalytic performance compared to their microporous parents. This is

Table 2.1 Benefits of mesoporous zeolites prepared by desilication in catalytic applications with respect to the purely microporous counterparts.

Zeolite	Reaction	Reported results	Reference
ZSM-5	Butene aromatization	Increased stability	[63]
ZSM-5	Liquid-phase HDPE degradation	Higher catalytic activity	[64]
		Higher yield to liquid products	
ZSM-5	N_2O-mediated benzene to phenol	Increased lifetime	[65]
		Higher phenol yield and selectivity	
ZSM-5	Methylation of 2-methylnaphthalene	Increased lifetime	[66]
		Higher activity	
ZSM-5	Synthesis of pyridine and pycolines	Increased stability	[67]
ZSM-5	Cracking of cumene and heavy oil	Higher cracking activity	[55]
		Higher yield to light olefins	
ZSM-5	Methanol to gasoline	Increased gasoline fraction (C5+)	[68]
		Higher propylene : ethylene ratio	
		Increased lifetime	
ZSM-5	Methanol to propylene	Higher propylene selectivity	[69]
		Higher propylene : ethylene ratio	
ZSM-5	Benzene alkylation with ethylene	Higher ethylbenzene yield	[41]
ZSM-5	Aromatization and isomerization of 1-hexene	Improved lifetime	[70]
ZSM-5	Cracking of *n*-octane	Higher activity	[71]
		Higher propylene selectivity	
		Reduced oligomerization	
ZSM-5	Oligomerization of styrene	Higher activity	[72]
		Higher selectivity	
ZSM-5	Isomerization of α-pinene	Higher activity and yield to products	[73]
LaZSM-5	Liquid-phase LDPE degradation	Higher activity	[74]
		Higher liquid quality	
MoZSM-5	Methane dehydroaromatization	Higher activity and selectivity to aromatics	[75]
		Higher tolerance to coking	
FeZSM-5	N_2O decomposition	Higher activity	[76]
ZnZSM-5	1-Hexene isomerization	Higher stability	[77]

Table 2.1 (*Continued*)

Zeolite	Reaction	Reported results	Reference
Mordenite	Benzene alkylation with ethylene	Increased lifetime	[45]
		Higher activity and ethylbenzene selectivity	
Mordenite	Isomerization of 2-methyl-2-pentene	Higher activity	[78]
Mordenite	Alkylation of benzene with benzyl alcohol	Higher activity	[78]
ZSM-12	Isomerization of α-pinene	Higher activity and yield to products	[73]
Octadecasil	LDPE pyrolysis	Higher degradation activity	[59]
Beta	LDPE pyrolysis	Higher degradation activity	[57]

due to the integration of the catalytic properties of the native micropores in the same material and the facilitated transport brought by interconnections with a mesopore network of inter- or intracrystalline nature. A number of templating and nontemplating routes are available to synthesize hierarchically structured zeolites in the form of nanocrystals, nanosheets, composites, and mesoporous crystals. However, the main short-term need in the industry is to have suitable methods that can be economically realized on a multi-ton scale, with the potential of enhancing the performance of commercial off-the-shelf zeolites. The selective extraction of silicon in an alkaline medium is an effective, versatile, simple, cheap, and scalable treatment to create accessible mesoporosity in zeolites. The improved catalytic performance of alkaline-treated zeolites compared to the parent purely microporous counterparts is a reality for a number of applications. However, more studies and protocols on diffusion and accessibility as well as detailed insight into the possible physicochemical and spatial alteration of the intrinsic zeolite characteristics are required. This understanding is of key importance for rational catalyst design since it bridges materials properties, transport, and catalytic performance. Therefore, in the next years, we expect more efforts in linking the synthesis and properties of these materials with their functions as diffusion medium, adsorbent, and catalyst.

References

1. Davis, M.E. (2002) *Nature*, **417**, 813–821.
2. Corma, A. (1997) *Chem. Rev.*, **97**, 2373–2420.
3. Taguchi, A. and Schüth, F. (2005) *Microporous Mesoporous Mater.*, **77**, 1–45.
4. Kresge, C.T., Leonowicz, M.E., Roth, W.J., Vartuli, J.C., and Beck, J.S. (1992) *Nature*, **359**, 710–712.
5. Zhao, D., Feng, J., Huo, Q., Melosh, N., Fredrickson, G.H., Chmelka, B.F., and Stucky, G.D. (1998) *Science*, **23**, 548–552.

6. Zhang, Z., Han, Y., Zhu, L., Wang, R., Yu, Y., Qiu, S., Zhao, D., and Xiao, F.-S. (2001) *Angew. Chem. Int. Ed.*, **40**, 1258–1262.
7. Mokaya, R. (2002) *ChemPhysChem*, **3**, 360–363.
8. Meynen, V., Cool, P., and Vansant, E.F. (2007) *Microporous Mesoporous Mater.*, **104**, 26–38.
9. Eejka, J. and Mintova, S. (2007) *Catal. Rev. - Sci. Eng.*, **49**, 457–509.
10. Davis, M.E., Saldarriaga, C., Montes, C., Garces, J., and Crowder, C. (1988) *Nature*, **331**, 698–699.
11. Freyhardt, C.C., Tsapatsis, M., Lobo, R.F., Balkus, K.J., and Davis, M.E.Jr. (1996) *Nature*, **381**, 295–298.
12. Strohmaier, K.G. and Vaughan, D.E.W. (2003) *J. Am. Chem. Soc.*, **125**, 16035–16039.
13. Corma, A., Díaz-Cabañas, M.J., Rey, F., Nicolopoulus, S., and Boulahyab, K. (2004) *Chem. Commun.*, 1356–1357.
14. Corma, A., Díaz-Cabañas, M.J., Martínez-Triguero, J., Rey, F., and Rius, J. (2002) *Nature*, **418**, 514–517.
15. Corma, A., Díaz-Cabañas, M.J., Jordá, J.L., Martínez, C., and Moliner, M. (2006) *Nature*, **443**, 842–845.
16. Lakes, R. (1993) *Nature*, **361**, 511–515.
17. Egeblad, K., Christensen, C.H., Kustova, M., and Christensen, C.H. (2008) *Chem. Mater.*, **20**, 946–960.
18. Tao, Y., Kanoh, H., Abrams, L., and Kaneko, K. (2006) *Chem. Rev.*, **106**, 896–910.
19. Pérez-Ramírez, J., Christensen, C.H., Egeblad, K., Christensen, C.H., and Groen, J.C. (2008) *Chem. Soc. Rev.*, **37**, 2530–2542.
20. Karge, H.G. and Weitkamp J. (eds) (2002) *Molecular Sieves 3 – Post-synthesis Modification I*, Springer-Verlag, Heidelberg, pp. 203–255.
21. van Donk, S., Janssen, A.H., Bitter, J.H., and de Jong, K.P. (2003) *Cat. Rev. - Sci. Eng.*, **45**, 297–319.
22. Janssen, A.H., Koster, A.J., and de Jong, K.P. (2001) *Angew. Chem. Int. Ed.*, **40**, 1102–1104.
23. Kortunov, P., Vasenkov, S., Kärger, J., Valiullin, R., Gottschalk, P., Elía, M.F., Perez, M., Stöcker, M., Drescher, B., McElhiney, G., Berger, C., Gläser, R., and Weitkamp, J. (2005) *J. Am. Chem. Soc.*, **127**, 13055–13059.
24. Groen, J.C., Bach, T., Ziese, U., Paulaime-van Donk, A.M., de Jong, K.P., Moulijn, J.A., and Pérez-Ramírez, J. (2005) *J. Am. Chem. Soc.*, **127**, 10792–10793.
25. Pavel, C.C., Park, S.-H., Dreier, A., Tesche, B., and Schmidt, W. (2006) *Chem. Mater.*, **18**, 3813–3820.
26. Pavel, C.C. and Schmidt, W. (2006) *Chem. Commun.*, 882–884.
27. Le Febre, R.A. (1989) "High-silica zeolites and their use as catalysts in organic chemistry", Ph.D. Thesis, Delft University of Technology.
28. Dessau, R.M., Valyocsik, E.W., and Goeke, N.H. (1992) *Zeolites*, **12**, 776–779.
29. von Ballmoos, R. and Meier, W.M. (1981) *Nature*, **289**, 782–783.
30. Althoff, R., Schulz-Dobrick, B., Schüth, F., and Unger, K.K. (1993) *Microporous Mater.*, **1**, 207–218.
31. Čižmek, A., Subotić, B., Subotić, I., Tonejc, A., Aiello, R., Crea, F., and Nastro, A. (1997) *Microporous Mater.*, **8**, 159–169.
32. Ogura, M., Shinomiya, S.Y., Tateno, J., Nara, Y., Kikuchi, E., and Matsukata, H. (2000) *Chem. Lett.*, 82–83.
33. Groen, J.C., Jansen, J.C., Moulijn, J.A., and Pérez-Ramírez, J. (2004) *J. Phys. Chem. B*, **108**, 13062–13065.
34. Groen, J.C., Peffer, L.A.A., Moulijn, J.A., and Pérez-Ramírez, J. (2005) *Chem. Eur. J.*, **11**, 4983–4994.
35. Groen, J.C., Peffer, L.A.A., Moulijn, J.A., and Pérez-Ramírez, J. (2004) *Colloids Surf., A*, **241**, 53–58.
36. Groen, J.C., Moulijn, J.A., and Pérez-Ramírez, J. (2007) *Ind. Eng. Chem. Res.*, **14**, 4193–4201.
37. Sano, T., Nakajima, Y., Wang, Z.B., Kawakami, Y., Soga, K., and Iwasaki, A. (1997) *Microporous Mater.*, **12**, 71–77.
38. Baes, C.F. and Mesmer, R.E. (1976) *The Hydrolysis of Cations*, John Wiley & Sons, Inc., New York, p. 121.

39. Le Page, J.F., Cosyns, J., and Courty, P. (1987) *Applied Heterogeneous Catalysis: Design, Manufacture, Use of Solid Catalysts*, Technip, Paris, p. 84.
40. Camblor, M.A., Corma, A., and Valencia, S. (1998) *Microporous Mesoporous Mater.*, **25**, 59–74.
41. Groen, J.C., Abelló, S., Villaescusa, L.A., and Pérez-Ramírez, J. (2008) *Microporous Mesoporous Mater.*, **114**, 93–102.
42. Lietz, G., Schnabel, K.H., Peuker, Ch., Gross, Th., Storek, W., and Völter, J. (1994) *J. Catal.*, **148**, 562–568.
43. Derouane, E.G., Detremmerie, S., Gabelica, Z., and Blom, N. (1981) *Appl. Catal.*, **1**, 201–224.
44. Groen, J.C., Zhu, W., Brouwer, S., Huynink, S.J., Kapteijn, F., Moulijn, J.A., and Pérez-Ramírez, J. (2007) *J. Am. Chem. Soc.*, **129**, 355–360.
45. Groen, J.C., Sano, T., Moulijn, J.A., and Pérez-Ramírez, J. (2007) *J. Catal.*, **251**, 21–27.
46. Wei, X. and Smirniotis, P.G. (2006) *Microporous Mesoporous Mater.*, **97**, 97–106.
47. Pirutko, L.V., Chernyavsky, V.S., Uriarte, A.K., and Panov, G.I. (2002) *Appl. Catal., A*, **227**, 143–157.
48. Pereira, M.S. and Nascimento, M.A.C. (2003) *Theor. Chem. Acc.*, **110**, 441–445.
49. Hensen, E.J.M., Pidko, E.A., Rane, N., and vanSanten, R.A. (2007) *Angew. Chem. Int. Ed.*, **46**, 7273–7276.
50. Zhu, Q., Kondo, J.N., Setoyama, T., Yamaguchi, M., Domen, K., and Tatsumi, T. (2008) *Chem. Commun.*, 5164–5166.
51. Selvaraj, M. and Kawi, S. (2008) *Catal. Today*, **131**, 82–89.
52. Srinivasu, P. and Vinu, A. (2008) *Chem. Eur. J.*, **14**, 3553–3561.
53. Groen, J.C., Caicedo-Realpe, R., Abelló, S., and Pérez-Ramírez, J. (2009) *Mater. Lett.*, **63**, 1037–1040.
54. Groen, J.C., Brouwer, S., Peffer, L.A.A., and Pérez-Ramírez, J. (2006) *Part. Part. Syst. Char.*, **23**, 101–106.
55. Zhao, L., Shen, B., Gao, J., and Xu, C. (2008) *J. Catal.*, **258**, 228–234.
56. Groen, J.C., Moulijn, J.A., and Pérez-Ramírez, J. (2005) *Microporous Mesoporous Mater.*, **87**, 153–161.
57. Pérez-Ramírez, J., Abelló, S., Bonilla, A., and Groen, J.C. (2009) *Adv. Funct. Mater.*, **19**, 164–172.
58. Gies, H., Marler, B., and Werthmann, U. (1998) Synthesis of porosils: Crystalline nanoporous silicas with cage- and channel-like void structures in *Molecular Sieves: Science and Technology*, vol. 1 (eds H.G.Karge and J. Weitkamp), Springer, Heidelberg, pp. 35–64.
59. Pérez-Ramírez, J., Abelló, S., Villaescusa, L.A., and Bonilla, A. (2008) *Angew. Chem. Int. Ed.*, **47**, 7913–7917.
60. Wang, Y. and Tuel, A. (2008) *Microporous Mesoporous Mater.*, **113**, 286–295.
61. Wang, Y., Lin, M., and Tuel, A. (2007) *Microporous Mesoporous Mater.*, **102**, 80–85.
62. Melián-Cabrera, I., Espinoza, S., Groen, J.C., van der Linden, B., Kapteijn, F., and Moulijn, J.A. (2006) *J. Catal.*, **238**, 250–259.
63. Song, Y., Zhu, X., Song, Y., and Wang, Q. (2006) *Appl. Catal., A*, **288**, 69–77.
64. Choi, D.H., Park, J.W., Kim, J.-H., and Sugi, Y. (2006) *Polym. Degrad. Stab.*, **91**, 2860–2866.
65. Gopalakrishnan, S., Zampieri, A., and Schwieger, W. (2008) *J. Catal.*, **260**, 193–197.
66. Jin, L., Zhou, X., Hu, H., and Ma, B. (2008) *Catal. Commun.*, **10**, 336–340.
67. Jin, F., Cui, Y., and Li, Y. (2008) *Appl. Catal., A*, **350**, 71–78.
68. Bjørgen, M., Joensen, F., Holm, M.S., Olsbye, U., Lillerud, K.-P., and Svelle, S. (2008) *Appl. Catal., A*, **345**, 43–50.
69. Mei, C., Wen, P., Liu, Z., Liu, H., Wang, Y., Yang, W., Xie, Z., Hua, W., and Gao, Z. (2008) *J. Catal.*, **258**, 243–249.
70. Li, Y., Liu, S., Zhang, Z., Xie, S., Zhu, X., and Xu, L. (2008) *Appl. Catal., A*, **338**, 100–113.
71. Jung, J.S., Park, J.W., and Seo, G. (2005) *Appl. Catal., A*, **288**, 149–157.
72. Kox, M.H.F., Stavitski, E., Groen, J.C., Pérez-Ramírez, J., Kapteijn, F., and

Weckhuysen, B.M. (2008) *Chem. Eur. J.*, **14**, 1718–1725.
73. Mokrzycki, Ł., Sulikowski, B., and Olejniczak, Z. (2009) *Catal. Lett.*, **127**, 296–303.
74. Zhou, Q., Wang, Y.-Z., Tang, C., and Zhang, Y.-H. (2003) *Polym. Degrad. Stab.*, **80**, 23–30.
75. Su, L., Liu, L., Zhuang, J., Wang, H., Li, Y., Shen, W., Xu, Y., and Bao, X. (2003) *Catal. Lett.*, **91**, 155–167.
76. Groen, J.C., Brückner, A., Berrier, E., Maldonado, L., Moulijn, J.A., and Pérez-Ramírez, J. (2006) *J. Catal.*, **243**, 212–216.
77. Li, Y., Liu, S., Xie, S., and Xu, L. (2009) *Appl. Catal., A*, doi: 10.1016/j.apcata.2009.02.039.
78. Li, X., Prins, R., and van Bokhoven, J.A. (2009) *J. Catal.*, **262**, 257–265.

3
Preparation of Nanosized Gold Catalysts and Oxidation at Room Temperature

Takashi Takei, Tamao Ishida, and Masatake Haruta

3.1
Introduction

In the history of heterogeneous catalysis covering more than a century, gold was regarded as being almost inactive as a catalyst. In the 1980s, the number of papers on this subject was around five a year. Since Haruta, in 1987, reported that gold nanoparticles (NPs) deposited on selected base metal oxides could catalyze CO oxidation even at a temperature as low as 200 K [1], catalysis by gold gradually attracted growing interest and was reported in several tens of papers a year in the mid-1990s. The discoveries of selective oxidation of hydrocarbons in gas phase [2] and of alcohols in liquid water [3] have stimulated many scientists and engineers to work on gold catalysts [4]. The number of scientific papers were 700 in 2005 and 600 in 2006.

There are three major streams in current research activities on gold catalysts: exploration of liquid-phase organic reactions [5, 6]; development of commercial applications, especially to room temperature air purification [7–9]; and discussion on the active states of gold [7, 10–13]. As for the active state of gold, most of the active gold catalysts are composed of metallic gold NPs larger than 2 nm in diameter and base metal oxide supports [7]. The junction perimeters between gold NPs and the supports, such as TiO_2, Fe_2O_3, Co_3O_4, NiO, and CeO_2, provide sites for the reaction of reactants with oxygen. It should be noted that gold NPs supported on carbons and polymers are not at all active for CO oxidation in gas phase, indicating the importance of the contribution from the supports to the reaction. Goodman reported that in surface science model catalysts prepared by depositing gold onto monolayer TiO_2 grown on Mo single crystal substrate or onto $TiO_2(110)$ single crystals, two atomic layer gold exhibited especially high catalytic activity [10]. Their hypothesis neglects the direct involvement of support materials in the catalytic functions. The cationic gold hypothesis has been proposed for Au/La_2O_3 [11], Au/Fe_2O_3 [12], and Au/CeO_2 [13] catalysts, which are moderately active but less active than the corresponding catalysts prepared under proper conditions [14].

The characteristic nature of gold as a catalyst can be depicted by four properties: low-temperature activity, unique selectivity, enhancement by water, and preference to basic conditions [7]. The main questions are why gold NPs are usually more

Novel Concepts in Catalysis and Chemical Reactors: Improving the Efficiency for the Future.
Edited by Andrzej Cybulski, Jacob A. Moulijn, and Andrzej Stankiewicz

ISBN: 978-3-527-32469-9

active than Pd and Pt catalysts at temperatures below 200 °C and why gold should be NPs. An answer to the first question is that the adsorption of reactants on the surfaces of gold NPs is weak, so that it is retarded at higher temperatures. In contrast, the adsorption of reactants is very strong over the surfaces of Pd and Pt due to the presence of vacancies in d-orbitals. Since the adsorption sites on the surfaces of gold NPs are likely to be corners or edges but not the terrace sites, only small gold NPs can adsorb reactant molecules. It is interesting to note that the apparent activation energies for many reactions catalyzed by gold NPs are relatively small, often from 20 to 40 kJ mol^{-1}.

In this chapter, we focus on the methods to deposit gold NPs on a number of materials and on gas-phase oxidation of methanol, its decomposed derivatives, and pollutants in ambient air at room temperature.

3.2 Preparation of Nanosized Gold Catalysts

3.2.1 Deposition of Gold Nanoparticles and Clusters onto Metal Oxides

Usually noble metal NPs highly dispersed on metal oxide supports are prepared by impregnation method. Metal oxide supports are suspended in the aqueous solution of nitrates or chlorides of the corresponding noble metals. After immersion for several hours to one day, water solvent is evaporated and dried overnight to obtain precursor (nitrates or chlorides) crystals fixed on the metal oxide support surfaces. Subsequently, the dried precursors are calcined in air to transform into noble metal oxides on the support surfaces. Finally, noble metal oxides are reduced in a stream containing hydrogen. This method is simple and reproducible in preparing supported noble metal catalysts.

However, this method of impregnation does not bring about active nanosized gold catalysts. During calcination, gold NPs aggregate with each other to form large particles, which show poor catalytic activity [15]. This is because gold salts are decomposed into metallic particles but not into oxides, which may have stronger interaction with the metal oxide supports. Chloride ion is found to markedly accelerate the coagulation of gold NPs within a few hours during the reaction. This section describes four major preparation methods that lead to active-oxide-supported gold catalysts.

3.2.1.1 Coprecipitation (CP)

The coprecipitation (CP) method was the first method that was found to be effective in depositing nanosized gold particles on base metal oxide surfaces [1, 16]. In this method, an aqueous solution of $HAuCl_4$ and the nitrate of a base metal is poured, under vigorous stirring, into an aqueous solution of Na_2CO_3 at 70 °C in a short period (within 3 min). The adequate concentration of both the solutions is 0.1–0.4 M. The pH of the mixed solution can be maintained at about 9 by using

Na_2CO_3 in excess by 1.2 times. Sodium carbonate is preferred as a neutralizing reagent because of its buffer function to adjust pH at about 9.

After stirring at 70 °C for 1 h, the coprecipitate of gold hydroxide and metal hydroxide or carbonate is washed several times until the pH of the supernatant reaches 7. Heating at 70 °C during neutralization and aging accelerates substitution of Au–Cl bonding in $[AuCl_4]^-$ complex by Au–OH bonding. In the case of Co_3O_4 support, low-temperature condition (0 °C ~ room temperature) during neutralization and aging is preferred, because the reduction of $[AuCl_n(OH)_{4-n}]^-$ complex by the oxidation of Co^{2+} to Co^{3+} occurs at higher temperatures. The coprecipitate is filtered and dried at 100 °C for over night. The subsequent calcination in air at 300 °C forms nanosized gold particles deposited on the oxides of base metals. Gold species, most likely cationic, may be excluded from a crystalline structure of the support material during calcination and results in the formation of nanosized gold particles on the surfaces (Figure 3.1).

This method is limited to base metals that can form hydroxides or carbonates from the corresponding metal ion solutions in the pH range from 6 to 10 where the solubility of $Au(OH)_3$ is low. This method is applicable in various supports, such as MgO, Fe_2O_3, Co_3O_4, NiO, ZnO, Al_2O_3, In_2O_3, SnO_2, and La_2O_3 [17]. Especially gold catalysts supported on Fe_2O_3, Co_3O_4, and NiO show excellent CO oxidation activity at a temperature as low as −70 °C. To deposit gold NPs on MgO by this method, addition of Mg citrate into the mixed solution soon after the completion of CP is effective. The CP method is a convenient and effective method to dope other metals into base metal oxide supports. In the cases of Fe_2O_3 [18] and ZnO [19], doping of Ti and Fe, respectively, improves not only the catalytic activity but also durability.

3.2.1.2 **Deposition–Precipitation of Hydroxide (DP)**

Metal oxide support in the form of powder, bead, honeycomb, or cloth is dispersed or immersed in an aqueous solution of $HAuCl_4$, the pH of which is adjusted

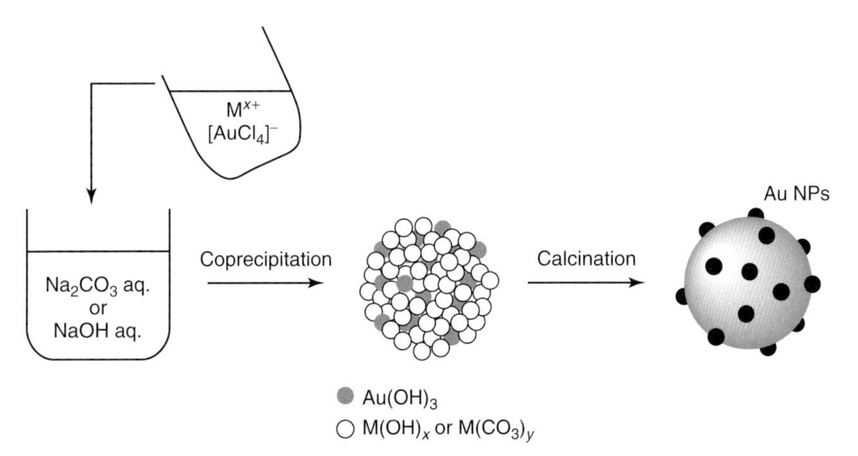

Figure 3.1 Schematic representation of coprecipitation (CP) method.

in the range of 7–10 at 70 °C [20]. The effect of heating at 70 °C is the same as in the case of CP method. Different from the case of CP, sodium hydroxide is preferred because the adjustment of pH can be done with smaller amount than sodium carbonate and, therefore, the ionic strength of the solution is kept low. The concentration of $HAuCl_4$ solution is important for the deposition–precipitation (DP) of gold hydroxide ($Au(OH)_3$) to take place exclusively on the oxide surfaces without precipitation in solution. The most acceptable concentration of $HAuCl_4$ solution, which depends on the surface area of a metal oxide support, is 1 mM. In general, a specific surface area of the support over $10\,m^2\,g^{-1}$ is desirable. After aging under agitation for 1 h at 70 °C, it is repeatedly washed by distilled water to remove sodium and chloride ions, the latter of which markedly accelerates aggregation of gold NPs. Washing by distilled water is continued until the pH of the supernatant reaches a constant value near 7. The support materials are filtrated and dried at 100 °C for half a day. The subsequent calcination in air at 300 °C reduces the gold precursors into metallic gold NPs without reduction in H_2 stream (Figure 3.2).

An important principle of the DP method is the electrostatic interaction between metal oxide surfaces and gold anion complexes such as $[AuCl(OH)_3]^-$ and $[Au(OH)_4]^-$. Metal oxide surfaces positively charged in water in the region of pH = 7–10 can electrostatically attract these gold anion complexes. Accordingly, this method is not applicable to the metal oxide supports, isoelectric point of which is below pH = 5, for example, SiO_2 and WO_3. Results of deposition of gold NPs on such acidic supports were hardly found in the literature.

Catalytic reactions over supported nanoparticulate gold catalysts are very sensitive to the types of supports. At present, base metal oxides are the most effective supports for various reactions. It is also interesting to explore new catalytic reactions using acidic supports. New preparation methods for acidic supports are required. Recently, for the deposition of gold sulfide in a strong acid solution, a new technique has been patented [21]. In the case of SiO_2 support, the surface is modified by silane coupling reagents containing amino groups. Consequently, a zeta potential shows positive charge in the wide range of pH = 3–9 and gold anion complexes are adsorbed on silica surface. In this procedure, gold NPs were successfully deposited on mesoporous silica (SBA-15) [22].

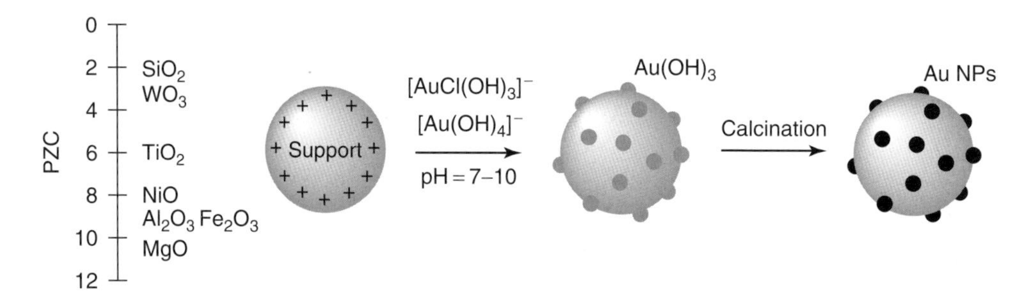

Figure 3.2 Schematic representation of deposition–precipitation (DP) method.

In the DP method, the kind of gold anion complexes and the control of surface charge in water (the control of pH) are key factors for obtaining good catalytic performance. The influence of pH on the preparation of Au/TiO_2 catalyst by the DP method has been reported in detail [23–26]. These results suggest that (i) the size of gold NPs markedly increases below pH = 6, (ii) the deposition efficiency of gold from solution is high in the range of pH = 4–7, and (iii) the samples prepared in the range of pH = 7–10 show high catalytic activity for CO oxidation at or below room temperature.

A modified DP method with urea as a precipitating agent, called the *DP urea method*, has been developed for TiO_2 support [27]. Deposition of gold NPs by this method was applied to Al_2O_3 [28], CeO_2 [28], and Mn_2O_3 [29] supports. Aqueous urea solution produces hydroxyl ions above 60 °C as follows:

$$CO(NH_2)_2 + 3H_2O \rightarrow CO_2 + 2NH_4^+ + 2OH^- \tag{3.1}$$

In this method, the pH of the solution gradually increases with increase in time. All of the gold content in a solution deposits on a support because of the formation of gold–NH_3 complexes during long time aging (~90 h), while the particle size of gold is controlled by the aging time.

3.2.1.3 Gas-Phase Grafting (GG)

This method is also called *chemical vapor deposition method*. Vapor of an organogold complex, especially $(CH_3)_2Au(CH_3COCHCOCH_3)$, abbreviated as $Me_2Au(acac)$, is adsorbed on the surfaces of support. The vapor pressure of $Me_2Au(acac)$ at room temperature is about 1.1 Pa. A simple vacuum line is used for the adsorption of $Me_2Au(acac)$ molecules on the support surfaces (Figure 3.3a). Before adsorption, support metal oxides are usually evacuated at 200 °C and then subsequently treated with oxygen at 200 °C to remove organic residue. These pretreatments are effective to improve the reproducibility of gold NPs deposition. After adsorption by evaporation at 33 °C, the sample is calcined at 300 °C in air. Gold NPs having diameters in the range 3.5–6.9 nm are deposited on Al_2O_3, TiO_2, SiO_2, and SiO_2–Al_2O_3 supports and even in the pores of mesoporous silica (MCM-41) [30, 31].

As previously described, it is difficult to deposit highly dispersed gold NPs on acidic metal oxides, such as SiO_2 and SiO_2–Al_2O_3, by liquid-phase method. Catalytic activity of Au/SiO_2 catalyst prepared by gas-phase grafting (GG) method for CO oxidation in the presence of moisture at 0 °C is close to that of Au/TiO_2, which is highly active for CO oxidation. This method enables gold NPs to deposit with strong interaction even on acidic metal oxide surfaces. As $Me_2Au(acac)$ molecule has a planer structure, strong interaction between oxygen atoms in acetylacetonate ring structure with the surface hydroxyl groups on metal oxide surfaces may prevent the gold NPs from aggregation during calcination (Figure 3.3b). The GG method can be applied to all kinds of supports including activated carbon (AC) and organic polymers.

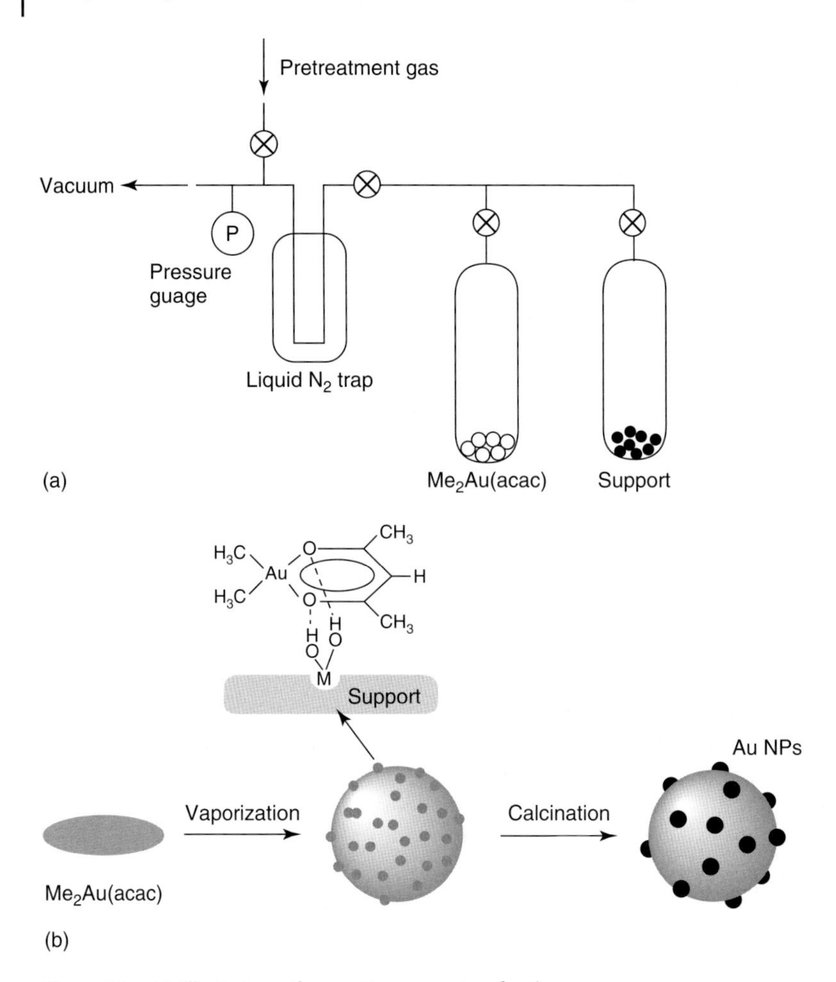

Figure 3.3 (a) Illustration of a reaction apparatus for the gas-phase grafting (GG) method and (b) its procedure.

3.2.1.4 Physical Vapor Deposition (PVD)

This method is one of the dry methods in which no chemical reaction is involved. Preparation of ultrafine particles by physical vapor deposition (PVD) dose not require washing and calcination, which are indispensable for chemical preparation such as in CP and DP methods. As waste water and waste gases are not by-produced, the arc plasma (AP) method is expected to grow in popularity as one of the industrial production methods for gold catalysts and as a clean preparation method.

Arc Plasma Method The principle of NPs synthesis in this method is based on evaporation by heating and condensation by cooling. The bulk metal is evaporated by heating with electrical resistance, electron beam, or high-frequency magnetics, and subsequently the vapor of metal atoms is condensed on a substrate as a solid film or particles. In the AP method, electrical charge filled in an external capacitor

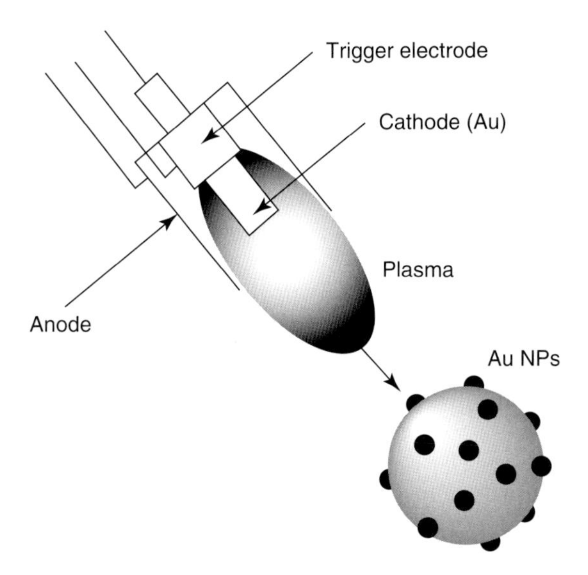

Figure 3.4 Schematic drawing for arc plasma (AP) method.

is discharged toward a cathode as target (gold) and a cathode material is melted and evaporated. Metal (gold) cations and electrons generated by ionization form plasma state successively. Metal cations fly and deposit on a substrate.

The advantages of the AP method over the conventional vacuum evaporation methods are as follows [32]: (i) the adjustment of the number and the interval of arc discharge enables us to control the particle size and surface coverage of gold, (ii) high purity of ultrafine gold particles can be maintained under ultra high vacuum (UHV), (iii) the adhesion of gold to the support materials is strong because of high kinetic energy of gold particles ($\sim$80 eV), and (iv) it can avoid a rise in the substrate temperature (Figure 3.4).

Sputtering Method In a sputtering method, atoms, molecules, or clusters evaporate from a metal target by the irradiation of ions such as argon ions to the target and the subsequent deposition of metal on a support material. Several kinds of sputtering methods have been developed: DC sputtering, radio-frequency (RF) sputtering, and magnetron sputtering. Veith *et al.* [33, 34] have studied the preparation of gold NPs on Al_2O_3, WO_3, and activated carbon by magnetron sputtering and measured catalytic activities for CO and glycerol oxidation. One of the remarkable features of the PVD method is that deposition of gold NPs on all kinds of support materials is possible, that is, even on WO_3, on which it is difficult to deposit gold NPs by the DP method.

Homogeneous deposition of ultrafine metal particles on the surfaces of fine powder is not easy using PVD. A device for stirring the powder support in a vacuum chamber is needed to avoid heterogeneous deposition. Sputter deposition units equipped with stirring powder supports have already been adapted for the industrial production of TiO_2 and carbon-supported gold catalysts by 3M [35].

3.2.2
Deposition of Gold Nanoparticles and Clusters onto Carbons and Polymers

3.2.2.1 Gold Colloid Immobilization (CI)

Gold colloidal immobilization (CI) method [36] is one of the most common techniques to deposit small gold NPs onto carbons and other inert supports. This method includes three steps: (i) the preparation of gold colloids, (ii) mixing gold colloids with the supports, and (iii) the removal of organic protecting agents by washing or thermal treatment (Figure 3.5). Citrate, poly(*N*-vinylpyrrolidone) (PVP), poly(vinyl alcohol) (PVA), and phosphine compounds have been frequently used as protecting agents for synthesizing gold colloids. Tetrahydroxymethylphosphonium chloride (THPC), which works both as a stabilizer and as a reducing agent, could also give small gold NPs around 3 nm on AC [37] and on TiO_2 [38] (Figure 3.5).

The important parameters for retaining the size of gold NPs after they are deposited are as follows: the selection of organic protecting agents, the ratio of stabilizers to gold, and post-treatments. Suitable organic protecting agents should be carefully chosen in particular for carbon supports, since gold NPs more easily aggregate on carbons than on metal oxides. For instance, PVA could retain the size of gold NPs after they were deposited onto AC, while PVP could not prevent gold NPs from aggregation even without calcination [39].

Since gold colloids are readily aggregated on carbons, Prati *et al.* [36, 39] and Mirescu *et al.* [40] avoided calcination after the deposition of colloidal gold and used Au/AC for catalytic reactions after washing with only water to remove chloride ions. The stabilizers might remain to some extent on the catalyst surfaces and affect the catalytic activity of gold. Önal *et al.* performed inert gas treatment after the deposition of colloidal gold stabilized by PVA and THPC on AC in He at 350 °C for 3 h to remove organic stabilizers from the catalyst surfaces [37].

The calcination conditions affect the size of gold particles when metal oxides are used as supports. The colloidal gold particles deposited on TiO_2 could retain their sizes after calcination at 200 °C, whereas gold NPs were sintered with an increase in calcination temperature over 200 °C [41]. Phosphine-compounds-stabilized Au_{55} clusters ($Au_{55}(PPh_3)_{12}Cl_6$) (1.4 nm) were successfully deposited onto inert supports such as SiO_2 and boron nitride (BN) [42]. Protected gold clusters were treated by calcination at 200 °C under vacuum for 2 h to completely remove phosphine ligands, keeping the size of gold particles at 1.5 nm. The size of gold particles after

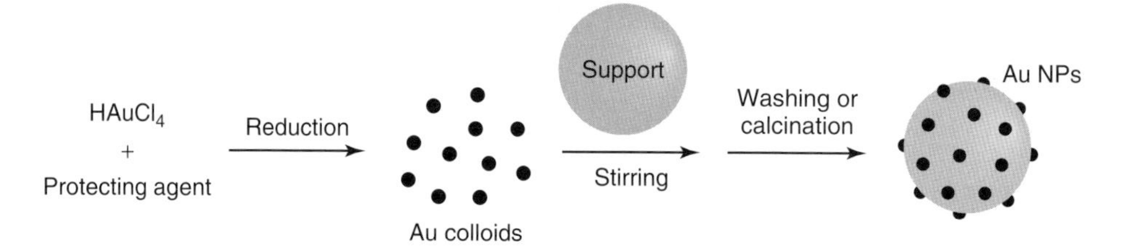

Figure 3.5 Schematic representation of colloid immobilization (CI) method.

the deposition was tunable by gold loading and gold clusters with a diameter of 1.4 nm were obtained for 0.6–3 wt% Au loadings. Higher gold loadings caused the aggregation of gold particles to form larger particles.

3.2.2.2 Deposition–Reduction (DR)

Most of the popular organic polymer beads such as polystyrene (PS) and poly(methyl methacrylate) (PMMA) have low points of zero charges so that DP is not applicable to them as in the case of carbon materials. The deposition of gold NPs onto polymer supports has also been generally performed by CI. The direct deposition of gold NPs is not feasible on most commercial organic polymers due to weak interactions between polymers and the gold complex ions such as $HAuCl_4$, and reduction of Au(III) usually takes place in the liquid phase. Even when CI is used, functionalization of polymer surfaces is favorable to improve the interactions of polymer surfaces with colloidal gold particles.

The deposition–reduction (DR) method is based on the weak electrostatic interactions of polymer surfaces with the oppositely charged Au(III) complex ions, leading to the reduction of Au(III) exclusively on the polymer surfaces. Appropriate anionic or cationic Au(III) precursors are chosen based on the zeta potentials of polymer supports (Figure 3.6) [43].

For instance, PS and PMMA have negative zeta potentials in a wide pH range, so that all loaded cationic Au(III) precursors, for example, $Au(en)_2Cl_3$ (en, ethylenediamine), are reduced to Au(0) only on the polymer surfaces when reducing agents are added, whereas the use of anionic $HAuCl_4$ causes the reduction in the liquid phase as well as on the polymer surfaces [43, 44]. Anionic Au(III) precursor, $HAuCl_4$, is chosen for polymers having cationic ammonium groups [45, 46]. In DR, gold NPs can be deposited without the addition of

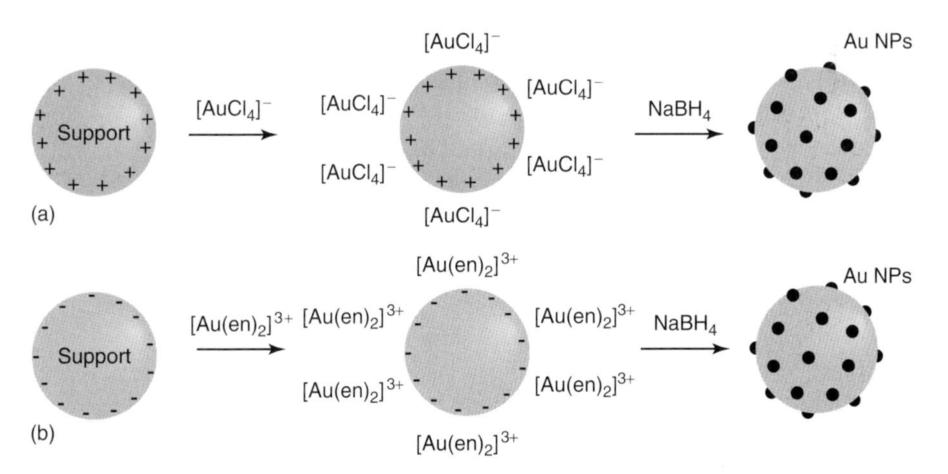

Figure 3.6 Schematic representation of deposition–reduction (DR) method: (a) positively charged supports and (b) negatively charged supports.

stabilizers or organic protecting groups which hinder substrate from contacting Au catalytic surfaces. Strong electrostatic interactions of polymer surfaces having cationic ammonium groups (NR_3^+) with $[AuCl_4]^-$ enable the direct deposition of gold NPs, yielding small gold clusters in the range of 1–2 nm [45] (Figure 3.6).

The reduction rate of Au(III) to Au(0) affects the degree of aggregation and the size of the gold particles. With the polymer supports that do not have cationic or anionic surface functional groups, dropwise addition of $NaBH_4$ gave smaller gold NPs than those prepared by rapid addition. It is suggested that Au(III) ions exist in the electrical double layer of polymer supports and gold NPs formed are deposited on polymer surfaces. While the rapid addition of $NaBH_4$ causes the aggregation of gold crystal seeds before they are deposited on polymer surfaces, the dropwise addition of $NaBH_4$ diminishes the numbers of the Au(0) crystal seeds formed at the outer layer at one time and prevents the coagulation of gold particles (Figure 3.7) [43].

In contrast, small gold NPs can be obtained by the rapid reduction of Au(III) for cationic or anionic polymers. It has been reported that gold NPs were directly deposited onto amine-functionalized polymer supports without using reducing agents by the reduction with surface amine groups [47–49]. However, gold NPs obtained in this manner are generally larger (e.g., >10 nm) than those obtained by $NaBH_4$ reduction due to the slow reduction rate [48–50], suggesting that Au(III) is gradually reduced on the Au(0) crystal seeds. When a large amount of amine groups in excess of gold was used, a large extent of adsorption sites for Au(III) was provided and it prevented gold particles from aggregation, resulting in the formation of small (<5 nm) gold NPs [46].

3.2.2.3 **Solid Grinding (SG)**

GG can be applied to all the kinds of supports including carbons and SiO_2. In the GG method, a volatile organogold complex such as dimethylacetylacetonate gold(III) ($Me_2Au(acac)$) is vaporized under vacuum at around 35–40 °C and is deposited onto the supports. Then the adsorbed Au(III) precursors are reduced by H_2 or calcination [30, 31]. GG gives gold NPs with diameters of around 5 nm that are deposited on acidic SiO_2–Al_2O_3, SiO_2, and AC. However, when powder support materials were used, GG hardly gave either narrow size distributions or uniform dispersion of gold NPs as compared to DP and CI. Efficient mixing of powder supports is necessary to ensure uniform adsorption of organogold complex.

Instead of vaporizing $Me_2Au(acac)$ under vacuum conditions, the grinding of $Me_2Au(acac)$ with supports can assist the vaporization under atmospheric pressure. In the solid grinding (SG) method, supports and $Me_2Au(acac)$ are ground in an agate mortar or in a ball mill in air, at room temperature for 20–60 min and then the adsorbed Au(III) precursors are decomposed or reduced by calcination or in a stream of H_2 in N_2 (Figure 3.8) [50, 51].

Grinding conditions are modified by the nature of supports. Gold clusters with a diameter of 1.5 nm were obtained on an aluminum-containing porous coordination polymer by grinding in an agate mortar in air for 20 min, followed by the reduction in a stream of H_2 in N_2 at 120 °C for 2 h [50]. In contrast, ball milling is favorable

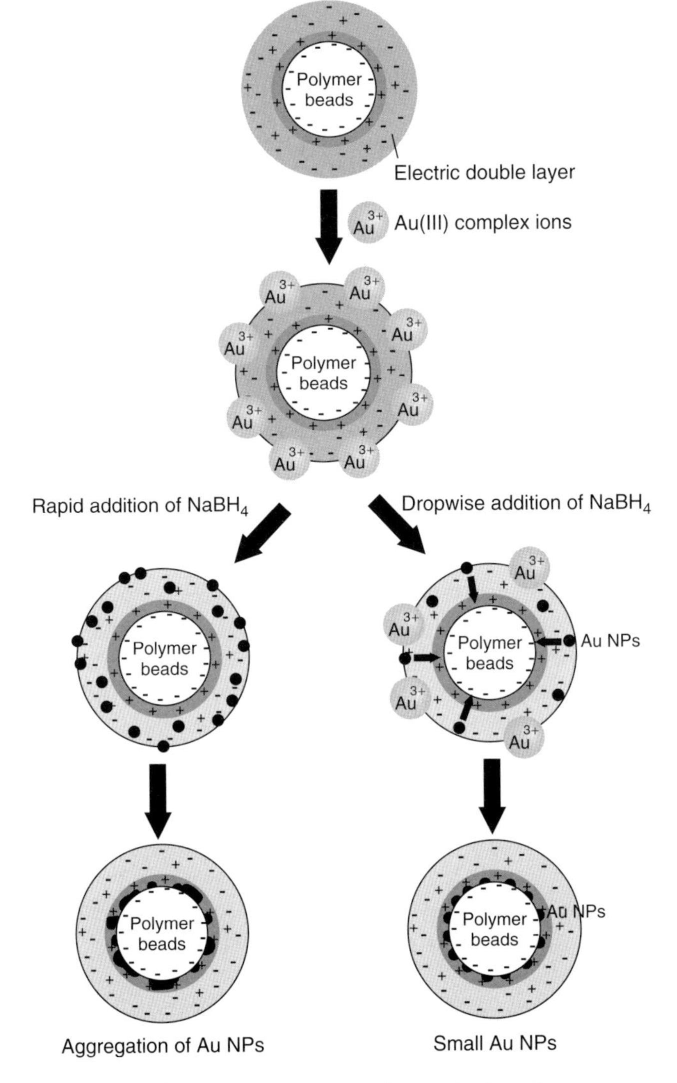

Figure 3.7 Schematic representation of the reduction of Au (III) ions at the outer layer of the electric double layer in aqueous media [43].

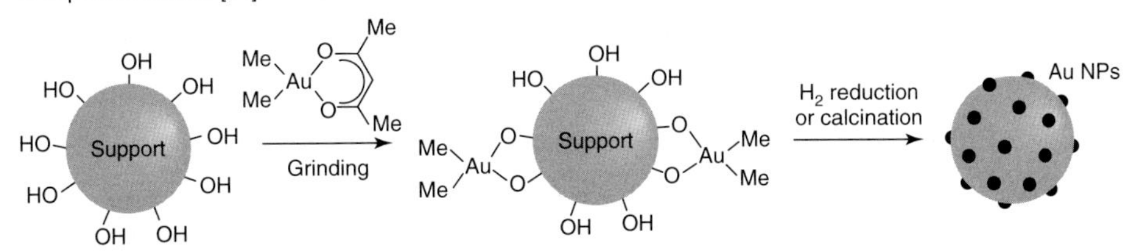

Figure 3.8 Schematic representation of solid grinding (SG) method.

for metal oxide supports. It is likely that harder metal oxide particles need stronger grinding than for soft carbon and polymer supports. The milling of metal oxides with $Me_2Au(acac)$ in organic solvent slurry enables the uniform mixing and high dispersion of gold clusters. Ball milling in air at 350 rpm for 1 h followed by calcination at 300 °C for 4 h gave gold NPs with a diameter of 2.6 nm on Al_2O_3 [51].

SG is an efficient method for chemically inert supports such as Al_2O_3, carbon materials, and polymers but is not applicable to semiconductive or reactive metal oxides (e.g., TiO_2). It can be suggested that Au(III) precursors adsorbed on the supports is reduced by M^{3+} cation radicals or by oxygen radical sites on the support surfaces to form Au(0) crystal seeds during grinding. Once Au(0) crystal seeds are formed on the supports, they enhance the reduction of Au(III) on their surface and thus cause the growth of large Au NPs [51].

3.3 Gas-Phase Oxidation Around Room Temperature

Owing to the strong adsorption of gaseous compounds on the surfaces of palladium and platinum, they are not catalytically active for oxidation at low temperatures except for H_2 oxidation. At temperatures above 200 °C, palladium and platinum catalysts tend to exhibit high catalytic activity in gas-phase oxidation of hydrocarbons. Organic compounds are completely oxidized to CO_2 and H_2O and the selectivity to organic oxygenates is poor. On the other hand, the oxidation power of gold is mild owing to the fulfilled d-band character, leading to high selectivity to partially oxidized organic compounds under mild reaction conditions (low temperature and pressure).

3.3.1 Oxidation of Methanol

Methanol is a major bulk chemical, and its global annual production exceeds 37 million tons. It is mainly used for the production of formaldehyde and methyl *t*-butyl ether (MTBE). Especially, formaldehyde is dominantly used for producing resins. At present, methanol and its decomposed derivatives can be oxidized to CO_2 and H_2O by the proper selection of supported noble metal catalysts such as palladium, platinum, and gold.

Figure 3.9 shows temperatures for 50% conversion (T_{50}) of CH_3OH and its decomposed derivatives over Pt/γ-Al_2O_3, Pd/γ-Al_2O_3, and Au/α-Fe_2O_3 catalysts [52]. For MeOH oxidation, palladium is more active than platinum, while gold lies in between. These three catalysts are similarly active for the oxidation of HCHO and HCOOH. Catalytic oxidation at temperatures below 0 °C can proceed over palladium and platinum for H_2 oxidation, while it happens over gold for CO oxidation.

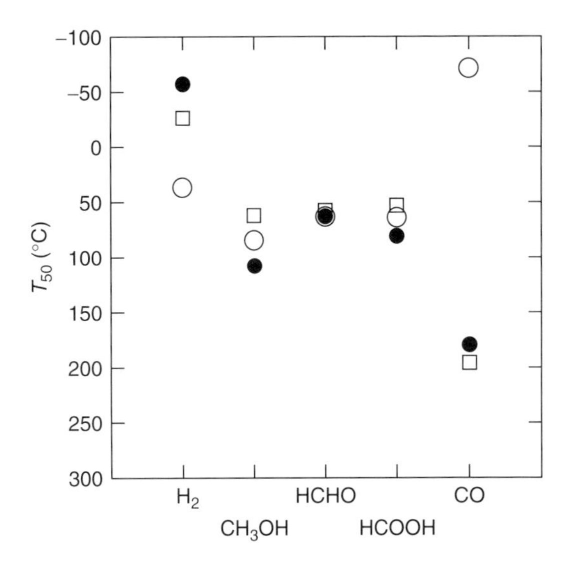

Figure 3.9 Temperature for 50% conversion in the oxidation of CH_3OH and its decomposed derivatives over supported Pd, Pt, and Au, catalysts [52]. ○ : Au/a-Fe_2O_3, ●: Pt/γ-Al_2O_3, □: Pd/γ-Al_2O_3 CH_3OH = 0.5 vol%, HCHO = 0.5 vol%, HCOOH = 0.5 vol%, H_2 = 1 vol%, CO = 1 vol%, space velocity (SV) = 20 000 h^{-1} ml g cat^{-1}.

Table 3.1 Catalytic activity of supported Ib metal catalysts for MeOH oxidation [53].

Catalyst	Metal loading (%)	Conversion (%)	Selectivity (%)		
			HCHO	CO	CO_2
Cu/Al_2O_3	1.5	47	20	–	80
Ag/Al_2O_3	2.5	40	50	–	50
Au/Al_2O_3	5	85	60	0	40
$Au/CeO_x/Al_2O_3$	5	36	75	5	20
$Au/Li_2O/Al_2O_3$	5	44	76	–	24

Reaction temperature: 250 °C, MeOH/O_2 = 1.

Concerning selective oxidation of MeOH to HCHO, Ag/γ-Al_2O_3 or Fe–Mo/γ-Al_2O_3 is used in the industry at temperatures above 600 and at 300 °C, respectively. Information on selective oxidation of MeOH to HCHO by gold catalysts is limited compared to the decomposition of MeOH. Recently, Nieuwenhuys *et al.* [53] showed that Au/Al_2O_3 catalyst exhibited a high selectivity of 60% to HCHO at 250 °C and the selectivity was improved by the addition of CeO_x or Li_2O. As shown in Table 3.1, gold exhibits higher selectivity to HCHO than the other coinage metals, Cu and Ag.

3.3.2
Oxidation of Formaldehyde

Formaldehyde is a harmful compound released from walls and furnitures in new houses because adhesives containing HCHO are often used in construction materials. In addition, HCHO is emitted by tobacco smoke and combustion exhaust gases. Long exposure to HCHO causes serious health problems called *sick house diseases*. In Japan, the concentration of HCHO in indoor air is regulated [54] to under 0.08 ppm based on the recommendation of the World Health Organization (WHO).

The elimination of HCHO by gas adsorption using activated carbon having high specific surface area is one of the effective removal methods. However, in this method, periodical replacement of carbon adsorbents, regeneration, and combustion after saturation with HCHO are required. An ideal method is the catalytic oxidation of HCHO to CO_2 at room temperature, using molecular oxygen in ambient atmosphere.

Zeolite doped with Fe^{3+} [55] or $MnO_x{\cdot}CeO_2$ [56] was reported to oxidize HCHO when heated at 100–190 °C. Some supported noble metal catalysts are active for HCHO oxidation at room temperature. For instance, Pt/TiO_2 catalyst prepared by an impregnation method with an aqueous solution of H_2PtCl_6 shows perfect oxidation of HCHO (100 ppm) at room temperature [57]. The demand for platinum will continue to increase in the long term because of its growing applications in automobile exhaust gas treatment and fuel cell electrodes. Moreover, the resources of platinum are limited and localized in the world. On the other hand, it is advantageous to use gold instead of platinum because of the lower cost and higher resistance to oxidation. Catalytic activity of gold NPs supported on Fe_2O_3 [52, 58], TiO_2 [52], CeO_2 [59], and ZrO_2 [60] for HCHO oxidation is summarized in Table 3.2. Although experimental conditions for HCHO oxidation are not the same, the catalytic activity of Au/ZrO_2 prepared by the DP method appears to be similar to that of Pt/TiO_2 prepared by an impregnation method and calcined at 400 °C for 2 h.

Table 3.2 Catalytic activity supported Au and Pt catalysts for HCHO oxidation.

Sample	Metal loading (wt%)	SV (h^{-1})	HCHO (ppm)	T_{50} (°C)	Reference
Au/ZrO_2	–	20 000	50	$<20^a$	[60]
Au/Fe_xO_y	7.1	54 000	5	38	[58]
Au/Fe_2O_3	1	20 000	5000	80	[52]
Au/CeO_2	0.85	32 000	600	90	[59]
Au/TiO_2	1	50 000	100	90	[57]
Pt/TiO_2	1	50 000	100	$<20^b$	[57]

[a] Conversion = 88%.
[b] Conversion = 100%.

3.3.3 Oxidation of Carbon Monoxide

The oxidation of CO at low temperatures was the first reaction discovered as an example of the highly active catalysis by gold [1]. Carbon monoxide is a very toxic gas and its concentration in indoor air is regulated to 10–50 ppm depending on the conditions [61]. An important point is that CO is the only gas that cannot be removed from indoor air by gas adsorption with activated carbon. On the other hand, metal oxides or noble metal catalysts can oxidize CO at room temperature. In the beginning of the twentieth century, mixed metal oxides such as $MnO_2{\cdot}CuO$ and $MnO_2{\cdot}Co_3O_4{\cdot}Ag_2O$ were developed for use in submarines and these are called *hopcalites*. A constraint is that the CO oxidation activity of these metal oxides is remarkably depressed by moisture. In contrast, Pt group metal catalysts are activated by moisture; however, usually they need to be heated to 100 °C or above. A palladium catalyst is less expensive, but is readily oxidized to PdO during exposure to air and during reaction. In contrast, gold NPs highly dispersed on semiconductive metal oxide supports lead to perfect CO oxidation at room temperature (Figure 3.10) and even at −80 °C. In addition, the catalytic activity of gold NPs supported on metal oxides is enhanced by moisture [62, 63]. This is advantageous in the practical use of gold catalysts for CO oxidation in atmosphere.

Several active sites for CO oxidation have been proposed. Among them, the major ones are as follows: (i) cationic gold [64], (ii) quantum-sized gold particles [65], and (iii) perimeter which means interface between gold NPs and the support surfaces [7]. In the majority of practically active gold catalysts, it can be assumed that CO molecules trapped on the surfaces of gold NPs move to the perimeter interfaces to react with oxygen molecules. Catalytic activity of supported gold catalysts for CO oxidation strongly depends on the contact structure between gold NPs and the

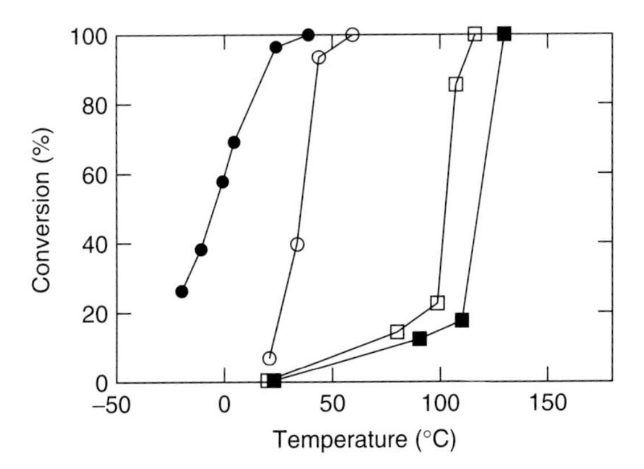

Figure 3.10 Conversion of CO in catalytic oxidation over supported PdO and Au catalysts as a function of catalyst temperature. ●: Au/TiO_2, ○: Au/CeO_2, □: PdO/CeO_2, ■: PdO/TiO_2 metal loading: 1.0 wt%, 1.0 vol% CO in air, SV = 20 000 h^{-1} ml g cat^{-1}.

supports. Oxygen vacancies at the support oxide on the perimeter appear to be a key to the oxidation of CO molecules at low temperatures.

Dominant factors for CO oxidation by gold catalysts are as follows: (i) the contact structure of gold particles, (ii) the kinds of support materials, (iii) the size of gold particles, and (iv) the water content in the reactant gases. Hemispherical gold NPs attached to the support are much more active for CO oxidation than spherical ones simply loaded on the support [15]. Gold NPs supported on carbon and organic polymer materials show no catalytic activity for CO oxidation. Only metal oxides or hydroxides are effective as supports for CO oxidation at room temperature. Especially, semiconductive metal oxides such as Fe_2O_3 and TiO_2 enable low-temperature CO oxidation. Although gold NPs on nonsemi-conductive metal oxide supports such as SiO_2 and Al_2O_3 exhibit lower catalytic activity than gold NPs on semiconductive metal oxides, moisture enhances CO oxidation.

The turnover frequency (TOF) based on surface-exposed atoms significantly increases with a decrease in the diameter of the gold particle from 5 nm [66]. This feature is unique to gold, because other noble metals usually show TOFs that decrease or remain the same with a decrease in the diameter [7]. The decrease in particle size gives rise to an increase in corner or edge and perimeter of NPs and change in electronic structure; however, the origin of size effects on catalytic activity for CO oxidation is not clear.

3.3.4
Oxidation of NH_3 and Trimethylamine

The catalytic oxidation of NH_3 proceeds by the following three main reactions:

$$4NH_3 + 5O_2 \rightarrow 4NO + 6H_2O \tag{3.2}$$

$$4NH_3 + 4O_2 \rightarrow 2N_2O + 6H_2O \tag{3.3}$$

$$4NH_3 + 3O_2 \rightarrow 2N_2 + 6H_2O \tag{3.4}$$

Reaction (3.2) is applied for the production of nitric acid through NO_2, which is produced from NO by the reaction with O_2. Pt–Rh alloy gauzes are used in the industry for producing NO. As these alloy catalysts operate at the high temperature of 850–900 °C, durability of the catalysts is 2–12 months. The product of N_2O (Reaction (3.3)) is useful as an oxidizing agent for hydrocarbons, for example, phenol from benzene over Fe–FMI–zeolite. Reaction (3.4) is an ideal one for decomposing NH_3 into a harmless substance, N_2. High selectivity to N_2 is required in catalysts used for treating ammonia pollution. A number of studies have been carried out on the selective oxidation of NH_3 by supported metal catalysts (Reaction (3.4)). Silver on Al_2O_3 [67] and Cu/Al_2O_3 [68] catalysts have been found to be highly active for the selective oxidation of NH_3. Gold catalyst is expected to be a favorable catalyst for this reaction because of the high activity for CO oxidation at low temperature. Table 3.3 summarizes the temperatures for 50% conversion (T_{50}) and the selectivities to N_2, N_2O, and NO in NH_3 oxidation over Ag/Al_2O_3,

Table 3.3 Catalytic activity of supported Ib metal catalysts for NH_3 oxidation [70].

Sample	T_{50} (°C)[a]	Selectivity (%)		
		N_2	N_2O	NO
Cu/Al_2O_3	361	97	–	–
Ag/Al_2O_3	341	95	–	–
Au/Al_2O_3	>400[b]	87	3	7
$Cu/CeO_x/Li_2O/Al_2O_3$	250	86	8	8
$Ag/CeO_x/Li_2O/Al_2O_3$	236	83	12	3
$Au/CeO_x/Li_2O/Al_2O_3$	239	66	8	0

[a]Temperature for 50% conversion.
[b]Conversion is 28% at 400 °C.
Reaction condition: $NH_3 = 4$ vol%, $NH_3{:}O_2 = 1:1$, $SV = 2500\,h^{-1}$.

Cu/Al_2O_3, and Au/Al_2O_3 catalysts and the effects of Li_2O and CeO_x addition [69]. However, the additives caused a decrease in the N_2 selectivity but remarkably improved the catalytic activity, in particular, a decrease in T_{50} over 200 °C in the case of gold. Gold catalysts have a potential for NH_3 oxidation at lower temperature if a proper kind of support metal oxides is selected.

Trimethylamine is one of the typical odorous organic compounds containing nitrogen. Supported noble metals, such as palladium and platinum, are useful in the combustion of various organic compounds. Gold NPs supported on $NiFeO_4$ can oxidize trimethylamine at lower temperatures compared to supported palladium and platinum catalysts [70]. Furthermore, $Au/NiFeO_4$ catalyst yields nitrogen and carbon dioxide, while Pt and Pd catalysts mainly produce nitrous and nitric oxides (N_2O and NO). The Au/Fe_2O_3 catalyst also exhibits high catalytic activity for the oxidation of trimethylamine at low temperature. Figure 3.11 shows conversion of trimethylamine to CO_2 as a function of catalyst temperature over single-component catalysts (Au/Fe_2O_3, Pt/SnO_2, and Ir/La_2O_3) and binary-component catalysts (Au/Fe_2O_3–Ir/La_2O_3 and Pt/SnO_2–Ir/La_2O_3) [71]. The catalytic activity of gold and palladium catalyst is greatly improved by combination with the iridium catalyst. Surprisingly, the iridium catalyst alone is much less active for this reaction. Such synergy effect is valid for improving not only catalytic activity but also durability.

3.4 Conclusions

Although bulk gold is poorly active as a catalyst, gold NPs attached to a variety of support materials exhibit unique catalytic properties in oxidation. The general features can be summarized as follows.

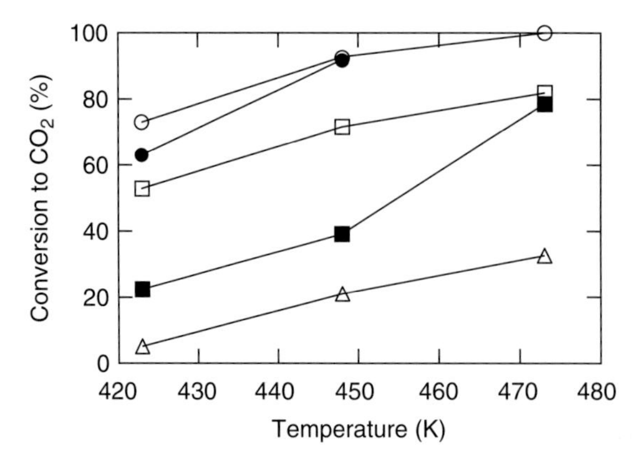

Figure 3.11 Yield of CO_2 as a function of catalyst temperature in the oxidation of trimethylamine over single-component and multicomponent noble metal catalysts [70]. ○: $Au/Fe_2O_3–Ir/La_2O_3$, ●: $Pt/SnO_2–Ir/La_2O_3$, □: Au/TiO_2, ■: Pt/SnO_2, □: Ir/La_2O_3. Metal loading: 2 atm% except for 2 wt% Pt/SnO_2–2 wt% Ir/La_2O_3. Preparation method: DP method trimethylamine 100 ppm in air, SV = 10 000 h^{-1} ml g cat^{-1}.

1) Highly active gold catalysts can be prepared by an appropriate selection of preparation methods such as CP, DP, DR, and SG with dimethyl Au(III) acetylacetonate, depending on the kind of support materials and reactions targeted.
2) There are four important conditions for gold to be active as a catalyst in oxidation: strong junction with reducible metal oxides, H_2O, OH^-, and the size of gold particles and films. At least two conditions should be fulfilled for gold to exhibit high catalytic activity. The catalysis by gold is often promoted by water and alkalies but killed by acids, which presents an interesting research topic of catalysis science.
3) For CO oxidation in gas phase at room temperature, gold NPs supported on reducible metal oxides such as TiO_2, Fe_2O_3, Co_3O_4, NiO, and CeO_2 are much more active than platinum group metal catalysts. They are active even at a temperature as low as $-70\,^{\circ}C$ when gold loadings are higher than 3 wt%.
4) In the copresence of H_2O, gold NPs supported on insulating metal oxides such as Al_2O_3 and SiO_2 can also exhibit catalytic activity at room temperature for CO oxidation.
5) Over most of reported gold catalysts, CO oxidation takes place at the junction perimeter between gold NPs and the metal oxide supports. Carbon monoxide is adsorbed on the edges, corners, and steps of gold NPs. Molecular oxygen is adsorbed on the support surfaces and may be activated at the oxygen defect sites at the perimeter interfaces, where the two adsorbates react with each other to form CO_2 in the gas phase. At the perimeter interfaces, gold is assumed to exist as Au^{3+}, which might be stabilized through bonding with OH^-.
6) Supported gold catalysts are, in general, less active than platinum group metal catalysts in the complete oxidation of hydrocarbons; however, by choosing

appropriate metal oxide supports, gold catalysts are useful for the removal of HCHO at room temperature.

References

1. Haruta, M., Kobayashi, T., Sano, H., and Yamada, N. (1987) *Chem. Lett.*, 405–408.
2. Hayashi, T., Tanaka, K., and Haruta, M. (1998) *J. Catal.*, **178**, 566–575.
3. Prati, L. and Rossi, M. (1998) *J. Catal.*, **176**, 552–560.
4. Haruta, M. (2005) *Nature*, **437**, 1098–1099.
5. Hashmi, A.S.K. and Hutchings, G.J. (2006) *Angew. Chem. Int. Ed.*, **45**, 7896–7936.
6. Ishida, T. and Haruta, M. (2007) *Angew. Chem. Int. Ed.*, **46**, 7154–7156.
7. Haruta, M. (2003) *Chem. Rec.*, **3**, 75–87.
8. Louis, C. (2007) Gold nanoparticles: Recent advances in CO oxidation, Chapter 15, in *Nanoparticles and Catalysis* (ed. D. Astruc), Wiley-VCH Verlag GmbH, Weinheim, pp. 475–502.
9. Saito, M. and Haruta, M. (2007) *Clean Technol.*, **9**, 58–61.
10. Chang, M.S. and Goodman, D.W. (2004) *Science*, **306**, 252–255.
11. Guzman, J. and Gates, B.C. (2004) *J. Am. Chem. Soc.*, **126**, 2672–2673.
12. Herzing, A.A., Kiely, C.J., Carley, A.F., Landon, P., and Hutchings, G.J. (2008) *Science*, **321**, 1331–1335.
13. Fu, O., Saltsburg, H., and Flitzani-Stephanopoulos, M. (2003) *Science*, **301**, 935–938.
14. Sakurai, H., Akita, T., Tsubota, S., Kiuchi, M., and Haruta, M. (2005) *Appl. Catal. A*, **291**, 179–187.
15. Bamwenda, G.R., Tsubota, S., Nakamura, T., and Haruta, M. (1997) *Catal. Lett.*, **44**, 83–87.
16. Haruta, M., Yamada, N., Kobayashi, T., and Iijima, S. (1989) *J. Catal.*, **115**, 301–309.
17. Haruta, M., Kageyama, H., Kamijo, N., Kobayashi, T., and Delannay, F. (1988) *Stud. Surf. Sci. Catal.*, **44**, 33–42.
18. Kobayashi, T., Haruta, M., Sano, H., Tsubota, S., and Nakane, M. (1988) *Sens. Actuators*, **13**, 339–349.
19. Takei, T., Suenaga, J., Kuwano, K., Ohashi, H., Ishida, T., and Haruta, M. (2008) JP 2008-179978.
20. Haruta, M., Tsubota, S., Kobayashi, T., Kageyama, H., Genet, M.J., and Delmon, B. (1993) *J. Catal.*, **144**, 175–192.
21. Ohashi, H., Haruta, M., Takei T., and Ishida, T. (2008) JP 2008-091587.
22. Yang, C.-M., Kalwei, M., Schuth, F., and Chao, K.J. (2003) *Appl. Catal. A: Gen.*, **254**, 289–296.
23. Tsubota, S., Cunningham, D.A.H., Bando, Y., and Haruta, M. (1995) *Stud. Surf. Sci. Catal.*, **91**, 227–235.
24. Wolf, A. and Schuth, F. (2002) *Appl. Catal. A: Gen.*, **226**, 1–13.
25. Moreau, F., Bond, G.C., and Taylor, A.O. (2005) *J. Catal.*, **231**, 105–114.
26. Moreau, F. and Bond, G.C. (2007) *Catal. Today*, **122**, 260–265.
27. Zanella, R., Giorgio, S., Henry, C.R., and Louis, C. (2002) *J. Phys. Chem. B*, **106**, 7634–7642.
28. Zanella, R., Delannoy, L., and Louis, C. (2005) *Appl. Catal. A: Gen.*, **291**, 62–72.
29. Wang, L.-C., Huang, X.-S., Liu, Q., Liu, Y.-M., Cao, Y., He, H.-Y., Fan, K.-N., and Zhuang, J.-H. (2008) *J. Catal.*, **259**, 66–74.
30. Okumura, M., Nakamura, S., Tsubota, S., Nakamura, T., Azuma, M., and Haruta, M. (1998) *Catal. Lett.*, **51**, 53–58.
31. Okumura, M., Tsubota, S., and Haruta, M. (2003) *J. Mol. Catal. A: Chem.*, **199**, 73–84.
32. Agawa, Y., Sakae, K., Saito, A., Yamaguchi, K., Matsuura, M., Suzuki, Y., Hara, Y., Nakano, Y., Tsukahara, N., and Murakami, H. (2006) *ULVAC Tech. J.*, **65**, 1–5.
33. Veith, G.M., Lupini, A.R., Pennycook, S.J., Ownby, G.W., and Dudney, N.J. (2005) *J. Catal.*, **231**, 151–158.

34. Veith, G.M., Lupini, A.R., Pennycook, S.J., Villa, A., Prati, L., and Dudney, N.J. (2007) *J. Catal. Today*, **122**, 248–253.
35. Brady, J.T., Jones, M.E., Brey, L.A., Buccellato, G.M., Chamberlain, C.S., Huberty, J.S., Siedle, A.R., Wood, T.E., Veeraraghavan, B., and Fansler, D.D. WO 2006/074126.
36. Prati, L. and Martra, G. (1999) *Gold Bull.*, **32**, 96–101.
37. Önal, Y., Schimpf, S., and Claus, P. (2004) *J. Catal.*, **223**, 122–133.
38. Grunwaldt, J.-D. and Baiker, A. (1999) *J. Phys. Chem. B*, **103**, 1002–1012.
39. Porta, F., Prati, L., Rossi, M., Coluccia, S., and Martra, G. (2000) *Catal. Today*, **61**, 165–172.
40. Mirescu, A., Berndt, H., Martin, A., and Prüße, U. (2007) *Appl. Catal. A: Gen.*, **317**, 204–209.
41. Tsubota, S., Nakamura, T., Tanaka, K., and Haruta, M. (1998) *Catal. Lett.*, **56**, 131–135.
42. Turner, M., Golovko, V.B., Vaughan, O.P.H., Abdulkin, P., Berenguer-Murcia, A., Tikhov, M.S., Johnson, B.F.G., and Lambert, R.M. (2008) *Nature*, **454**, 981–984.
43. Ishida, T., Kuroda, K., Kinoshita, N., Minagawa, W., and Haruta, M. (2008) *J. Colloid Interface Sci.*, **323**, 105–111.
44. Kuroda, K., Ishida, T., and Haruta, M. (2009) *J. Mol. Catal. A: Chem.*, **298**, 7–11.
45. Schrinner, M., Polzer, F., Mei, Y., Lu, Y., Haupt, B., Ballauf, M., Göldel, A., Drechsler, M., Preussner, J., and Glatzel, U. (2007) *Macromol. Chem. Phys.*, **208**, 1542–1547.
46. Ishida, T., Okamoto, S., Makiyama, R., and Haruta, M. (2009) *Appl. Catal. A: Gen.*, **353**, 243–248.
47. Biffis, A. and Prati, L. (2007) *J. Catal.*, **251**, 1–6.
48. Biffis, A. and Minati, L. (2005) *J. Catal.*, **236**, 405–409.
49. Zhang, M., Liu, L., Wu, C., Fu, G., Zhao, H., and He, B. (2007) *Polymer*, **48**, 1989–1997.
50. Ishida, T., Nagaoka, M., Akita, T., and Haruta, M. (2008) *Chem. Eur. J.*, **14**, 8456–8460.
51. Ishida, T., Kinoshita, N., Okatsu, H., Akita, T., Takei, T., and Haruta, M. (2008) *Angew. Chem. Int. Ed.*, **47**, 9265–9268.
52. Haruta, M., Ueda, A., Tsubota, S., and Sanchez, R.M.T. (1996) *Catal. Today*, **29**, 443–447.
53. Lippits, M.J., Iwema, R.R.H.B., and Nieuwenhuys, B.E. (2008) *Catal. Today*, **145**, 27–33.
54. Ministry of Health, Labour and Welfare, Japan (2000) Guide to concentration of volatile chemicals in room.
55. Xuzhuang, Y., Yuenian, S., Zhangfu, Y., and Huaiyong, Z. (2005) *J. Mol. Catal. A: Chem.*, **237**, 224–231.
56. Tang, X., Li, Y., Huang, X., Xu, Y., Zhu, H., Wang, J., and Shen, W. (2006) *Appl. Catal. B: Environ.*, **62**, 265–273.
57. Zhang, C., He, H., and Tanaka, K. (2006) *Appl. Catal. B: Environ.*, **65**, 37–43.
58. Li, C., Shen, Y., Jia, M., Sheng, S., Adebajo, M.O., and Zhu, H. (2008) *Catal. Commun.*, **9**, 355–361.
59. Shen, Y., Yang, X., Wang, Y., Zhang, Y., Zhu, H., Gao, L., and Jia, M. (2008) *Appl. Catal. B: Environ.*, **79**, 142–148.
60. Okumura, M., Akita, T., Ueda, A., Tsubota, S., and Haruta, M. (1992) Abstract of the 90th Meeting of Catalysis Society of Japan, 33.
61. Ministry of Health, Labour and Welfare, Japan (2004) Revision of Standard Regulation of Hygiene in Office.
62. Date, M., Okamura, M., Tsubota, S., and Haruta, M. (2004) *Angew. Chem. Int. Ed.*, **43**, 2129–2132.
63. Costello, C.K., Kung, M.C., Oh, H.-S., Wang, Y., and Kung, H.H. (2002) *Appl. Catal. A: Gen.*, **232**, 159–168.
64. Fierro-Gonzalez, J.C., Bhirud, V.A., and Gates, B.C. (2005) *Chem. Commun.*, 5275–5277.
65. Valden, M., Lai, X., and Goodman, D.W. (1998) *Science*, **281**, 1647–1650.
66. Haruta, M. (1997) *Catal. Today*, **36**, 153–166.
67. Gang, L., Anderson, B.G., van Grondelle, J., and van Santen, R.A. (2003) *Appl. Catal. B: Environ.*, **40**, 101–110.
68. Gang, L., Anderson, B.G., van Grondelle, J., and van Santen, R.A. (1999) *J. Catal.*, **186**, 100–109.

69. Lippits, M.J., Gluhoi, A.C., and Nieuwenhuys, B.E. (2008) *Catal. Today*, **137**, 446–452.
70. Ueda, A. and Haruta, M. (1992) *J. Resour. Environ.*, **28**, 1035–1038.
71. Okumura, M., Akita, T., Haruta, M., Wang, X., Kajikawa, O., and Okada, O. (2003) *Appl. Catal. B: Environ.*, **41**, 43–52.

4
The Fascinating Structure and the Potential of Metal–Organic Frameworks

Luc Alaerts and Dirk E. De Vos

4.1
Introduction

The chemical industry extensively uses microporous materials. The superior performance of these materials as catalysts and adsorbents is directly related to the host–guest interactions in their nanosized pores, which are a tunable yet well-defined environment for interactions on a molecular scale. Microporous materials, for instance, zeolites, are frequently used as heterogeneous catalysts for conversion of base chemicals to valuable intermediates, or as adsorbents in separation and purification processes [1, 2]. An exciting class of novel materials is the metal–organic frameworks (MOFs). MOFs can be considered to be complementary to zeolites due to their mainly organic interior and the presence of structural metal–organic complexes instead of single exchanged cations. Literature reports are still focusing mainly on MOF material synthesis, while the search for applications is just emerging. This chapter highlights the structural characteristics of MOFs, and their potential for heterogeneous catalysis and adsorptive separations.

4.2
Preparation and Structure

4.2.1
Definition, Synthesis, and Properties of MOFs

The term *metal–organic framework* was first used in 1999 by the group of Yaghi and coworkers and later adopted by many authors [3, 4]. Other terminology is also in use, for instance, *metal–organic coordination network* [5] or just *coordination polymer* [6]. The abbreviation MOF will be consistently used throughout this text. MOFs are considered as a subclass in the family of coordination polymers. A strict definition describes them as relatively rigid, crystalline, and porous three-dimensional lattices consisting of metal ions connected by multidentate organic ligands, often polycarboxylic acids, via coordinative bonds [3, 7]. A broader definition also

Novel Concepts in Catalysis and Chemical Reactors: Improving the Efficiency for the Future.
Edited by Andrzej Cybulski, Jacob A. Moulijn, and Andrzej Stankiewicz

ISBN: 978-3-527-32469-9

includes crystalline materials with coordinative metal-organic bonds in only one or two dimensions [5]. Further subdivisions are regularly introduced, for instance, the covalent organic frameworks (COFs), with a fully organic interior generated by condensation of organic boronic acids [8, 9], or the zeolitic imidazolate frameworks (ZIFs), claimed to have a relatively higher stability [10, 11]. The overview of definitions and terminology presented here should rather be seen as a snapshot of a field evolving at high pace.

Most MOFs are synthesized under hydrothermal or solvothermal conditions, favoring the dissolution of the organic ligands and the subsequent exchange with solvent ligands coordinated to the metal ions. New synthesis methods are being developed continuously, for instance, using microwaves, spectacularly shortening the duration of a synthesis [12], or using an electrical current via metal electrodes immersed in a solution containing the ligand [4]. Recently, the use of high-throughput experimentation (HTE) techniques has emerged as an efficient approach for optimization of synthesis procedures or for the search for new structures or trends via parallelization and miniaturization of synthesis reactions. In this way, the influence of many synthesis parameters can be quickly monitored [13, 14]. During a synthesis, solvent or unreacted ligand molecules often play the role of a weak template and obstruct the pores. An activation treatment is therefore required, for instance, a thermal and/or vacuum treatment [15, 16]. This has to be carried out with care owing to the relatively low thermal stability of MOFs (see below) [17].

Figure 4.1 schematically shows the crystal structure of MOF-5, a zinc terephthalate MOF. Coordinative bonds are established between the carboxylate groups of the terephthalate ligands and the Zn^{II} ions. In this way, elementary zinc carboxylate clusters based on a tetrahedron of Zn^{II} ions with a central oxygen atom are formed. This and other structural units are well known from coordination chemistry and function as secondary building units (SBUs) of the crystal lattices [3, 18]. A MOF lattice is, in fact, a familiar metal–organic complex extended into three dimensions by the use of polyfunctional ligands. A wide variety of MOFs has already been developed, incorporating not only transition metals but also s- or p-block elements [19, 20], and ligands ranging from simple, commercially available polycarboxylic acids like terephthalic acid (1,4-benzenedicarboxylic acid (BDC)) or trimesic acid (1,3,5-benzenetricarboxylic acid (BTC)) to synthetically challenging polyaromatic molecules like 4,4′,4″-s-triazine-2,4,6-triyltribenzoic acid (TATB) [20–24]. The building scheme of MOFs also seems to offer many more possibilities for the synthesis of chiral lattices in comparison with inorganic porous materials (see Sections 4.3.1.2 and 4.3.1.3) [25].

MOFs can be considered as organic zeolite analogs, as their pore architectures are often reminiscent of those of zeolites; a comparison of the physical properties of a series of MOFs and of zeolite NaY has been provided in Table 4.1. Although such coordinative bonds are obviously weaker than the strong covalent Si–O and Al–O bonds in zeolites, the stability of MOF lattices is remarkable, especially when their mainly organic composition is taken into account. Thermal decomposition generally does not start at temperatures below 300 °C [3, 21], and, in some cases,

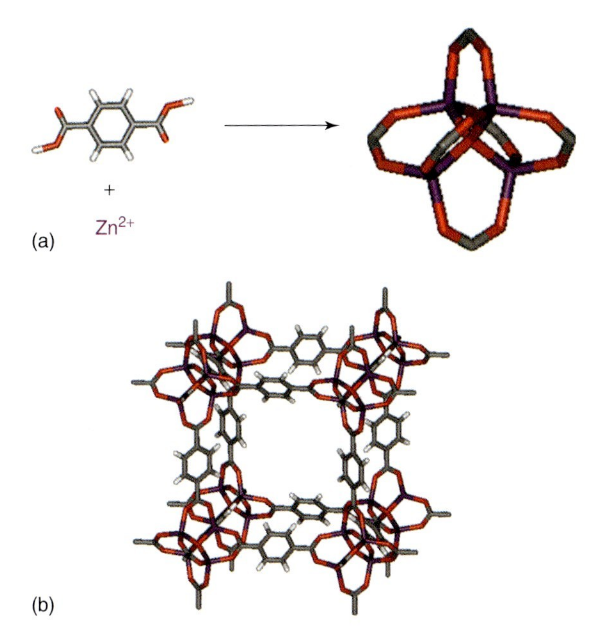

Figure 4.1 Representation of the structure of MOF-5: (a) six carboxylic acid groups from terephthalic acid are connected to four Zn^{II} cations to form a metal–organic elementary zinc carboxylate cluster (only the carboxylate moieties are shown; red = oxygen, gray = carbon, purple = zinc, white = hydrogen) and (b) the porous three-dimensional lattice of MOF-5.

Table 4.1 Comparison of the physical properties of MOF-5, $[Cu_3(BTC)_2]$, MIL-101E, and Zeolite NaY (3, 21, 32, 33).

	MOF-5	$[Cu_3(BTC)_2]$	MIL-101	Zeolite NaY
Specific Langmuir surface ($m^2\ g^{-1}$)	2900	1600	5900	890
Thermal stability (°C)	350	275	275	800
Window diameter (Å)	8	9	12–16	7
Cage diameter (Å)	11	12	29–34	12
Pore volume ($ml\ g^{-1}$)	1.0	0.6	2.0	0.3
Density ($g\ cm^{-3}$)	0.6	1.2	0.6	2.3

even not below 500 °C [20, 26]. The chemical stability of MOFs deserves special attention, as a number of MOFs has been found to be unstable toward water in solvent concentrations, or even to atmospheric moisture [27–29]. However, considering the high research intensity in the field of MOF synthesis, which has already resulted in a large number of stable and cheap structures that are presently available, early claims that MOFs would be too instable or impractical for industrial applications have definitely been proven invalid [30, 31].

The degree of robustness of MOFs and the ordering of their lattices are remarkable compared to those of coordination polymers, which have identical building blocks but lack the long-range ordering characteristic of MOFs [34]. This robustness is a consequence of the relative rigidity and the particular coordination mode of the ligands. In most MOFs, the ligands are resonance-stabilized structures, containing aromatic rings in many cases, and rotation around internal bonds is often severely impeded. Carboxylate complexes are most frequently encountered in MOF lattices because of their capability to lock several metal ions together into rigid clusters. Such SBUs contain much more directional information in comparison with, for instance, metal complexes consisting of a single metal ion connected to ligands with monocipital functional groups like amino groups. Moreover, many MOF lattices have a high resistance to shear due to the staggered implantation of the carboxylate groups on opposite sites of an SBU [18, 21]. However, in some MOFs, particular linkages may still have a considerable degree of rotational freedom. This results in lattices that may reversibly contract or expand as a response to external stimuli, for instance, adsorption or desorption of particular compounds (see Section 4.3.2). An example of such behavior can be found in MIL-53. In this lattice, the connections of the carboxylate groups with the metal clusters function as kneecaps, allowing breathing of the structure [35]. In the family of isoreticular MIL-88 frameworks, a similar mechanism is responsible for an increase in pore volume of more than 300%, involving atomic displacements larger than 10 Å [36, 37]. The simultaneous use of the terms *flexible* and *rigid* for MOFs may be perceived as contradictory here, but while rigidity clearly points to a relative stability that allows the existence of permanent micropores, flexibility refers to the dynamic properties of MOFs in comparison with zeolites rather than to a lack of structural cohesion or stability.

The pore structure of most MOFs remains intact after removal of the synthesis solvent [3]. As the pore walls are only one organic molecule thick (see Figure 4.1), MOFs are open structures with low densities, resulting in world records for specific surface per mass unit with values much higher than 1000 $m^2\ g^{-1}$ for many materials (Table 4.1) [3, 21, 32]. The inner walls of MOFs are largely organic in composition and contain aromatic nuclei in many cases. The effective pore diameters of MOFs are often difficult to extract from the literature, as they may be expressed as crystalline distances neglecting the van der Waals volumes of the surrounding atoms, hence overestimating the true diameter; moreover, the pores in MOFs can sometimes have irregular shapes. Effective pore diameters can, for instance, be estimated by probing the adsorption of a series of voluminous adsorbates [38].

4.2.2
Discovery Instead of Design

Figure 4.1 clearly illustrates the modular construction of MOFs, with ligands used as pillars and metal clusters as node points. One unique feature of MOFs seems to be the possibility to systematically modify their lattice. For instance, by replacing the terephthalate ligands of MOF-5 (Figure 4.1) with different dicarboxylate ligands, a family of isoreticular metal–organic frameworks (known in literature as the

IRMOF-series) with the same topology has been produced on the basis of the zinc carboxylate SBU [18, 39]. This and a small number of similar examples at first seemed to convince the scientific community that a rationalization of the synthesis of MOFs is possible, as the large amount of directional information already present in the building blocks would imply that new MOFs can be easily devised. This view seemed to be supported by the possibility to describe MOFs as *augmented topological networks* [18, 40, 41]. Terminology like *tailor-made MOFs* and *MOFs by design* has therefore often been used, but this is, in a sense, misleading. In theory, it is possible to systematically modify pore size and functionality by a deliberate choice of linkers and metal ions. However, the number of structures is, in practice, obviously limited by theoretical constraints, like enthalpies and energetic minima, which have fixed values for the possible structures to be synthesized, and which cannot be controlled or chosen. Furthermore, it appears that rather unpredictable parameters like crystallization rates, solubility constants, and nucleation and growth rates also critically determine the outcome of a synthesis. From this viewpoint, new MOF structures can be discovered or suggested, but they can neither be created, nor designed [42].

Recently, a more balanced view has come into use in the literature, which more adequately describes the possibilities for rationalization of MOF synthesis and also incorporates the limitations mentioned above. For those cases in which the synthetic conditions of the particular metal–organic SBUs of an existing MOF are well known, or for MOFs constructed from SBUs that can be formed in a rather wide range of synthetic conditions, it has been demonstrated that it is possible to apply modifications to their lattice. The synthesis of the IRMOF series can, for instance, be viewed from this perspective, as this achievement was mainly based on the mastering of the conditions for the formation of the elementary zinc carboxylate SBU (Figure 4.1). Other examples are the synthesis of $[Cu_3(BTC)_2]$, $[Cu_3(BTB)_2]$, and $[Cu_3(TATB)_2]$ each based on the same Cu paddlewheel complex [21, 23, 43], or the trioctahedral bipyramidal Fe^{III} cluster found in MIL-88, MIL-89, and MIL-101 (Figure 4.2) [44, 45]. Taking advantage of this degree of predictability, structure determination from XRD data can be much simplified [32]. Substitution of metal

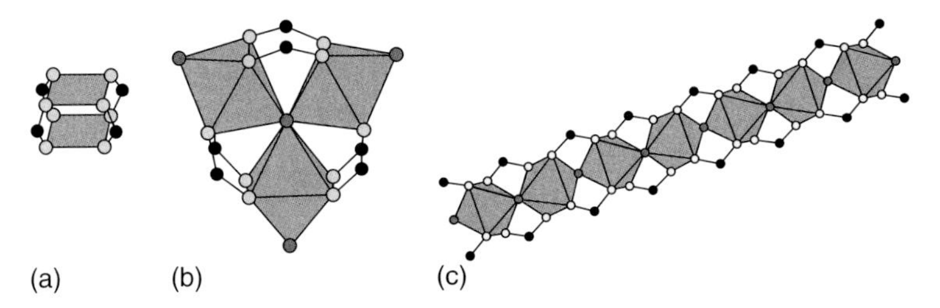

Figure 4.2 A number of secondary building units encountered in MOFs: (a) the paddlewheel complex; (b) the trioctahedral bipyramidal complex; and (c) the infinite chain of octahedra. The squares and octahedra represent the metals within their coordination sphere; carbon and oxygen atoms are in black and light gray, respectively; the medium gray atoms represent either an oxygen atom derived from a water molecule, a hydroxyl group, or a fluorine atom.

ions is also possible, for instance, in MIL-53 with Al^{III}, Cr^{III}, or Fe^{III} (Figure 4.2) [20, 46, 47], or in MIL-101 with Cr^{III} or Fe^{III} [44, 48]. In the case of $[Cu_3(BTC)_2]$ and $[Mo_3(BTC)_2]$, the existence of both structures was anticipated from the existence of the same type of carboxylate paddlewheel complex of Cu^{II} and Mo^{II} [21, 49]. A thorough knowledge of the mechanisms governing MOF synthesis is clearly at the base of a more rational search for new MOFs. HTE techniques are particularly helpful in developing these insights, as outlined above.

4.3 Applications

As of 2009 no industrial applications were using MOFs. The probing of their potential is at present largely the domain of academic research and their suitability for real applications should still be assessed in large-scale setups. As can be seen in Figure 4.3, there has been a strong belief in the catalytic potential of MOFs right from the onset of their development. Their potential for separations in liquid and gas phases has only been realized much more recently, with the first reports appearing in the literature starting from 2006.

In the following sections, we discuss a number of important achievements in catalysis and separation using MOFs, without trying to be exhaustive. Wherever possible, stability and reusability of MOFs are assessed as well, and their performance is compared to that of homogeneous or heterogeneous reference systems.

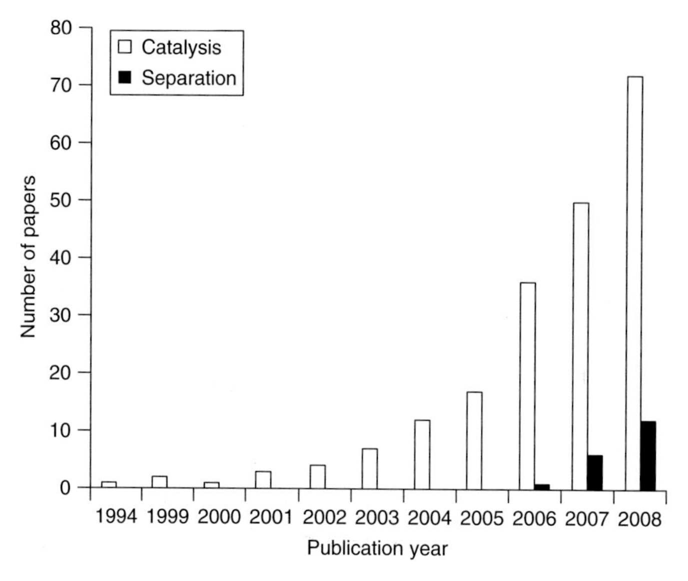

Figure 4.3 Number of published papers dealing with the use of MOFs for catalysis or separation [50].

4.3.1 Catalysis

Both of the components of MOFs, metal ions and ligands, can be involved in heterogeneous catalysis. The framework as a whole can also serve as a heterogeneous carrier that hosts catalytic functionalities. The central question of this section is whether it is possible to introduce reactive centers that are designed to transform organic molecules and that will simultaneously not affect the organic components of the frameworks themselves.

4.3.1.1 Active Metal Centers

Most MOFs are constructed from transition metal ions. The particular catalytic activities of many of these metal ions are well known and they are therefore the active ingredient of many homogeneous Lewis acid and oxidation catalysts. Their incorporation in a solid MOF lattice can be seen as a strategy to generate self-supporting catalysts. The fact that the metal sites are integral parts of the structure not only results in a high concentration of active sites but is also an excellent guarantee for easy catalyst separation and against leaching. The high stability and rigidity of MOFs are to a large extent a consequence of the high coordinative saturation of the metal ions by their ligands; however, the metal ions are inaccessible in most cases. Only in a limited number of structures do the metal ions in the SBUs still have solvent ligands in their coordination sphere, for instance, water or DMF. Upon removal of these ancillary ligands via a thermal or vacuum treatment, open metal sites are created.

Open metal sites can accept electron pairs from reagent molecules, which is a prerequisite for Lewis acid-catalyzed conversions. A typical example of such a site in MOFs is the Cu paddlewheel cluster (Figure 4.2). The most well-known MOF constructed from this SBU is $[Cu_3(BTC)_2]$; it has an open structure with easy access to the open metal Cu sites inside the large pores [21]. The Lewis acid properties of these Cu sites have been first exemplified using cyanosilylation reactions [15] and have thoroughly been investigated with two organic transformations: the isomerization of α-pinene oxide to campholenic aldehyde and the cyclization of (+)-citronellal to (−)-isopulegol, a precursor of (−)-menthol (Figure 4.4). In these reactions, the selectivity obtained with $[Cu_3(BTC)_2]$ as catalyst, respectively 84 and 66%, was equal or higher than those of a series of Cu- and Zn-based reference catalysts. The resistance of $[Cu_3(BTC)_2]$ toward leaching of Cu into the solution proves its heterogeneous character, facilitating an easy product separation after reaction. Another reaction, the rearrangement of the ethylene ketal of 2-bromopropiophenone (Figure 4.4), allows to obtain a detailed view on the acid nature of a catalyst, as the obtained product distribution can be directly related to the activity of particular acid sites. On Brönsted acid sites, the ketal is converted back to the ketone, while with soft Lewis acids, a dioxene compound is obtained and with hard Lewis acids, an ester compound. The yields (ketone, 0%; ester, 75%; dioxene, 25%) not only show that $[Cu_3(BTC)_2]$ is a performant catalyst for this reaction, but also prove that the active sites are hard Lewis acids and that Brönsted acid sites are absent [51]. These

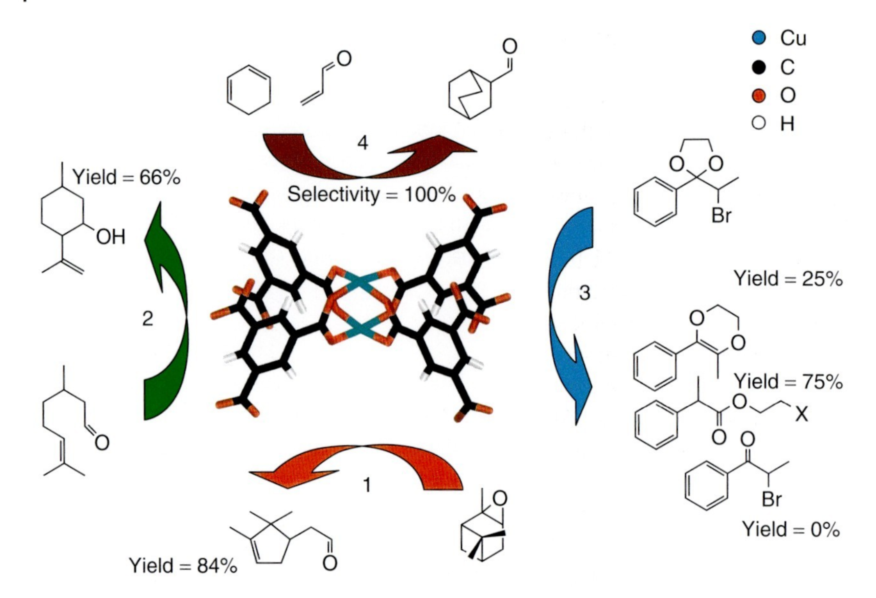

Figure 4.4 Reactions catalyzed by $[Cu_3(BTC)_2]$ (only the active site is shown). Yields or selectivities are given for each reaction product.

findings were confirmed by IR spectroscopy on adsorbed carbon monoxide and acetonitrile [38, 51]. As hard Lewis acids are generally performant Diels–Alder catalysts, $[Cu_3(BTC)_2]$ was tested for the cyclization of cyclohexadiene and acrolein; only cyclization products were obtained [38]. The activity of $[Cu_3(BTC)_2]$ was rather low as a consequence of a strong shielding effect of the four oxygen atoms surrounding the active sites, decreasing their degree of coordinative unsaturation (Figure 4.2) [52]. Formation of deposits in the pores caused a slow but progressive deactivation of the $[Cu_3(BTC)_2]$ catalysts. While this phenomenon is often encountered for these kinds of reactions, the fact that a MOF cannot withstand a calcination treatment urges the search for alternative reactivation treatments. In this particular case, a washing treatment in polar solvents was sufficient to restore the activity to close to its initial value [51].

Another MOF constructed from Cu paddlewheel SBUs and with 5-nitro-1,3-benzenedicarboxylate as the ligand can be used for the acetylation of methyl 4-hydroxybenzoate with acetic acid anhydride. The framework did not remain intact upon exposure to acetic acid, one of the reaction products [53].

Acetalization of benzaldehyde with trimethyl orthoformate can be carried out with a series of MOFs constructed from In and BDC or BTC ligands with open In sites. The catalysts are even stable in aqueous medium and can be reused without loss of activity. Owing to the small pores of these MOFs, the reaction only takes place at the outer surface of the crystals [54]. In another MOF constructed from In and 4,4′-(hexafluoroisopropylidene)bis(benzoic acid), the same reaction takes place inside the pores [55].

Another SBU with open metal sites is the tri-μ-oxo carboxylate cluster (see Section 4.2.2 and Figure 4.2). The tri-μ-oxo Fe^{III} clusters in MIL-100 are able to catalyze Friedel–Crafts benzylation reactions [44]. The tri-μ-oxo Cr^{III} clusters of MIL-101 are active for the cyanosilylation of benzaldehyde. This reaction is a popular test reaction in the MOF literature as a probe for catalytic activity; an example has already been given above for $[Cu_3(BTC)_2]$ [15]. In fact, the very first demonstration of the catalytic potential of MOFs had already been given in 1994 for a two-dimensional Cd bipyridine lattice that catalyzes the cyanosilylation of aldehydes [56]. A continuation of this work in 2004 for reactions with imines showed that the hydrophobic surroundings of the framework enhance the reaction in comparison with homogeneous Cd(pyridine) complexes [57]. The activity of MIL-101(Cr) is much higher than that of the Cd lattices, but in subsequent reaction runs the activity decreases [58]. A MOF with two different types of open Mn^{II} sites with pores of 7 and 10 Å catalyzes the cyanosilylation of aromatic aldehydes and ketones with a remarkable reactant shape selectivity. This MOF also catalyzes the more demanding Mukaiyama-aldol reaction [59].

The role of structural defects in MOFs has been probed as well. For instance, although the Zn atoms in intact MOF-5 are inaccessible for ligation, catalytic activities have been reported for this material, for instance, for esterification reactions or for para alkylation of large polyaromatic compounds [4, 60]. It is most probable that Zn–OH defects are created inside the pores as a consequence of adsorption of moisture [28].

For open metal sites in MOFs to be able to catalyze oxidation reactions, the framework must generally be able to accommodate the modified coordinative demand of the metal ions undergoing a change of the oxidation number. For MOFs constructed from Cu paddlewheel SBUs, this does not need to be problematic, as the molecular species ligated to the Cu sites may change its oxidation state instead of the metal atoms themselves. For instance, the incorporation of this complex into a two-dimensional MOF with *trans*-1,4-cyclohexanedicarboxylate ligands generates an active heterogeneous and reusable oxidation catalyst for the conversion of various aliphatic and aromatic alcohols to aldehydes or ketones with H_2O_2 with selectivities above 99%. This MOF consists of sheets connected by Cu-O bonds between adjacent Cu paddlewheels. Upon oxidation, the Cu paddlewheels are disconnected, rotated, and again linked to each other by peroxo bridges (Figure 4.5) [61, 62]. The intrinsic potential of Cu paddlewheel clusters to catalyze oxidation reactions is also illustrated by a MOF constructed from Cu^{II} and 5-methylisophthalate ligands with pores of about 8 Å used for CO oxidation. Activities are similar to or higher than those of CuO and CuO/Al_2O_3 reference catalysts, with full conversion of CO at 200 °C. The porous lattice has a stable activity and remains intact after catalytic reaction; no CuO is formed [63].

Mixed-valence $Ru^{II}-Ru^{III}$ paddlewheel carboxylate complexes also have potential for oxidation reactions after incorporation in a microporous lattice with porphyrinic ligands. This MOF can be used for oxidation of alcohols and for hydrogenation of ethylene. Both the porosity of the lattice and the ability of the diruthenium centers to chemisorb dioxygen are essential for the performance of the catalyst [62, 64].

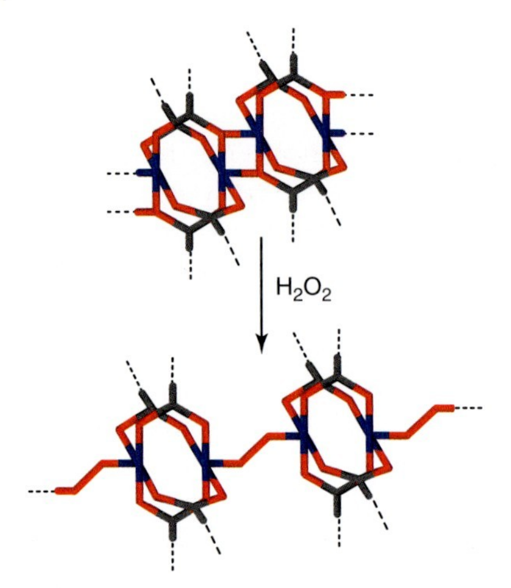

Figure 4.5 A MOF oxidation catalyst based on Cu paddlewheels connected by 1,4-*trans*-cyclohexane-dicarboxylate ligands (only carboxylate groups are explicitly shown). After oxidation by H_2O_2, paddlewheels are linked by peroxo bridges (Cu = blue; O = red; C = gray) [34]. (Reproduced by permission of the Royal Society of Chemistry.)

A study comparing the activities of a series of MOFs containing Sc, Y, or La has shown that the ability of their lattices to allow modifications of the coordination environment of the metals is crucial for their performance in the oxidation of sulfides with H_2O_2. Y and La more easily assume a high coordination number than Sc; the accommodation of an extra oxygen atom in the coordination sphere is necessary to allow the ligation of the proposed peroxo derivative. All catalysts are fully heterogeneous and can be reused without loss of selectivity. In comparison with the corresponding metal oxides, all four catalysts present a similar selectivity, but a higher activity; large, almost stoichiometric amounts of the catalysts were needed [26, 65, 66].

A MOF constructed from Pd and 2-hydroxypyrimidinolate ligands, $[Pd(2\text{-pymo})_2]_n$, was shown to be a versatile catalyst due to the fact that the framework is able to accommodate the change in the coordinative demand of Pd from tetra- to dicoordination, for instance, in the Suzuki–Miyaura coupling between phenylboronic acid and 4-bromoanisole. $[Pd(2\text{-pymo})_2]_n$ can also catalyze the aerobic oxidation of cinnamyl alcohol to cinnamylaldehyde and the hydrogenation of alkenes. Reactant shape selectivity is observed in the reaction of cyclododecene compared to the reaction of 1-octene. This observation also excludes a possible homogeneous hydrogenation activity of leached Pd species [67].

4.3.1.2 **Functionalized Organic Ligands**

In a few MOFs, the ligands incidentally show weak catalytic activity. For instance, the OH ligands connecting the metal ions in a number of structures, like in MIL-100, may have weak Brønsted acidity, but strength of these ligands is not comparable with the Brønsted acidity of zeolite OH groups [68]. On the other hand, protonated terminal carboxylic groups may also be Brønsted acid sites, although activity of such sites has not been detected yet [51]. The OH groups in MIL-53(Cr) have electron-donating capabilities that are involved in adsorption of CO_2 [69]. Basic sites capable of proton abstraction are found in the ligands of a Cu-pyrazine-2,3-carboxylate MOF. Their activity is sufficiently high to initiate anionic polymerization of monosubstituted acetylenes [70]. Basic sites capable of Knoevenagel condensation reactions are found in MOFs with aminated ligands, for instance, IRMOF-3 and amino-MIL-53(Al) [71].

Catalytically active ligands can also be deliberately introduced after synthesis. An example of this approach is found in MIL-100(Cr): its open metal sites can be grafted with different alcohols and also water, generating Brønsted acid sites with different strengths [72]. The open metal sites of MIL-101(Cr) can be grafted with ethylenediamine or diethylenetriamine for use in Knoevenagel reactions with benzaldehyde and ethyl cyanoacetate with selectivities above 99%. Reactant shape selectivity shows that the reaction occurs inside the pores [73]. Several examples of postsynthetic modification of MOF ligands using traditional methods of organic synthesis have been reported, but these efforts have not yet been directed toward devising catalytically active systems [74–76].

An original strategy to prepare catalytically active MOFs starts from particular ligands that themselves contain metal–organic complexes [77]. This approach is, for instance, encountered in the preparation of a Zn-based MOF with Mn(salen) ligands [78]. These ligands are anchored to sheets composed of Zn paddlewheels connected by 1,4′-biphenyldicarboxylate ligands. These sheets have only a structural role and are not involved in catalysis. The resulting crystalline structure has channels with dimensions of 6.2×15.7 and 6.2×6.2 Å (Figure 4.6). This material was applied in the asymmetric epoxidation of 2,2-dimethyl-2H-chromene. The reaction was shown to occur within the pores by reactant shape selectivity studies. A comparison with the activity of free Mn(salen) ligands shows that the activity of the MOF catalyst is lower, but more stable over time, as the immobilization of the Mn(salen) complexes prohibits reactive encounters between different complexes. The enantioselectivity of the MOF was only slightly lower than that of the homogeneous Schiff base complex (ee of 82 and 88%, respectively), possibly due to the higher rigidity of the ligands as a consequence of their immobilization. Upon reuse of the MOF, a small loss (4–7%) of catalytically inactive manganese was noticed after each reaction run, while enantioselectivity remained constant.

A similar system is encountered in a MOF constructed from Cd^{II} and (*R*)-6,6-dichloro-2,2-dihydroxy-1,1-binaphthyl-4,4-bipyridine ligands (Scheme 4.1). This MOF has a specific surface of 600 $m^2\,g^{-1}$ and a pore volume of 0.26 $ml\,g^{-1}$. After reaction of $Ti(OiPr)_4$ with the OH groups of the ligands, a heterogeneous Ti–BINOL catalyst is formed that is able to catalyze the addition of $ZnEt_2$ to aromatic aldehydes to afford chiral secondary alcohols with ee values higher than 80% for most aromatic substrates tested. The inactivity of a mixture of $Cd(pyridine)_2(H_2O)_2$ and $Ti(OiPr)_4$ illustrates the heterogeneous nature of the MOF catalyst [79, 80].

A MOF constructed from rhodium paddlewheel clusters linked to porphyrinic ligands already discussed in Section 4.3.1.1 shows an interesting synergetic behavior when the porphyrinic rings are loaded with metals like Cu^{II}, Ni^{II}, or Pd^{II}. In the hydrogenation of olefins, the hydride species at the rhodium center is transferred to the coordinated olefin adsorbed on a metal ion in the center of the porphyrin ring to form an alkyl species, and next this alkyl species reacts with a hydride species activated at the rhodium center to form the alkane [81].

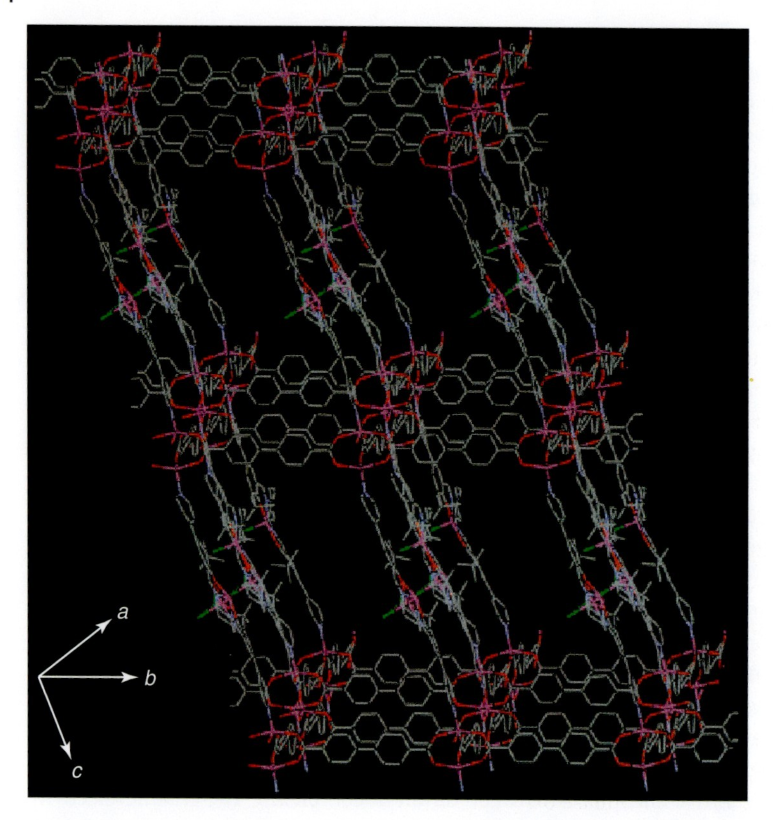

Figure 4.6 An enantioselective MOF epoxidation catalyst. Zn-(1,4′-biphenyldicarboxylate) sheets in the *ab* direction are connected by functionalized Mn–salen ligands to form a doubly interpenetrated porous structure (Zn = purple; Cl = green; C = gray; O = red; N = blue; Mn = pink; hydrogen atoms have been omitted for clarity) [34]. (Reproduced by permission of the Royal Society of Chemistry.)

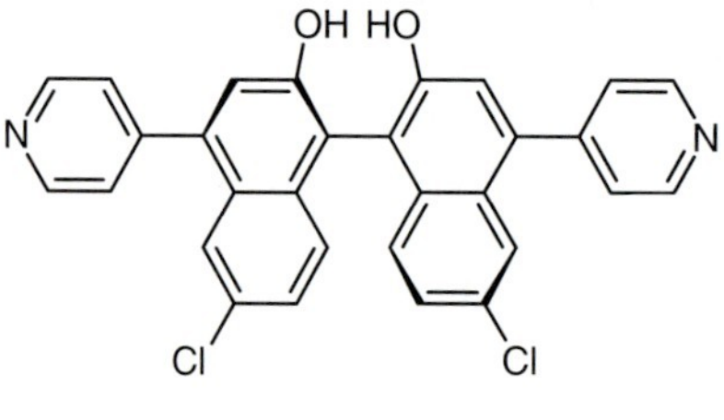

OH HO
N N
Cl Cl

Scheme 4.1 Structure of (*R*)-6,6-dichloro-2,2-dihydroxy-1,1-binaphthyl-4,4-bipyridine.

4.3.1.3 **Chiral MOF Catalysts**

Chiral MOFs have to cope with two kinds of challenges when they are used as catalysts. First, in those cases where chirality is induced via sp^3 carbon centers, the incorporation of such centers may confer a degree of flexibility to the framework that eventually impairs the stability after desolvation. Such problems can be circumvented, for instance, using mixed ligand systems where a rigid nonchiral backbone ensures the stability of the structure [82]. MOFs with large and flexible chiral ligands may also suffer from a lack of rigidity. A MOF constructed from catalytically active Cu paddlewheels and chiral 2,2′-dihydroxy-1,1′-binaphthalene-5,5′-dicarboxylic acid ligands is active for the ring opening of different epoxides with amines (ee = 50%), but its three-dimensional structure is reversibly destroyed upon desolvation of the lattice [83]. A strategy to counter such a lack of rigidity has already been encountered in the Mn–salen MOF discussed in Section 4.3.1.2.

Secondly, for a chiral lattice to be active as a chiral catalyst, the chiral groups should be located as close as possible to the active sites. This is, for instance, clear in the enantioselective D-POST-1 framework, constructed from Zn ions and chiral carboxylate ligands with catalytically active pyridyl groups positioned quite far from the chiral tartrate ligands. This MOF can be used for the transesterification of 2,4-dinitrophenyl acetate with 1-phenyl-2-propanol to 2,4-dinitrophenol and (S)-(2-phenyl)-isopropoxyacetate with a yield of 77% but a rather low ee value of 8% [84]. Another MOF constructed from Zn, L-lactic acid and terephthalic acid is able to adsorb sulfoxides inside its pores with modest ee values between 20 and 27%, but despite the presence of chiral centers inside the pores, no asymmetric induction was found for the oxidation of sulfides to sulfoxides [82]. In this regard, one of the advantages of the chiral metalloligand catalysts presented in Section 4.3.1.2 is that the active center itself is embedded in a chiral coordination sphere.

4.3.1.4 **MOF Lattices as Hosts for Catalytic Functionalities**

In spite of its low chemical stability, impregnated MOF-5 surprisingly has been intensively studied for this purpose. For instance, after impregnation with $AgNO_3$, it was used for the gas-phase oxidation of propylene to propylene oxide with O_2 at 220 °C, with a selectivity of 10.3% at a propylene conversion of 4.3% [85]. MOF-5 has also been loaded with metals by chemical vapor deposition for use in CO oxidation. When Pd or Cu is used, the obtained MOF samples show some catalytic activity, but their stability is problematic. A MOF-5 sample loaded in the same manner with Au particles with diameters of 5–20 nm remains intact, but is catalytically inactive, apparently owing to the rather large size of the particles or the lack of a strong metal–support interaction or promotion [86]. Incipient wetness impregnation of $Pd(acac)_2$ in MOF-5 leads to an active catalyst for hydrogenation of styrene, with an activity comparable to that of Pd on activated carbon [87]. The same authors also reported a coprecipitation method for the preparation of Pd-loaded MOF-5. Ethyl cinnamate can be hydrogenated with this catalyst with an activity twice as high as that of commercial Pd/C catalysts, but the active Pd finally ends up at the outer surface of the degraded catalyst [88].

The much more stable MIL-100(Cr) lattice can also be impregnated with $Pd(acac)_2$ via incipient wetness impregnation; the loaded catalyst is active for the hydrogenation of styrene and the hydrogenation of acetylene and acetylene–ethene mixtures to ethane [58]. MIL-101(Cr) has been loaded with Pd using a complex multistep procedure involving an addition of ethylene diamine on the open Cr sites of the framework. The Pd-loaded MIL-101(Cr) is an active heterogeneous Heck catalyst for the reaction of acrylic acid with iodobenzene [73].

MOFs with a pore system consisting of cages connected via relatively narrow windows can be used for occlusion of voluminous catalytically active species by physically preventing leaching (ship-in-a-bottle catalysts). The relatively mild synthesis conditions of many MOFs also permit a one-step construction and encapsulation of the occluded species. For instance, Mo/W-containing Keggin-type polyoxometalates can be placed inside the large pores of [$Cu_3(BTC)_2$] [89]. Such systems are heterogeneous and reusable acid catalysts, for instance, for the hydrolysis of esters. Conglomeration and deactivation of the polyoxometallate clusters is prohibited by their physical separation in the [$Cu_3(BTC)_2$] cages [90]. Another example of a ship-in-a-bottle MOF catalyst is the inclusion of porphyrinic compounds inside the matrix of an indium–imidazoledicarboxylate-based MOF, used for the oxidation of cyclohexane. Owing to the size of its cages, one porphyrin molecule is isolated per cage, preventing their oxidative self-degradation [91].

4.3.2
Separation

The search for adsorptive applications of MOFs has up to now mainly focused on the storage of small molecules in gas phase, for instance, H_2, CO_2, CH_4, or NO [46, 92, 93]. This section focuses on the application of MOFs for adsorptive separation of larger molecules and in the liquid phase, a domain in which their potential only recently has been recognized (Figure 4.3).

MOFs are promising materials for the separation of alkylaromatics, an industrial challenge due to the similarity in boiling points and dimensions of these compounds and the enormous markets for their derivatives, like PET or polystyrene plastics. In MIL-47 and MIL-53(Al), two MOFs with one-dimensional diamond-shaped pores and constructed from terephthalic acid and V^{IV} and Al^{III}, respectively, the adsorption selectivities measured between alkylaromatics are equal to or higher than those measured on industrially used zeolites; moreover, their uptake capacities are about twice as high. The potential of both materials for real separations in the liquid phase has been demonstrated by breakthrough experiments. Regeneration can be easily accomplished using a flow of aliphatic solvent [94, 95].

For both MOFs, the selectivity is a consequence of packing effects occurring at high pore loadings, and the adsorption preference is determined by enthalpic interactions. In MIL-47, the framework is not involved in these interactions: the adsorbed xylenes interact among themselves via $\pi-\pi$ interactions between their aromatic rings (Figure 4.7). For *ortho-* and *para-*xylene, a perfectly parallel orientation results in a high affinity. The lower uptakes of *meta-*xylene and ethylbenzene are due to an increasing sterical hindrance, which, in the case of ethylbenzene, even impedes the formation of $\pi-\pi$ pairs [94, 96]. For MIL-53, interactions between the alkyl side chains of the adsorbates and the framework determine the adsorption preference. Here, the geometrical implantation of the side chains is crucial: *ortho-*xylene can benefit the most from these interactions due to the nearby implantation of its alkyl groups (Figure 4.8). An identical mechanism was found in the selective adsorption of the larger ethyltoluene and cymene isomers [95].

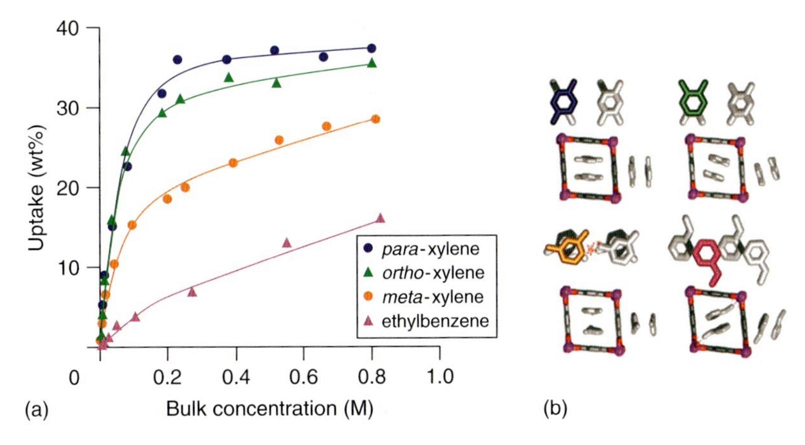

Figure 4.7 (a) Single-compound adsorption isotherms and (b) packing in the pores of C_8 alkylaromatics in MIL-47.

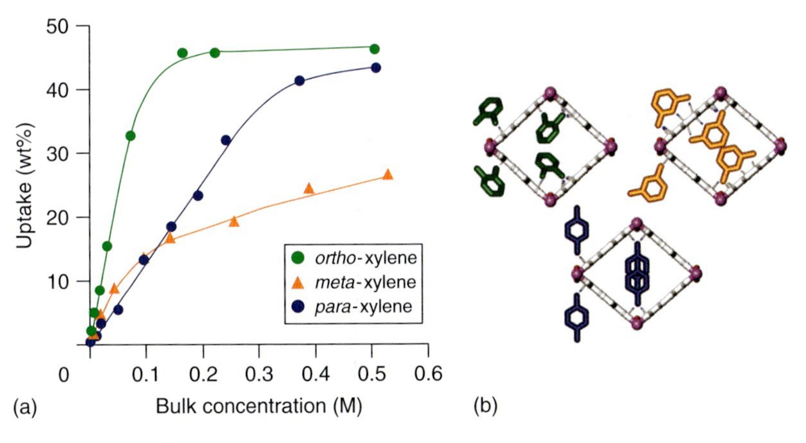

Figure 4.8 (a) Single-compound adsorption isotherms and (b) packing in the pores of C_8 alkylaromatics in MIL-53(Al).

Although MIL-47, and especially MIL-53(Al), had been found on many occasions to dynamically respond to adsorption of particular compounds, referred to as *breathing* [35] in the literature, in these liquid phase conditions, only minor changes of the lattice parameters have been observed. A study of xylene separations in vapor phase on MIL-53(Al) shows that breathing profoundly influences the shape of the obtained breakthrough profiles as a function of adsorbate concentration [97].

The potential of MIL-47 and MIL-53(Al) for adsorption of other types of aromatic adsorbates has also been explored, for instance, of dichlorobenzene, cresol, or alkylnaphthalene isomers [17, 98]. The removal of sulfur-containing aromatics from fuels via physisorption on MOFs has been investigated on several instances in literature, for instance, via the selective removal of thiophene from a stream of methane gas by MIL-47 [99], the removal of tetrahydrothiophene from methane by

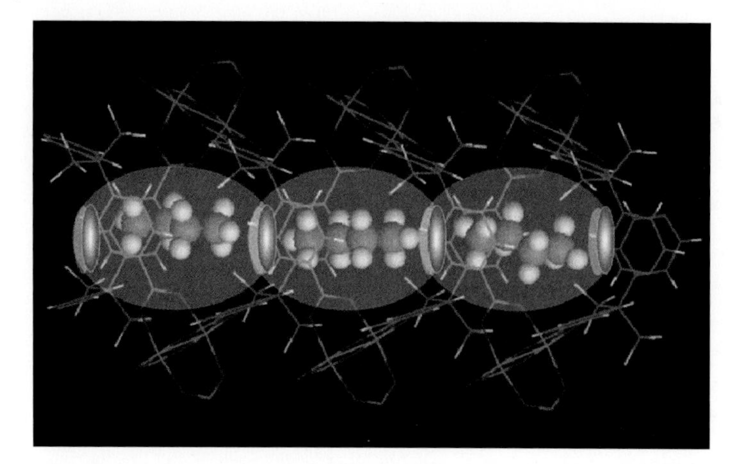

Figure 4.9 Arrangement of butane molecules in a microchannel of a Cu-hfipbb MOF (H_2hfipbb = 4,4′-(hexafluoroisopropylidene)bis(benzoic acid)); one molecule is adsorbed per cage. Copyright Wiley-VCH Verlag GmbH & Co. KGaA. Reproduced with permission from [103].

$[Cu_3(BTC)_2]$ [4] and the high uptake capacity of, for instance, MOF-5, $[Cu_3(BTC)_2]$, or MOF-177 for refractory sulfur compounds like dimethyldibenzothiophene [100].

Additional adsorption sites are provided on open metal sites, when available. $[Cu_3(BTC)_2]$ is performant in the selective adsorption and separation of olefinic compounds. The highly relevant separations of propene from propane and of isobutene from isobutane have been accomplished with separation factors of 2.0 and 2.1, respectively [101, 102]. $[Cu_3(BTC)_2]$ also selectively takes up pentene isomers from aliphatic solvent in liquid phase, and even discriminates between a series of *cis*- and *trans*-olefin isomer mixtures with varying chain length, always preferring a double bond in cis-position. This behavior is ascribed to π-complexation with the open Cu^{II} sites [100].

A peculiar shape-selective behavior has been observed in a Cu-based MOF containing one-dimensional tunnels with narrow necks at regular distances of 7.3 Å (Figure 4.9). The diameter of such a neck, 2.3 Å, is sufficiently large for a linear C-C chain to pass, but too small to also be an equilibrium adsorption position. The largest compound allowed inside the pores is a linear molecule limited in length to four carbon atoms due to the distance between two subsequent necks [103]. Another example of shape-selective behavior is found in a Zn-based MOF able to encapsulate linear hexane while branched hexanes are blocked [104].

4.4 Conclusion

MOFs entered the scientific scene almost 10 years ago. Since then, some have considered them to be another hype in materials science, while others have remained convinced that this is the onset of a new branch of high-potential materials. Today, this question is has not yet been fully resolved. This chapter convincingly shows that the research focusing on the applicability of MOFs in catalysis and separation has already resulted in a number of remarkable realizations and breakthroughs on the academic level in these 10 years.

For catalysis, a MOF can be considered as a novel way for immobilization of metal ions, and the high selectivities obtained in a series of Lewis acid and redox reactions confirm their potential as heterogeneous catalysts. The activity of open metal sites in the current generation of MOFs is, in many cases, low compared to the activity of exchanged transition metal cations in zeolites. Indeed, MOFs are neutral lattices and the metal ions have an important structural role as node points of the lattice, which partly or fully impedes their availability for catalysis. This viewpoint may also suggest an inverse relationship between activity of the metal ions and stability of the framework. For oxidation reactions, the performance and stability of MOFs can be hampered by the difficulty for the metal ions to modify their coordination environment, again because they are integral parts of the solid lattice. Besides the fact that the current efforts to understand the synthesis mechanisms of MOFs will result in the development of an increasing number of structures with accessible metal ions, there is clearly a need for strategies to increase the degree of exposure of the metal sites. The deliberate introduction of structural defects may prove to be an efficient active site engineering strategy.

The issue of inactive metal sites can be circumvented via two approaches. A first approach exploits the organic ligands of the lattice. While they offer, in theory, a variety of opportunities for introducing catalytic functionalities, only in the case of metal-loaded chiral Salen and binaphthol ligands has this approach led to the production of a series of active and selective MOF catalysts; however, the high cost of such ligands is problematic for eventual applications. The high activity of these metal ions without a structural role seems to confirm the hypothesis forwarded in the previous section. Focusing on chiral MOF catalysis, the examples also illustrate that chirality can only be efficiently induced in the reaction products when the immediate environment of the active site is chiral. In a second approach, MOFs can be loaded with catalytically active phases, for instance, metal nanospheres or occluded polyoxometallate ions.

While the lack of or the low activity of open metal sites is often seen as the major bottleneck presently hampering a wider application of MOFs in catalysis, it can be considered as advantageous in the field of separation, where side reactions of the adsorbates are undesirable. The potential of MOFs in this area lies in the large number of new pore topologies available and the mainly organic nature of the pore walls. The excellent performance of MOFs for separations of alkylaromatic isomers illustrates that they provide ideal environments for selective molecular confinement, a property that has also become apparent in a number of examples of peculiar shape-selective behavior. In all examples, two major advantages of MOFs compared to traditional zeolitic adsorbents are clear. Their uptake capacities per unit of mass are about twice those of traditional inorganic adsorbents, allowing lower amounts of material to be used. Their easy regeneration is an important advantage for the energetics of future adsorption processes.

One of the challenges in MOF research is certainly the search for a more systematic understanding of their behavior, to discern trends in the present collection of papers describing active or selective MOFs, which could presently be considered as rather isolated reports. Another challenge is the translation of the concepts of the modular construction of MOFs, now already explored in the synthesis of isoreticular series, in deliberate modifications of Lewis acid or oxidative properties of the metals, or in variations of the inner surface polarity or pore size. Beyond activity and selectivity, the questions of time-on-stream behavior and leaching stability will also be crucial parameters eventually determining whether MOFs will be used in industrial processes. The cost of MOF ingredients will certainly decrease with a scale-up of their syntheses and with the development of novel synthesis pathways for the organic ligands. A trend toward developing MOFs with ingredients of low toxicity has already been discerned in the presently increasing number of MOFs containing Al, Mg, or Fe.

Acknowledgments

L.A. is grateful to FWO (Research Foundation – Flanders) for a position as a postdoctoral fellow. We thank K.U. Leuven for support in the frame of the CECAT excellence and Methusalem projects.

Abbreviations

MOF	metal–organic framework
SBU	secondary building unit
BDC	1,4-benzenedicarboxylate
BTC	1,3,5-benzenetricarboxylate
MIL	matérial Institut Lavoisier
TATB	4,4′,4″,-s-triazine-2,4,6-triyltribenzoate
BTB	4,4′,4″-benzene-1,3,5-triyl-tribenzoate
DMF	dimethylformamide
BINOL	1,1′-binaphthyl-2,2′-diol

References

1. Chen, Y., Kam, A., Kennedy, C., Ketkar, A., Nace, D., and Ware, R. (1988) US Patent No. 4740292.
2. Kulprathipanja, S. (1995) US Patent No. 5453560.
3. Li, H., Eddaoudi, M., O'Keeffe, M., and Yaghi, O. (1999) *Nature*, **402**, 276.
4. Müller, U., Schubert, M., Teich, F., Pütter, H., Schierle-Arndt, K., and Pastré, J. (2006) *J. Mater. Chem.*, **16**, 626.
5. Janiak, C. (2003) *Dalton Trans.*, **14**, 2781.
6. (a) Kitagawa, S., Kitaura, R., and Noro, S. (2004) *Angew. Chem.*, **116**, 2388; (b) Kitagawa, S., Kitaura, R., and Noro, S. (2004) *Angew. Chem. Int. Ed.*, **43**, 2334.

7. Rowsell, J. and Yaghi, O. (2004) *Microporous Mesoporous Mater.*, **73**, 3.
8. Cote, A., Benin, A., Ockwig, N., O'Keeffe, M., Matzger, A., and Yaghi, O. (2005) *Science*, **310**, 1166.
9. (a) Mastalerz, M. (2008) *Angew. Chem.* **120**, 453; (b) Mastalerz, M. (2008) *Angew. Chem. Int. Ed.*, **47**, 445.
10. Sung Park, K., Ni, Z., Cote, A., Yong Choi, J., Huang, R., Uribe-Romo, F., Chae, H., O'Keeffe, M., and Yaghi, O. (2006) *Proc. Natl. Acad. Sci.*, **103**, 10186.
11. Banerjee, R., Phan, A., Wang, B., Knobler, C., Furukawa, H., O'Keeffe, M., and Yaghi, O. (2008) *Science*, **319**, 939.
12. Jhung, S., Lee, J., Yoon, J., Serre, C., Férey, G., and Chang, J. (2007) *Adv. Mater.*, **19**, 121.
13. Bauer, S., Serre, C., Devic, T., Horcajada, P., Marrot, J., Ferey, G., and Stock, N. (2008) *Inorg. Chem.*, **47**, 7568.
14. Biemmi, E., Christian, S., Stock, N., and Bein, T. (2009) *Microporous Mesoporous Mater.*, **117**, 111.
15. Schlichte, K., Kratzke, T., and Kaskel, S. (2004) *Microporous Mesoporous Mater.*, **73**, 81.
16. Dietzel, P., Johnsen, R., Blom, R., and Fjellvag, H. (2008) *Chem. Eur. J.*, **14**, 2389.
17. Alaerts, L., Maes, M., Jacobs, P., Denayer, J., and De Vos, D. (2008) *Phys. Chem. Chem. Phys.*, **10**, 2979.
18. Yaghi, O., O'Keeffe, M., Ockwig, N., Chae, H., Eddaoudi, M., and Kim, J. (2003) *Nature*, **423**, 705.
19. Senkovska, I. and Kaskel, S. (2006) *Eur. J. Inorg. Chem.*, **22**, 4564.
20. Loiseau, T., Serre, C., Huguenard, C., Fink, G., Taulelle, F., Henry, M., Bataille, T., and Férey, G. (2004) *Chem. Eur. J.*, **10**, 1373.
21. Chui, S., Lo, S., Charmant, J., Orpen, A., and Williams, I. (1999) *Science*, **283**, 1148.
22. (a) Latroche, M., Surblé, S., Serre, C., Mellot-Draznieks, C., Llewellyn, P., Lee, J., Chang, J., Jhung, S., and Ferey, G. (2006) *Angew. Chem.*, **118**, 8407; (b) Latroche, M., Surblé, S., Serre, C., Mellot-Draznieks, C., Llewellyn, P., Lee, J., Chang, J., Jhung, S., and Ferey, G. (2006) *Angew. Chem. Int. Ed.*, **45**, 8227.
23. Sun, D., Ke, M., Collins, D., and Zhou, H. (2006) *J. Am. Chem. Soc.*, **128**, 3896.
24. Ma, S. and Zhou, H. (2006) *J. Am. Chem. Soc.*, **128**, 11734.
25. Lin, W. (2005) *J. Solid State Chem.*, **178**, 2486.
26. Miller, S., Wright, P., Serre, C., Loiseau, T., Marrot, J., and Férey, G. (2005) *Chem. Commun.*, 3850.
27. Lin, C., Chui, S., Lo, S., Shek, F., Wu, M., Suwinska, K., Lipkowski, J., and Williams, I. (2002) *Chem. Commun.*, 1642.
28. Huang, L., Wang, H., Chen, J., Wang, Z., Sun, J., Zhao, D., and Yan, Y. (2003) *Microporous Mesoporous Mater.*, **58**, 105.
29. (a) Reineke, T., Eddaoudi, M., O'Keeffe, M., and Yaghi, O. (1999) *Angew. Chem.*, **111**, 2721; (b) Reineke, T., Eddaoudi, M., O'Keeffe, M., and Yaghi, O. (1999) *Angew. Chem. Int. Ed.*, **38**, 2590.
30. Millange, F., Serre, C., and Ferey, G. (2002) *Chem. Commun.*, 822.
31. (a) Millange, F., Serre, C., Guillou, N., Ferey, G., and Walton, R. (2008) *Angew. Chem.*, **120**, 4168; (b) Millange, F., Serre, C., Guillou, N., Ferey, G., and Walton, R. (2008) *Angew. Chem. Int. Ed.*, **47**, 4100.
32. Ferey, G., Mellot-Draznieks, C., Serre, C., Millange, F., Dutour, J., Surble, S., and Margiolaki, I. (2005) *Science*, **309**, 2040.
33. Online product catalog, directory Catalysis and Inorganic Chemistry. Information obtained from Sigma-Aldrich® available online at *http://www.sigmaaldrich.com/sigma-aldrich/home.html* (accessed May 2010).
34. Alaerts, L., Wahlen, J., Jacobs, P., and De Vos, D. (2008) *Chem. Commun.*, 1727.
35. Serre, C., Bourrelly, S., Vimont, A., Ramsahye, N., Maurin, G., Llewellyn, P., Daturi, M., Filinchuk, Y., Leynaud, O., Barnes, P., and Férey, G. (2007) *Adv. Mater.*, **19**, 2246.
36. Ferey, G. (2007) *Stud. Surf. Sci. Catal.*, **170**, 66.

37. Mellot-Draznieks, C., Serre, C., Surble, S., Audebrand, N., and Ferey, G. (2005) *J. Am. Chem. Soc.*, **127**, 16273.
38. Alaerts, L., Thibault-Starzyk, F., Séguin, E., Denayer, J., Jacobs, P., and De Vos, D. (2007) *Stud. Surf. Sci. Catal.*, **170**, 1996.
39. Eddaoudi, M., Kim, J., Rosi, N., Vodak, D., Wachter, J., O'Keeffe, M., and Yaghi, O. (2002) *Science*, **295**, 469.
40. Eddaoudi, M., Moler, D., Li, H., Chen, B., Reineke, T., O'Keeffe, M., and Yaghi, O. (2001) *Acc. Chem. Res.*, **34**, 319.
41. Kim, J., Chen, B., Reineke, T., Li, H., Eddaoudi, M., Moler, D., O'Keeffe, M., and Yaghi, O. (2001) *J. Am. Chem. Soc.*, **123**, 8239.
42. (a) Jansen, M. and Schön, J. (2006) *Angew. Chem.*, **18**, 3484; (b) Jansen, M. and Schön, J. (2006) *Angew. Chem. Int. Ed.*, **45**, 3406.
43. Chen, B., Eddaoudi, M., Hyde, S., O'Keeffe, M., and Yaghi, O. (2001) *Science*, **291**, 1021.
44. Horcajada, P., Surblé, S., Serre, C., Hong, D., Seo, Y., Chang, J., Grenèche, J., Margiolaki, I., and Férey, G. (2007) *Chem. Commun.*, 2820.
45. (a) Serre, C., Millange, F., Surblé, S., and Férey, G. (2004) *Angew. Chem.*, **116**, 6446; (b) Serre, C., Millange, F., Surblé, S., and Férey, G. (2004) *Angew. Chem. Int. Ed.*, **43**, 6285.
46. Férey, G., Latroche, M., Serre, C., Millange, F., Loiseau, T., and Percheon-Guégan, A. (2003) *Chem. Commun.*, 2976.
47. Whitfield, T., Wang, X., Liu, L., and Jacobson, A. (2005) *Solid State Sci.*, **7**, 1096.
48. (a) Férey, G., Serre, C., Mellot-Draznieks, C., Millange, F., Surblé, S., Dutour, J., and Margiolaki, I. (2004) *Angew. Chem.*, **116**, 6456; (b) Férey, G., Serre, C., Mellot-Draznieks, C., Millange, F., Surblé, S., Dutour, J., and Margiolaki, I. (2004) *Angew. Chem. Int. Ed.*, **43**, 6296.
49. Kramer, M., Schwarz, U., and Kaskel, S. (2006) *J. Mater. Chem.*, **16**, 2245.
50. ACS SciFinder Scholar search engine.
51. Alaerts, L., Seguin, E., Poelman, H., Thibault-Starzyk, F., Jacobs, P., and De Vos, D. (2006) *Chem. Eur. J.*, **12**, 7353.
52. Prestipino, C., Regli, L., Vitillo, J., Bonino, F., Damin, A., Lamberti, C., Zecchina, A., Solari, P., Kongshaug, K., and Bordiga, S. (2006) *Chem. Mater.*, **18**, 1337.
53. Burrows, A., Frost, C., Mahon, M., Winsper, M., Richardson, C., Attfield, J., and Rodgers, J. (2008) *Dalton Trans.*, **47**, 6788.
54. Gomez-Lor, B., Gutierrez-Puebla, E., Iglesias, M., Monge, M., Ruiz-Valero, C., and Snejko, N. (2005) *Chem. Mater.*, **17**, 2568.
55. Gandara, F., Gomez-Lor, B., Gutierrez-Puebla, E., Iglesias, M., Monge, M., Proserpio, D., and Snejko, N. (2008) *Chem. Mater.*, **20**, 72.
56. Fujita, M., Kwon, Y., Washizu, S., and Ogura, K. (1994) *J. Am. Chem. Soc.*, **116**, 1151.
57. Ohmori, O., and Fujita, M. (2004) *Chem. Commun.*, 1586.
58. Henschel, A., Gedrich, K., Kraehnert, R., and Kaskel, S. (2008) *Chem. Commun.*, 4192.
59. Horike, S., Dinca, M., Tamaki, K., and Long, J. (2008) *J. Am. Chem. Soc.*, **130**, 5854.
60. Ravon, U., Domine, M., Gaudillere, C., Desmartin-Chomel, A., and Farrusseng, D. (2008) *New J. Chem.*, **32**, 937.
61. Kato, C., Hasegawa, M., Sato, T., Yoshizawa, A., Inoue, T., and Mori, W. (2005) *J. Catal.*, **230**, 226.
62. Kato, C. and Mori, W. (2007) *C. R. Chim.*, **10**, 284.
63. Zou, R., Sakurai, H., Han, S., Zhong, R., and Xu, Q. (2007) *J. Am. Chem. Soc.*, **129**, 8402.
64. Mori, W., Takamizawa, S., Kato, C., Ohmura, T., and Sato, T. (2004) *Microporous Mesoporous Mater.*, **73**, 31.
65. Perles, J., Iglesias, M., Ruiz-Valero, C., and Snejko, N. (2004) *J. Mater. Chem.*, **14**, 2683.
66. Perles, J., Iglesias, M., Martin-Luengo, M., Monge, M., Ruiz-Valero, C., and Snejko, N. (2005) *Chem. Mater.*, **17**, 5837.

67. Llabres i Xamena, F., Abad, A., Corma, A., and Garcia, H. (2007) *J. Catal.*, **250**, 294.
68. Vimont, A., Goupil, J., Lavalley, J., Daturi, M., Surblé, S., Serre, C., Millange, F., Férey, G., and Audebrand, N. (2006) *J. Am. Chem. Soc.*, **128**, 3218.
69. Vimont, A., Travert, A., Bazin, P., Lavalley, J., Daturi, M., Serre, C., Ferey, G., Bourrelly, S., and Llewellyn, P. (2007) *Chem. Commun.*, 3291.
70. (a) Uemura, T., Kitaura, R., Ohta, Y., Nagaoka, M., and Kitagawa, S. (2006) *Angew. Chem.*, **118**, 4218; (b) Uemura, T., Kitaura, R., Ohta, Y., Nagaoka, M., and Kitagawa, S. (2006) *Angew. Chem. Int. Ed.*, **45**, 4112.
71. Gascon, J., Aktay, U., Hernandez-Alonso, M., Van Klink, G., and Kapteijn, F. (2009) *J. Catal.*, **261**, 75.
72. Vimont, A., Leclerc, H., Mauge, F., Daturi, M., Lavalley, J., Surble, S., Serre, C., and Ferey, G. (2007) *J. Phys. Chem. C*, **111**, 383.
73. (a) Hwang, Y., Hong, D., Chang, J., Jhung, S., Seo, Y., Kim, J., Vimont, A., Daturi, M., Serre, C., and Ferey, G. (2008) *Angew. Chem.*, **120**, 4212; (b) Hwang, Y., Hong, D., Chang, J., Jhung, S., Seo, Y., Kim, J., Vimont, A., Daturi, M., Serre, C., and Ferey, G. (2008) *Angew. Chem. Int. Ed.*, **47**, 4144.
74. (a) Wang, Z. and Cohen, S. (2008) *Angew. Chem.*, **120**, 4777; (b) Wang, Z., Cohen, S. (2008) *Angew. Chem. Int. Ed.*, **47**, 4699.
75. Wang, Z., Tanabe, K., and Cohen, S. (2009) *Inorg. Chem.*, **48**, 296.
76. (a) Burrows, A., Frost, C., Mahon, M., and Richardson, C. (2008) *Angew. Chem.*, **120**, 8610; (b) Burrows, A., Frost, C., Mahon, M., and Richardson, C. (2008) *Angew. Chem. Int. Ed.*, **47**, 8482.
77. Kitagawa, S., Norob, S., and Nakamura, T. (2006) *Chem. Commun.*, 701.
78. Cho, S., Ma, B., Nguyen, S., Hupp, J., and Albrecht-Schmitt, T. (2006) *Chem. Commun.*, 2563.
79. Wu, C., Hu, A., Zhang, L., and Lin, W. (2005) *J. Am. Chem. Soc.*, **127**, 8940.
80. (a) Wu, C. and Lin, W. (2007) *Angew. Chem.*, **119**, 1093; (b) Wu, C. and Lin, W. (2007) *Angew. Chem. Int. Ed.*, **46**, 1075.
81. Sato, T., Mori, W., Kato, C., Yanaoka, E., Kuribayashi, T., Ohtera, R., and Shiraishi, Y. (2005) *J. Catal.*, **232**, 186.
82. (a) Dybtsev, D., Nuzhdin, A., Chun, H., Bryliakov, K., Talsi, E., Fedin, V., and Kim, K. (2006) *Angew. Chem.*, **118**, 930; (b) Dybtsev, D., Nuzhdin, A., Chun, H., Bryliakov, K., Talsi, E., Fedin, V., and Kim, K. (2006) *Angew. Chem. Int. Ed.*, **45**, 916.
83. Tanaka, K., Odaa, S., and Shiro, M. (2008) *Chem. Commun.*, 820.
84. Seo, J., Whang, D., Lee, H., Jun, S., Oh, J., Jeon, Y., and Kim, K. (2000) *Nature*, **404**, 982.
85. Muller, U., Lobree, L., Hesse, M., Yaghi, O.M., and Eddaoudi, M. (2003) US Patent No. 6624318.
86. (a) Hermes, S., Schröter, M., Schmid, R., Khodeir, L., Muhler, M., Tissler, A., Fisher, R., and Fisher, R. (2005) *Angew. Chem.*, **117**, 6394; (b) Hermes, S., Schröter, M., Schmid, R., Khodeir, L., Muhler, M., Tissler, A., Fisher, R., and Fisher, R. (2005) *Angew. Chem. Int. Ed.*, **44**, 6237.
87. Sabo, M., Henschel, A., Froede, H., Klemm, E., and Kaskel, S. (2007) *J. Mater. Chem.*, **17**, 3827.
88. Opelt, S., Tuerk, S., Dietzsch, E., Henschel, A., Kaskel, S., and Klemm, E. (2008) *Catal. Commun.*, **9**, 1286.
89. Yang, L., Naruke, H., and Yamase, T. (2003) *Inorg. Chem. Commun.*, **6**, 1020.
90. Sun, C., Liu, S., Liang, D., Shao, K., Ren, Y., and Su, Z. (2009) *J. Am. Chem. Soc.*, **131**, 1883.
91. Alkordi, M., Liu, Y., Larsen, R., Eubank, J., and Eddaoudi, M. (2008) *J. Am. Chem. Soc.*, **130**, 12639.
92. Bourrelly, S., Llewellyn, P., Serre, C., Millange, F., Loiseau, T., and Ferey, G. (2005) *J. Am. Chem. Soc.*, **127**, 13519.
93. McKinlay, A., Xiao, B., Wragg, D., Wheatley, P., Megson, I., and Morris, R. (2008) *J. Am. Chem. Soc.*, **130**, 10440.
94. (a) Alaerts, L., Kirschhock, C., Maes, M., van der Veen, M., Finsy, V., Depla, A., Martens, J., Baron, G., Jacobs, P., Denayer, J., and De Vos, D. (2007) *Angew.*

Chem., **119**, 4371; (b) Alaerts, L., Kirschhock, C., Maes, M., van der Veen, M., Finsy, V., Depla, A., Martens, J., Baron, G., Jacobs, P., Denayer, J., and De Vos, D. (2007) *Angew. Chem. Int. Ed.*, **46** (23), 4293.

95. Alaerts, L., Maes, M., Giebeler, L., Jacobs, P., Martens, J., Denayer, J., Kirschhock, C., and De Vos, D. (2008) *J. Am. Chem. Soc.*, **130**, 14170.

96. Finsy, V., Verelst, H., Alaerts, L., De Vos, D., Jacobs, P., Baron, G., and Denayer, J. (2008) *J. Am. Chem. Soc.*, **130**, 7110.

97. Finsy, V., Kirschhock, C.E.A., Vedts, G., Maes, M., Alaerts, L., De Vos, D.E., Baron, G.V., and Denayer, J.F.M. (2009) *Chem. Eur. J.*, **15**, 7724.

98. Alaerts, L., Maes, M., van der Veen, M., Jacobs, P., and De Vos, D. (2009) *Phys. Chem. Chem. Phys.*, **11**, 2903.

99. (a) Wang, X., Liu, L., and Jacobson, A.J. (2006) *Angew. Chem.*, **118**, 6649; (b) Wang, X., Liu, L., and Jacobson, A.J. (2006) *Angew. Chem. Int. Ed.*, **45**, 6499.

100. Cychosz, K., Wong-Foy, A., and Matzger, A. (2008) *J. Am. Chem. Soc.*, **130**, 6938.

101. Wagener, A., Schindler, M., Rudolphi, F., and Ernst, S. (2007) *Chem. Ing. Tech.*, **79**, 851.

102. Hartmann, M., Kunz, S., Himsl, D., and Tangermann, O. (2008) *Langmuir*, **24**, 8634.

103. (a) Pan, L., Olson, D., Ciemnolonski, L., Heddy, R., and Li, J. (2006) *Angew. Chem.*, **118**, 632; (b) Pan, L., Olson, D., Ciemnolonski, L., Heddy, R., and Li, J. (2006) *Angew. Chem. Int. Ed.*, **45**, 616.

104. Barcia, P., Zapata, F., Silva, J., Rodrigues, A., and Chen, B. (2007) *J. Phys. Chem. B*, **111**, 6101.

5
Enzymatic Catalysis Today and Tomorrow

Piotr Kiełbasiński, Ryszard Ostaszewski, and Wiktor Szymański

5.1
Introduction

Enzymes, which are extremely efficient catalysts developed by nature to catalyze practically all the chemical reactions that take place in living organisms, have also been found capable of accepting and transforming unnatural, man-made substrates [1], also among them the heteroatom-containing ones [2]. Moreover, enzymes can work in a nonaqueous environment – in organic solvents and other types of unusual media (see Sections 5.2.3.1 and 5.2.3.2). Hence, there is a continuously growing interest in their use as catalysts in organic synthesis, including industrial processes [3, 4]. The methodology, generally called *biocatalysis,* includes fermentation, biotransformations by whole cells, and catalysis by isolated enzymes. Enzymes, which are involved in each of the above procedures, are proteins[1] consisting of 40–1000 amino acid residues and, as such, have all the features characteristic of these classes of compounds.

Compared to synthetic catalysts, enzymes have many advantages. First of all, being natural products, they are environmentally benign and therefore their use does not meet public opposition. Enzymes act at atmospheric pressure, ambient temperature, and at pH between 4 and 9, thus avoiding extreme conditions, which might result in undesired side reactions. Enzymes are extremely selective (see below). There are also, of course, some drawbacks of biocatalysts. For example, enzymes are known in only one enantiomeric form, as they consist of natural enantiomeric (homochiral) amino acids; their possible modifications are difficult to achieve (see Section 5.3.2); they are prone to deactivation owing to inappropriate operation parameters and to inhibition phenomena.

However, the particular usefulness of enzymes stems from their selectivity. There are three types of selectivity exerted by enzymes: chemoselectivity, regioselectivity, and stereoselectivity.

1) An exception being ribozymes and DNA-zymes – polynucleotides – whose application in organic synthesis is, however, scarce.

Novel Concepts in Catalysis and Chemical Reactors: Improving the Efficiency for the Future.
Edited by Andrzej Cybulski, Jacob A. Moulijn, and Andrzej Stankiewicz

ISBN: 978-3-527-32469-9

5.1.1 Chemoselectivity

Chemoselectivity of enzymatic reaction is the enzyme-catalyzed transformation of one type of functional group in the presence of other sensitive groups present in the substrate molecule. As a result, reactions catalyzed by enzymes generally tend to be "cleaner" and purification of product(s) from impurities can largely be omitted. Therefore, all enzymatic reactions generate less by-products and waste compared to chemical transformations. This simplifies all operations and reduces costs of transformation, which is of great importance for the industry.

5.1.2 Regioselectivity

The enzyme catalytic center (active site) has a complicated three-dimensional structure that easily distinguishes between two or more identical functional groups located in different sites of the substrate molecule. As a result, only one group participates in the reaction, which is far better than the possibilities with chemical catalysts. A good example, which illustrates the regioselectivity in enzyme-catalyzed reactions, is deacylation of sugar derivatives, which is hardly available using other methods [5, 6]. The deacylation of methyl 2,3,4,6-tetra-*O*-acyl-D-hexopyranosides **1** possessing four different hydroxyl group protected with pentanoyl groups proceeds smoothly in the presence of the lipase from *Candida cylindracea* to give the 6-OH derivatives **2** in high yields (80–90%, Scheme 5.1) [7].

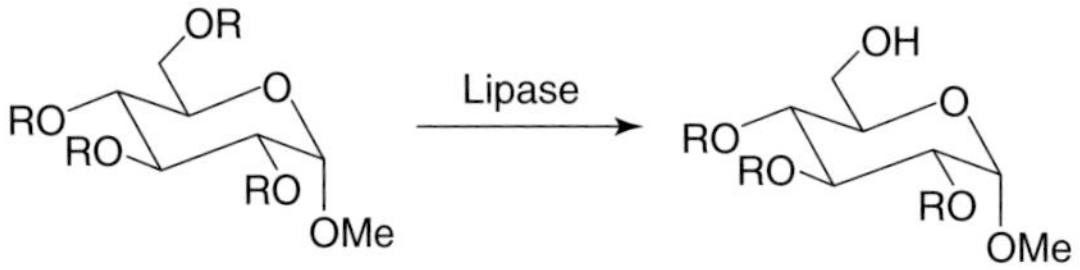

1 —Lipase→ **2**

$R = COC_4H_9$

Scheme 5.1 Regioselective deacylation of polyacylated sugars.

5.1.3 Stereoselectivity

Stereoselectivity, in particular, enantioselectivity, is the most important feature of enzymes. It should be stressed that enzymes are capable of recognizing any type of chirality of the substrates. It does not seem necessary to prove here how important the synthesis of sterically defined products is, because the differences in biological activity of particular stereoisomers of a given substance are well known. There are three approaches to the synthesis of enantiomerically enriched

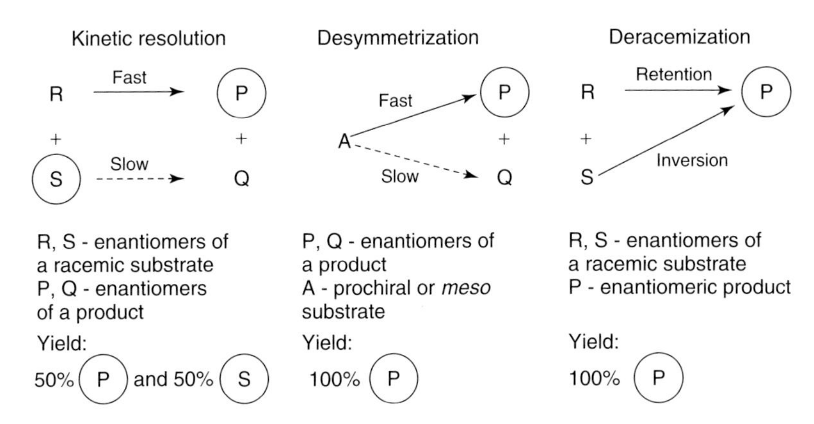

Scheme 5.2 Methods of synthesis of enantiomerically enriched (pure) compounds [8]. (Reproduced with permission from Kluwer Academic Publishers.)

(or pure) compounds in which enzymes are particularly useful (Scheme 5.2) [8]. Kinetic resolution of a racemic mixture rests upon chiral discrimination of enantiomers – each enantiomer is transformed into a product at a different rate. The theoretical yield is then 50% of the unreacted substrate and 50% of the product, each having opposite absolute configuration. In the desymmetrization process, a prochiral or *meso* substrate is preferentially transformed into one enantiomer of the product; in an ideal case, the yield of the product may be 100%. Finally, deracemization is a process in which a racemic substrate is transformed into one enantiomer of the product (for more details see Section 5.2.2).

All the transformations described above may be realized using various classes of enzymes. However, the importance for practical applications in organic synthesis is not the same for each class: of the six enzyme classes, two are most commonly used: hydrolases and oxidoreductases (altogether >85% of the total applications) [1, 3].

It should be added that in the last few years, considerable progress has been made in biocatalysis, which brings us closer to thinking of enzymatic process optimization in terms used so far only for synthetic catalysts. In certain cases, the specific features of enzymes allow even to explore areas of optimization that have no analogy in synthetic systems. The aim of this chapter is to signal these recent achievements. Section 5.2 focuses on the achievements in industrial processes (novel immobilization processes, new reaction media, and recent concepts in deracemization processes). Selected examples of simultaneous use of enzymes and organometallic catalysts for the dynamic kinetic resolution (DKR) processes will also be shown. Section 5.3 aims at the introduction of new optimization protocols, which show the future possibilities in biocatalysis. Among these, special attention is paid to directed evolution of enzymes, formation of bioconjugates, incorporation of nonnatural amino acids into enzyme molecules, and advantage taken from a "catalytic promiscuity" of enzymes.

5.2 Enzymatic Catalysis Today

5.2.1 Recent Developments in Heterogeneous Biocatalysis

During evolution, enzymes have been designed to work in water under physiological conditions and their activity is often regulated by substrate(s) or product(s). However, when they are used as catalysts in organic synthesis, they must accept different substrates at a high concentration. Moreover, for organic reactions, water is not the solvent of choice since the solubility of most organic compounds in this medium is very low. For large-scale operations, the removal of water is more tedious and expensive than of organic solvents. Therefore, application of the latter in industrial processes is of great importance. Fortunately, enzymes have been found to retain their catalytic activity in organic solvents as well, which has made it possible to develop a "nonaqueous" modification of enzymatic reactions. Since enzymes are generally insoluble in organic solvents, they act in this case as heterogeneous catalysts. Organic solvents have numerous advantages over water for biocatalytic transformations [9]. Thus, enzymes can easily be filtered off after the reaction and reused. Nonpolar substrates are better soluble in organic solvents, separation and purification of products is easy, side reactions are suppressed, and, in many cases, thermodynamic equilibrium is shifted to favor synthesis over hydrolysis. Therefore, this approach has been widely used in organic synthesis and there are a huge number of applications reported in the chemical literature. Hence, only one representative example is shown here.

The enantiomerically pure 3-arylglutaric ester are precursors for the synthesis of (−)-paroxetine [10], a selective serotonin reuptake inhibitor used in the treatment of depression, obsessive compulsive disorder, and panic, and (*R*)-Baclofen [11], a $GABA_B$ receptor agonist, which is used clinically in the treatment of spasticity (Chart 5.1).

The 3-arylglutaric esters **3** required as substrates were obtained by condensation of 2 equiv. of ethyl acetylacetate with respective benzaldehydes, followed by hydrolysis and decarboxylation of the resulting product under basic conditions. After esterification, ester **3** was obtained in 75% overall yield ($R^1 = Cl$, $R^2 = Me$) [10].

Chart 5.1

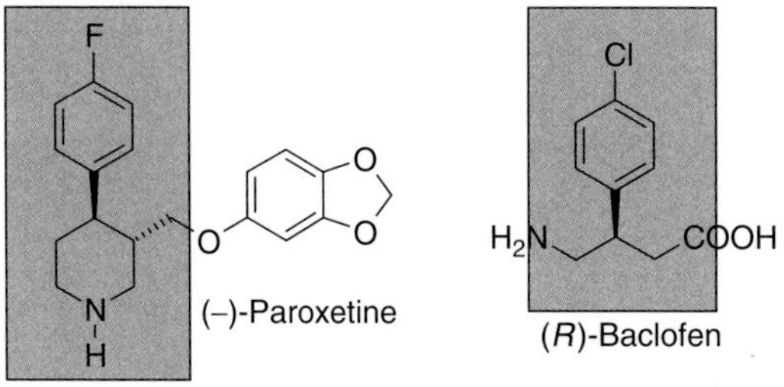

(–)-Paroxetine

(*R*)-Baclofen

For desymmetrization of diesters **3** via their hydrolysis in water, pig liver esterase [12], α-chymotrypsin [12, 13a], and *Candida antarctica* lipase (CAL-B) [14] were successfully used. However, further studies showed that respective anhydrides **5** can be used as substrates for enzyme-catalyzed desymmetrization in organic solvents [15]. The desired monoesters **4** were obtained in high yield in this way, using immobilized enzymes: Novozym 435 or Chirazyme L-2 (Scheme 5.3). After the reaction, enzymes were filtered off, organic solvents were evaporated, and the crude products were crystallized. This was a much simpler experimental procedure in which control of the reaction progress was not necessary, and all problems associated with extraction of products from aqueous phase and their further purification were omitted [15].

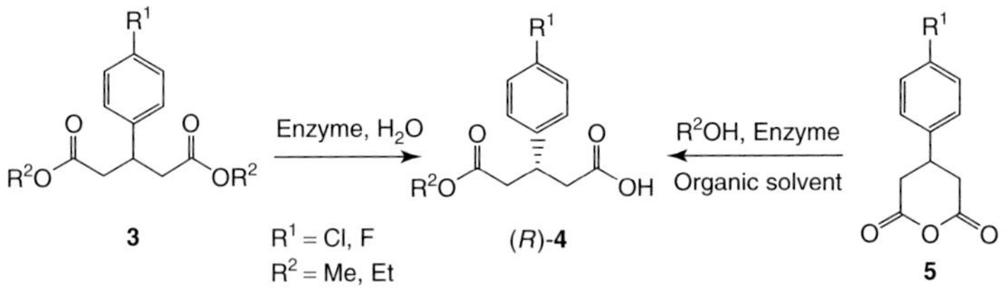

Enzyme, H_2O → ; ← R^2OH, Enzyme, Organic solvent

3 ; (*R*)-**4** ; **5**

R^1 = Cl, F
R^2 = Me, Et

Scheme 5.3 Enzymatic desymmetrization of diesters **3** and anhydrides **5**.

Another consequence of application of organic solvents as a reaction medium is associated with the mechanism of enzyme-catalyzed transformations. According to the commonly accepted mechanism, the first product of the interaction of a hydrolytic (serine) enzyme with an ester is an O-acylenzyme (Scheme 5.4). When the reaction is performed in an aqueous solution, water acts as a nucleophile in the next step, to give acid **B**. If more nucleophilic hydrogen peroxide is present in the reaction mixture, peroxycarboxylic acids **C** are formed. However, in organic solvents the O-acylenzyme also reacts readily with other nucleophiles, such as

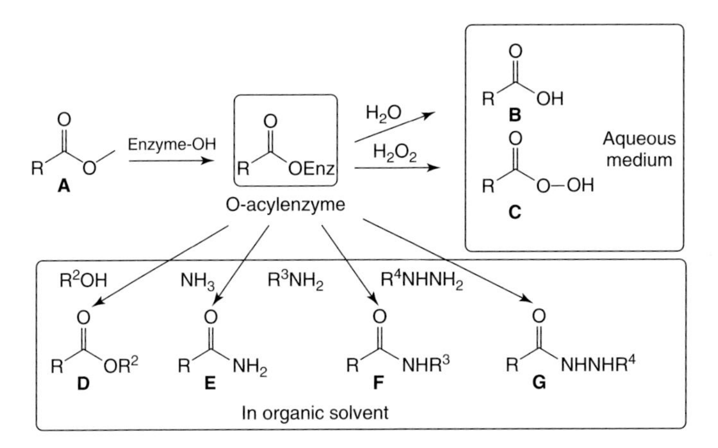

Scheme 5.4 Biotransformation of esters in water and in organic solvents.

alcohols, ammonia, amines, and hydrazines, leading to the formation of esters **D**, amides **E** and **F**, and hydrazones **G**, respectively. One can expect that other types of nuleophiles can also participate in this reaction, which will give access to completely new types of products that are not available using the currently known synthetic procedures.

For most applications, enzymes are purified after isolation from various types of organisms and microorganisms. Unfortunately, for process application, they are then usually quite unstable and highly sensitive to reaction conditions, which results in their short operational lifetimes. Moreover, while used in chemical transformations performed in water, most enzymes operate under homogeneous catalysis conditions and, as a rule, cannot be recovered in the active form from reaction mixtures for reuse. A common approach to overcome these limitations is based on immobilization of enzymes on solid supports. As a result of such an operation, heterogeneous biocatalysts, both for the aqueous and nonaqueous procedures, are obtained.

Enzymes are immobilized by a variety of methods. Two general types of immobilization procedures are used. The first-type procedures are based on weak interactions between the support and the enzyme and are classified as *physical methods*. The second-type procedures rest upon the formation of covalent bonds between the enzyme and the support and are classified as *chemical methods*.

The most important physical methods are physical and ionic adsorption on a water-insoluble matrix, inclusion and gel entrapment, and microencapsulation with a liquid or a solid membrane. The most important chemical methods include covalent attachment to a water-insoluble matrix, cross-linking with the use of a multifunctional, low-molecular weight reagent, and co-cross-linking with other neutral substances, for example proteins.

As a general rule, the optimal immobilization method is found empirically by a process of trial and error, where the selectivity, activity, and operational stability of the enzyme after immobilization are taken into account. The immobilization process is very sensitive to different parameters and is treated as a kind of art [16].

The importance of proper immobilization of enzymes can be shown in the kinetic resolution of racemic α-acetoxyamides. This group of compounds is an important class of chemicals since they can be readily transformed into α-amino acids [17], N-methylated amino acids, and tripeptide mimetics [18], amino alcohols [19], 1,2-diols [20], 1,2-diamines [21], and enantiopure 1,4-dihydro-4-phenyl isoquinolinones [22].

For the kinetic resolution of racemic α-acetoxyamide **6** several native enzymes were used (Scheme 5.5). The native lipases from *Pseudomonas cepacia* (PCL) and porcine pancreas (PPL) showed the highest, although still unsatisfactory, enantioselectivity ($E = 5.1$ and 3.5, respectively). Upon immobilization into a solgel matrix, the enantioselectivity of PCL was improved significantly to 30.5. The covalent immobilization on Eupergit® increased the enantioselectivity even more (34.0) [23].

Recently, an interesting example of the enzymatic kinetic resolution of α-acetoxyamide **8** was demonstrated using native wheat germ lipase and immobilized lipase PS (AMANO) (Scheme 5.6).

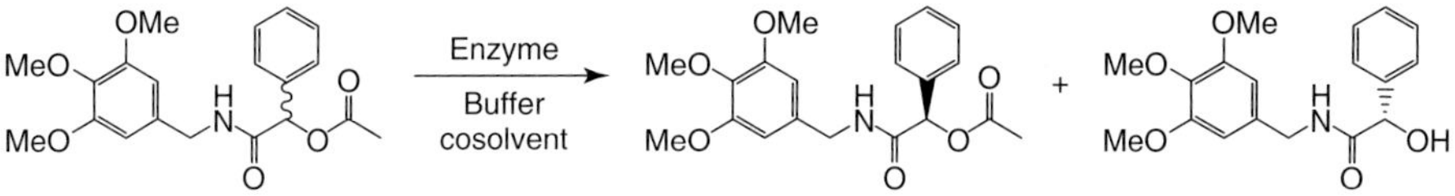

Scheme 5.5 Enzyme-promoted kinetic resolution of an α-acetoxyamide.

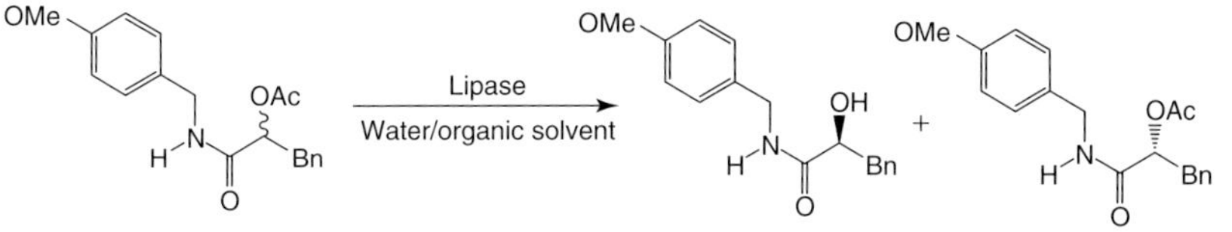

Scheme 5.6 Influence of a cosolvent on enzymatic kinetic resolution.

An intriguing influence of a cosolvent immiscible with water on the enantioselectivity of the enzyme-catalyzed hydrolysis was observed. It was proven that enzyme enantioselectivity is directly correlated with the cosolvent hydrophobicity. In the best example, for ethyl ether as cosolvent, the reaction proceeded with $E = 55$, and the target compound was obtained in 33% yield with 92.7% ee. This finding may be of great practical importance, particularly in industrial processes [24], since it will enable better optimization of enzyme-catalyzed processes. It is clear that, in future, immobilized enzymes, as heterogeneous catalysts, will be widely used in most industrial transformations, especially in the preparation of pharmaceuticals [25].

5.2.2 Deracemization Processes

Of the two former processes shown in Scheme 5.2, the kinetic resolution of racemates has found a much greater number of applications than the desymmetrization of prochiral or *meso* compounds. This is due to the fact that racemic substrates are much more common than prochiral ones. However, kinetic resolution suffers from a number of drawbacks, the main being the following:

- The theoretical yield of each enantiomer cannot be higher than 50%.
- At the most preferable conversion of 50%, the enantiomeric ratios of a substrate and a product exhibit the lowest values.
- Separation of the unreacted substrate from the resulting product may in certain cases be difficult or troublesome.
- If only one of the enantiomers is desired, which is often the case, half of the starting material is lost.

To avoid the problems shown above, a new approach has been developed. It is called *deracemization* and consists in the transformation of a racemic substrate into one enantiomer of the product. In other words, it means that one enantiomer of the racemic mixture is transformed into the product, with retention, and, the other one, with inversion of configuration, at the stereogenic center. Several processes have been elaborated to achieve this. The most important of these are discussed below.

5.2.2.1 **Dynamic Processes (Combination of Enzymes and Chemical Catalysts, DKR)**

For most chemical transformations, especially for industrial applications, the yield of 50% cannot be accepted. Since each enantiomer constitutes only 50% of the racemic mixture, the best way to increase the yield of the desired enantiomer is racemization of the unwanted one (Scheme 5.7). This reaction must proceed simultaneously with the enzymatic kinetic resolution. In order to indicate the dynamic character of such processes, the term *dynamic kinetic resolution* has been introduced.

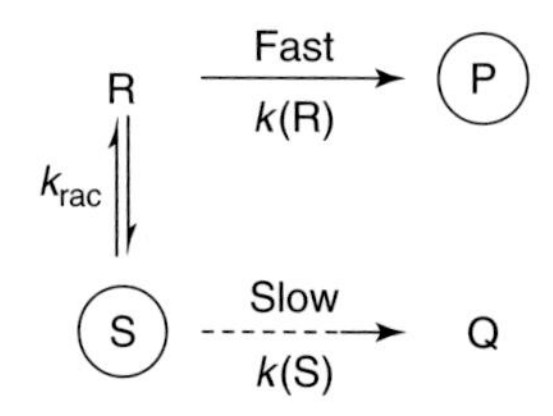

Scheme 5.7 The idea of dynamic kinetic resolution.

For successful DKR two reactions: an *in situ* racemization (k_{rac}) and kinetic resolution [$k(R) \gg k(S)$] must be carefully chosen. The detailed description of all parameters can be found in the literature [26], but in all cases, the racemization reaction must be much faster than the kinetic resolution. It is also important to note that both reactions must proceed under identical conditions. This methodology is highly attractive because the enantiomeric excess of the product is often higher than in the original kinetic resolution. Moreover, the work-up of the reaction is simpler since in an ideal case only the desired enantiomeric product is present in the reaction mixture. This concept is used for preparation of many important classes of organic compounds like natural and nonnatural α-amino acids, α-substituted nitriles and esters, cyanohydrins, 5-alkyl hydantoins, and thiazolin-5-ones.

For practical reasons, the best scenario is associated with spontaneous substrate racemization. This is often the case for compounds possessing α-stereogenic center bearing an acidic proton adjacent to an activating carbonyl group. When this process is too slow, the substrate must be modified or racemization can be achieved by a reversible cleavage of covalent bonds such as the cleavage of cyanohydrins.

In some cases, substrate racemization can be achieved by an appropriate change of a functional group. A good example of this strategy is the synthesis of acid **10**, which is the key intermediate for roxifiban – a potent selective antagonist of the platelet plycoprotein IIb/IIa receptor (Scheme 5.8) [27, 28].

Thus, racemic acid **12** (R = H) was obtained by [3+2] cycloaddition in 90–95% yield (Scheme 5.9) [28]. Its resolution into enantiomers could be achieved either by chiral preparative HPLC, or by fractional crystallization of its cinchonidine salts. Better results were obtained upon enzymatic kinetic resolution of its *iso*-butyl ester **12** (R = *i*-Bu) [29]. However, further work showed that racemic thiolester **13**, which

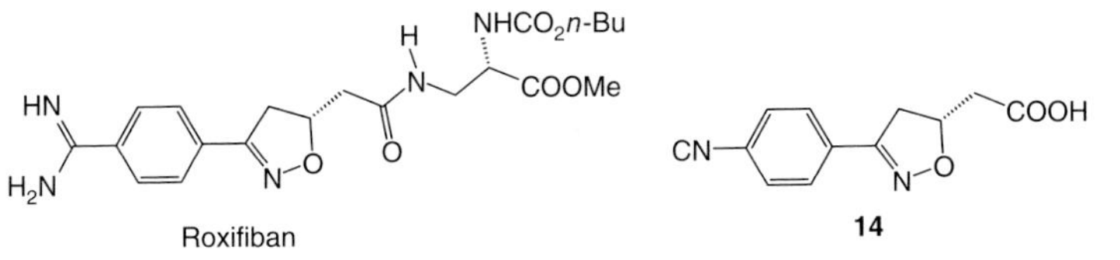

Scheme 5.8 Roxifiban and compound **14**.

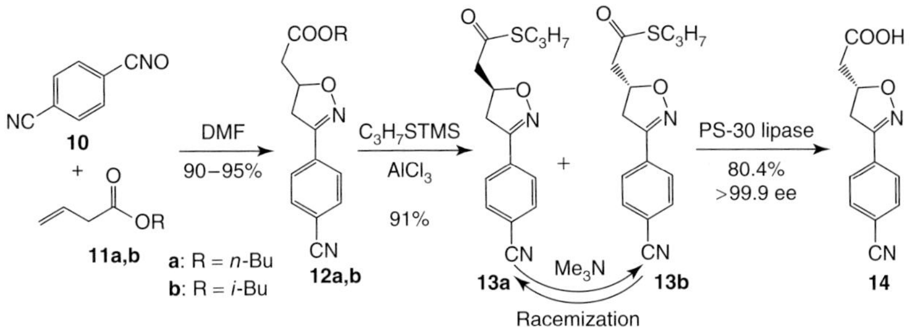

Scheme 5.9 Base-promoted racemization in dynamic kinetic resolution of substrate **13**.

could be readily obtained from ester **12** upon treatment with propyl trimethylsilyl sulfide in the presence of aluminum chloride, was a better substrate for enzymatic reaction [30]. While racemization of ester **12** was difficult to achieve, the thiolester **13** readily underwent retro-Michael–Michael racemization in the presence of a base. When this process was combined with enzymatic hydrolysis, the acid **14** was obtained in 88.7% yield and more than 99.5% ee on a kilogram scale [30].

In many cases, the racemization of a substrate required for DKR is difficult. As an example, the production of optically pure α-amino acids, which are used as intermediates for pharmaceuticals, cosmetics, and as chiral synthons in organic chemistry [31], may be discussed. One of the important methods of the synthesis of amino acids is the hydrolysis of the appropriate hydantoins. Racemic 5-substituted hydantoins **15** are easily available from aldehydes using a commonly known synthetic procedure (Scheme 5.10) [32]. In the next step, they are enantioselectively hydrolyzed by D- or L-specific hydantoinase and the resulting *N*-carbamoyl amino acids **16** are hydrolyzed to optically pure α-amino acid **17** by other enzymes, namely, L- or D-specific carbamoylase. This process was introduced in the 1970s for the production of L-amino acids **17** [33]. For many substrates, the racemization process is too slow and in order to increase its rate enzymes called *racemases* are used. In processes the three enzymes, racemase, hydantoinase, and carbamoylase, can be used simultaneously; this enables the production of α-amino acids without isolation of intermediates and increases the yield and productivity. Unfortunately, the commercial application of this process is limited because it is based on L-selective hydantoin-hydrolyzing enzymes [34, 35]. For production of D-amino acid the enzymes of opposite stereoselectivity are required. A recent study indicates that the inversion of enantioselectivity of hydantoinase, the key enzyme in the

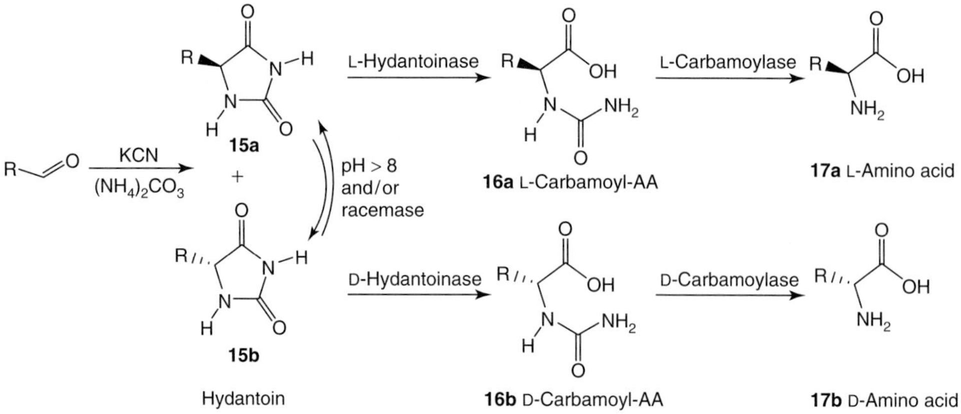

Scheme 5.10 Enzyme-catalyzed synthesis of α-amino acids from hydantoins.

process, can be achieved by directed evolution [36], which can elaborate the synthesis potential of this method.

Recently, recombinant biocatalysts obtained using *Escherichia coli* cells were designed for this process. The overexpression of all enzymes required for the process, namely, hydantoinase, carbamoylase, and hydantoin racemase from *Arthrobacter* sp. DSM 9771 was achieved. These cells were used for production of α-amino acids at the concentration of above $50\,g\,l^{-1}$ dry cell weight [37]. This is an excellent example presenting the power of biocatalysis with respect to classical catalysis, since a simultaneous use of three different biocatalysts originated from one microorganism can be easily achieved.

The DKR of secondary alcohols can be efficiently performed via enzymatic acylation coupled with simultaneous racemization of the substrates. This method was first used by Bäckvall for the resolution of 1-phenylethanol and 1-indanol [38]. Racemization of substrate **18** by a ruthenium catalyst (Scheme 5.11) was combined with transesterification using various acyl donors and catalyzed by *C.antarctica* B lipase. From all the acyl donors studied, 4-chlorophenyl acetate was found to be the best. The desired product **19** was obtained in 80% yield and over 99% ee.

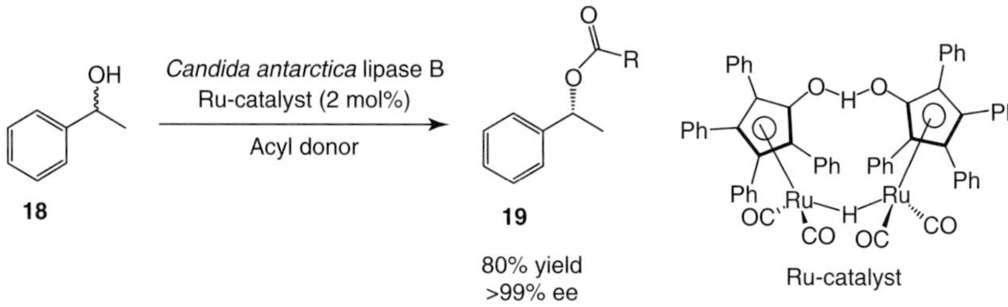

Scheme 5.11 Dynamic kinetic resolution of alcohol **18** by combination of enzymatic transesterification and ruthenium-catalyzed racemization.

The method is not restricted to secondary aryl alcohols and very good results were also obtained for secondary diols [39], α- and β-hydroxyalkylphosphonates [40], 2-hydroxyalkyl sulfones [41], allylic alcohols [42], β-halo alcohols [43], aromatic chlorohydrins [44], functionalized γ-hydroxy amides [45], 1,2-diarylethanols [46], and primary amines [47]. Recently, the synthetic potential of this method was expanded by application of an air-stable and recyclable racemization catalyst that is applicable to alcohol DKR at room temperature [48]. The catalyst type is not limited to organometallic ruthenium compounds. Recent report indicates that the *in situ* racemization of amines with thiyl radicals can also be combined with enzymatic acylation of amines [49]. It is clear that, in the future, other types of catalytic racemization processes will be used together with enzymatic processes.

5.2.2.2 **Other Deracemization Processes**

Stereoinversion Stereoinversion can be achieved either using a chemoenzymatic approach or a purely biocatalytic method. As an example of the former case, deracemization of secondary alcohols via enzymatic hydrolysis of their acetates may be mentioned. Thus, after the first step, kinetic resolution of a racemate, the enantiomeric alcohol resulting from hydrolysis of the fast reacting enantiomer of the substrate is chemically transformed into an activated ester, for example, by mesylation. The mixture of both esters is then subjected to basic hydrolysis. Each hydrolysis proceeds with different stereochemistry – the acetate is hydrolyzed with retention of configuration due to the attack of the hydroxy anion on the carbonyl carbon, and the mesylate – with inversion as a result of the attack of the hydroxy anion on the stereogenic carbon atom. As a result, a single enantiomer of the secondary alcohol is obtained (Scheme 5.12) [8, 50a].

Deracemization via the biocatalytic stereoinversion is usually achieved by employing whole cells. In the case of secondary alcohols, it is believed that microbial stereoinversion occurs by an oxidation–reduction sequence

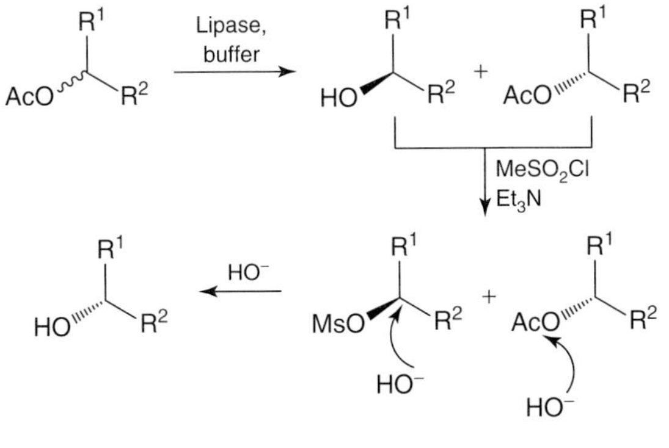

Scheme 5.12 Deracemization of secondary alcohols via resolution followed by chemical stereoinversion.

using two different enzymes: a dehydrogenase and a redox enzyme. One enantiomer of a racemate is selectively oxidised to a ketone by a dehydrogenase while the other enantiomer remains unchanged. The resulting ketone is reduced by another enzyme exhibiting opposite stereopreference. The redox cofactor such as NAD(P)H is recycled internally by using the whole cell system (Scheme 5.13). It is important that one of the redox reactions is irreversible [26].

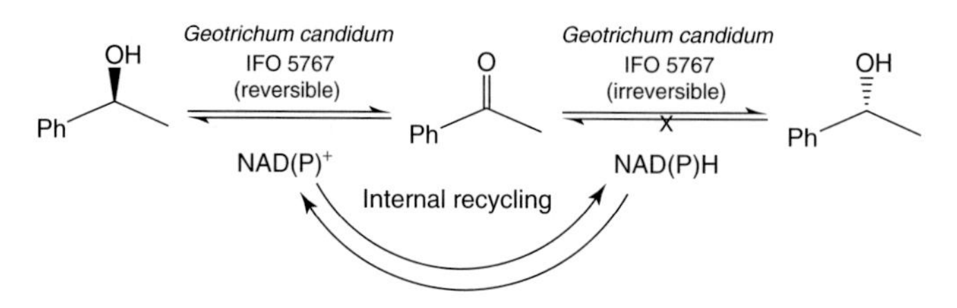

Scheme 5.13 Deracemization of secondary alcohols based on biocatalytic stereoinversion [26, 50b].

Enantioconvergence A process is considered as *enantioconvergent* when each enantiomer of a racemic substrate is converted into the same enantiomer of the product via independent pathways [26]. Three different approaches have been developed: using two biocatalysts, using a single biocatalyst, and using combined bio- and chemocatalysis. The two former methods are of limited applicability and have been applied mainly in deracemization of epoxides [51, 52]. The latter method has been successfully used in deracemization of secondary alcohols via hydrolysis of their sulfate esters. Thus, certain sulfatases hydrolyze secondary alcohol sulfates with inversion of configuration at the stereogenic carbon atom. In this way, after kinetic resolution, both the unreacted sulfate and the resulting alcohol possess the same absolute configuration. Appropriately chosen conditions of chemical hydrolysis of the unreacted sulfate, which must proceed without involvement of the stereogenic center (thus with retention of configuration), lead to the formation of a single enantiomer of the product (Scheme 5.14) [53, 54].

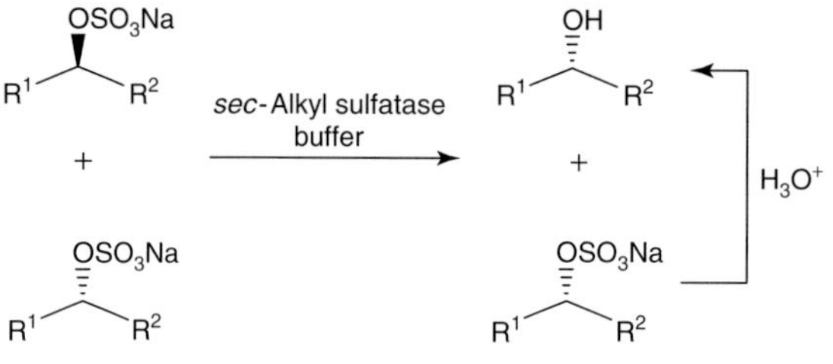

Scheme 5.14 Chemoenzymatic enantioconvergent deracemization of secondary alcohols via hydrolysis of their sulfate esters.

5.2.3 Novel Reaction Media

5.2.3.1 Ionic Liquids as Designer Solvents for Biotransformations

Ionic liquids (ILs) are substances that are completely composed of ions and are liquid at temperatures below 100 °C. Since the first reports on the biotransformations in ILs in 2000 [55], considerable development has been achieved in this area [56]. Growing interest in these techniques can be attributed to special features of ILs as solvents (very small vapor pressures [57], thermal stability [58]), and the convenient possibility to tune their properties – polarity, hydrophobicity, and solvent miscibility) by changing the nature of their cationic and anionic components. This latter feature adds another dimension to optimization studies on enzyme activity and selectivity. With respect to their general solvent properties, ILs resemble polar organic solvents, such as acetonitrile or methanol [59], which makes them good solvents for many polar organic compounds. However, the use of ILs in large-scale processes is still limited by their toxicity [60] and negligible biodegradability [61].

In biotransformation reactions, ILs can act as tunable solvents or immobilizing agents or additives. They can also be coupled to substrates or other reagents (e.g., acylating agents in lipase-catalyzed transesterifications [62]). Recent examples of chosen applications are presented below.

The outstanding influence of the anionic component on the activity and selectivity of enzymes was demonstrated in the *Candida rugosa* lipase-catalyzed kinetic resolution of ibuprofen, a nonsteroidal antiinflammatory drug with sales of USD 183 million in 2006 (Scheme 5.15) [63].

The change of the anion results in alteration of enzymatic activity and also allows improvement in the enantioselectivity from an unacceptable level (1.1) to a synthetically useful value of 24. The cationic component of the IL also affects the activity and selectivity of the biocatalyst. Scheme 5.16 presents the study on the kinetic resolution of adrenaline-type aminoethanol in ILs [64].

The introduction of different cations in the IL, which served as a solvent, resulted in considerable changes in enantioselectivity of the reaction. The best results both in terms of selectivity and activity were obtained for [EMIM] [BF_4]. However, in the case of heteroorganic substrates, namely hydroxymethylphosphinates, the best results were obtained using [BMIM] [PF_6], while [BMIM] [BF_4] proved to be completely useless [65c].

The use of ILs as solvents for biotransformation reactions is, of course, not limited to kinetic resolutions and the use of hydrolytic enzymes. Numerous other

COOH $\xrightarrow{C_3H_7OH}$ $COOC_3H_7$ (X^-)

X^-	Conversion (%)	E
$OctSO_4^-$	62	1.1
$MeSO_4^-$	50	1.2
BF_4^-	33	6.4
PF_6^-	30	24

Scheme 5.15 Kinetic resolution of ibuprofen in ionic liquids.

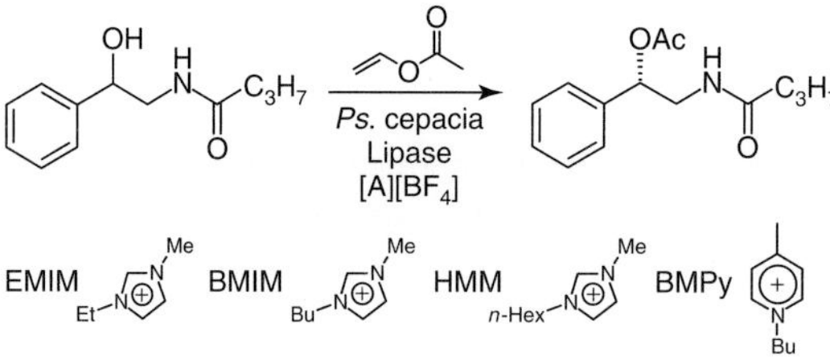

A	Conversion (%)	*E*
BMIM	23	80
HMIM	19	100
BMPy	29	140
EMIM	31	160

Scheme 5.16 Kinetic resolution of adrenaline-type aminoethanol in ionic liquids.

applications include glycosidation reactions catalyzed by glycosidases [65], redox reactions catalyzed mainly by whole cell systems [66], and addition reactions catalyzed by lyases [67]. It should be noted, however, that it is still impossible to predict the best medium for a given reaction without optimization study, and there is no "best" IL for all the reactions [56].

In biphasic water/IL mixtures, the latter can be used as immobilization systems. This idea was used for the synthesis of conducting polyaniline by IL-immobilized horseradish peroxidase [68]. Tuning the IL hydrophobicity by changing the anionic component allowed the increase in the yield of the product by altering the affinity of the product to the IL. After completion of the reaction, the IL phase was separated, facilitating the recovery of the enzyme.

Apart from being employed solely as solvents, ILs can also be used as reagent carriers (TSILs, task-specific ionic liquids). An efficient system that uses ILs as solvents and anchoring/acylating reagents for the CAL B-catalyzed separation of racemic alcohols is shown in Scheme 5.17 [69].

The TSIL used in this study has an ester moiety in its structure, which enables it to react in enzyme-catalyzed transesterification reaction. In the first part of the cycle, one of the enantiomers of the racemic alcohol is preferentially transformed into an ester of the IL-coupled acid. The other, unreacted enantiomer of the alcohol is then extracted, while the newly formed IL ester is treated with an excess of ethanol in the presence of the same enzyme. This process is accompanied by the regeneration of TSIL in the active form. The main advantage of the presented

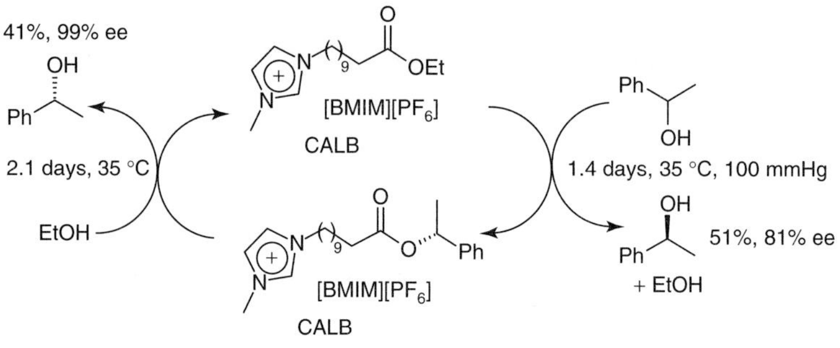

Scheme 5.17 The use of TSILs as reagent carriers.

system is the possibility of cyclic reuse, as no significant loss of efficiency was observed over four process cycles.

5.2.3.2 Supercritical Fluids as Green Solvents for Biotransformations

Supercritical fluids (SCFs) are compounds that exist at a temperature and pressure that are above their corresponding critical values [70, 71]. They exhibit the properties of both gases and liquids. With gases, they share the properties of low surface tension, low viscosity, and high diffusivity. Their main liquid-like feature is the density, which results in enhanced solubility of solutes compared with the solubility of gases. Furthermore, the solubility of solutes can be manipulated by changes in pressure and temperature near the critical point [72].

Among the SCFs, supercritical carbon dioxide ($scCO_2$) provides additional benefits [73], since it is environmentally benign, inexpensive, available in large quantities, nonflammable, and exhibits low toxicity. Its critical pressure is relatively low (73.4 bar) and it has an ambient critical temperature (31.3 °C). CO_2 can be easily removed from reaction mixtures by depressurization [74].

Reactors used for biotransformations in $scCO_2$ offer the possibility of changing temperature and pressure of the reaction. These parameters were shown to have a crucial influence on both activity and enantioselectivity of the biocatalysts [75]. It is suggested that large changes in the SCF density as a result of changing the pressure and temperature alter the interaction between CO_2 and the enzyme to progressively affect the enzyme conformation and, therefore, the enantioselectivity of the reaction [76]. Changing the temperature and the pressure of biotransformation reactions influences also protein stability [77] and mass transfer within the system [78].

In several studies, $scCO_2$ has shown to have adverse effects on enzymatic activity [79]. This fact results in the demand for novel methods of enzyme stabilization. The most effective of these include immobilization [80], lipid coating [81], surfactant coating [82], use of cross-linked enzyme crystals [83], cross-linked enzyme aggregates [84], and membrane reactors [85].

5.3 Enzymatic Catalysts of Tomorrow

5.3.1 Bioconjugates of Compatible Enzymes as Functional Catalysts for Multistep Processes

Most enzymes are catalytically active under very similar conditions (temperatures in the range of 20–40 °C, normal pressure, pH close to neutral). This results in the compatibility of enzymes, and suggests the possibility of performing many enzyme-catalyzed reactions in "one pot," thus shifting the equilibria of selected steps and increasing the efficiency of the multistep processes. Additional possibilities are offered by the construction of enzyme bioconjugates – biocatalysts connected to each other (usually via short linkers). Similar polyenzymatic clusters are well known in nature and are utilized by living organisms for the catalysis of subsequent reactions. Close proximity of enzymes in space results in the tunneling of intermediates between the active sites and increases the efficiency of the processes.

Several attempts to mimic this phenomenon have been undertaken in recent years. The conjugates obtained were used in sugar synthesis (UDP-Gal 4-epimerase + α−1,3-galactosyltransferase) [86], sugar determination (P-galactosidase + galactose dehydrogenase) [87], resveratrol synthesis (4-coumarate-CoA ligase + stilbene synthase) [88], and Bayer–Villiger type oxidations of ketones (Bayer–Villiger monooxygenase + cofactor-regenerating enzyme [89].

In one of the above-mentioned cases [86], a significant improvement of reaction rates was observed when compared to the reactions carried out by uncoupled biocatalysts. This fact suggests that owing to the close proximity of the enzymes the local concentration of the intermediate is higher around the fused biocatalyst.

5.3.2
Directed Evolution – Enzyme Tuning Toward Higher Selectivity

The activity and selectivity of enzymes is a function of their structure, which depends on the sequence of amino acids. The use of genetic engineering, which aims at influencing the amino acid sequence, is a fairly young concept in biotransformations, which arises from the specific nature and origin of biocatalysts. It has no analog in organic chemistry (although some analogies may be drawn toward combinatorial approach to catalyst synthesis), being a possibility offered distinctively by biological systems.

The use of genetic engineering for the preparation of more active and/or more selective enzymes consists in the possibility of a parallel synthesis of a large number of similar catalysts (by expression in bacterial cells) that can be screened for desired abilities. Since methods for random or single-site mutations within chosen gene are well established, and effective expression systems have been developed, the main bottleneck in this approach is the need for high-throughput screening (HTS) assays for activity and enantioselectivity. The fast probing of the latter feature is especially challenging, as the differences between the concentrations of enantiomeric substrates or products are difficult to measure rapidly. One has to bear in mind, that even a small library of clones derived from randomization of two amino acids in the native enzyme needs to have around 2000 mutants screened (400 possible amino acid combinations multiplied by 5 in order to cover over 95% of the possibilities). Even when the chiral separation can be found using chromatography methods (chiral GC, HPLC), the assay rarely takes less than 10 min per sample. This gives a total time for the analysis of this small library to be 20 000 min (almost 14 days). Such assay duration is unacceptable.

An important breakthrough in HTS ee assays came from the group of Reetz in late 1990, with the introduction of mass spectroscopy (MS)-based procedures [90]. These methods use special asymmetrically isotope-labeled compounds. Enzymatic transformations of these compounds usually lead to two pseudoenantiomeric compounds whose relative concentration can be estimated using MS techniques.

Consider, for example, a hydrolase-based desymmetrization of a *pseudo-meso* compound **20** (Scheme 5.18) [91].

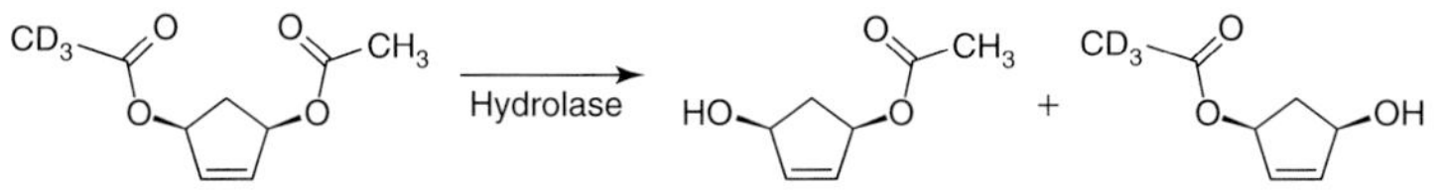

Scheme 5.18 Enzymatic hydrolysis of a pseudo-*meso* compound **20**.

Products **21** and **22** obtained in this reaction differ in their ESI-MS spectra, and the difference in the abundance of respective signals can be expressed quantitatively. Studies have shown that the pseudo-enantiomeric-excess values obtained in this way are in agreement ±5% with the data obtained by chromatographic methods, which is sufficient for studying relative values and choosing most selective mutants.

Initial approaches to directed evolution of enzymes rested upon the introduction of random mutations in random sites of the enzyme by the use of the error-prone PCR technique [92] or on the DNA-shuffling method [93]. Extensive research has also been reported in which every amino acid site in an enzyme was systematically subjected to saturation mutagenesis [94].

An example [95] of the use of above-mentioned modifications is depicted in Scheme 5.19.

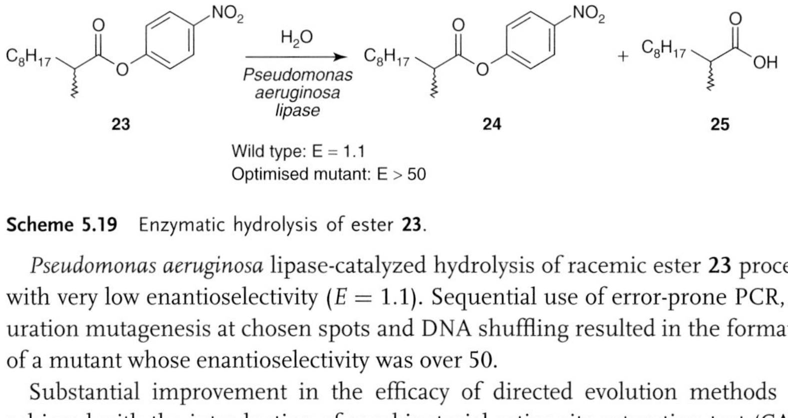

Wild type: E = 1.1
Optimised mutant: E > 50

Scheme 5.19 Enzymatic hydrolysis of ester **23**.

Pseudomonas aeruginosa lipase-catalyzed hydrolysis of racemic ester **23** proceeds with very low enantioselectivity ($E = 1.1$). Sequential use of error-prone PCR, saturation mutagenesis at chosen spots and DNA shuffling resulted in the formation of a mutant whose enantioselectivity was over 50.

Substantial improvement in the efficacy of directed evolution methods was achieved with the introduction of combinatorial active site saturation test (CAST) by Reetz and coworkers in 2005 [96]. CASTing consists in systematic design and screening of focused libraries around the complete binding pocket. On the basis of the 3D structure of the enzyme, a two to three amino acid sites that reside close to the active site are chosen. Following this, all amino acids in the chosen site are randomized simultaneously, which results in the formation of small libraries of mutants. After finding the best mutant, another site is addressed using the same approach. The success of CASTing depends on the proper choice of the order in which the sites should be optimized, and usually the trial–error method is used to determine it [97]. The first enzymes that were optimized using the CASTing method were epoxide hydrolases (E improved from 4.6 to 115).

5.3.3
Introduction of Nonproteinogenic Amino Acids – Toward More Selective, Stable, and Easily Handled Biocatalysts

The repertoire of α-amino acids used by nature in the biosynthesis of proteins is limited to about 20 structures. Since the diversity of synthetically accessible α-amino acids is enormous, initial studies have been undertaken to investigate the possibilities offered by incorporation of noncoded amino acids into enzymes. Some selected examples of successful modifications are presented below.

Nitroreductase, used for prodrug activation in cancer treatment, has been shown to exhibit insufficient activity in native form, and active site residue Phe124 has emerged as critical for substrate binding. Further studies have suggested that Phe124 interacts with the prodrug or NADH through π-stacking when the active site is occupied. Mutant enzymes possessing all 20 natural amino acids in position 124 were prepared [98] and the enzyme possessing lysine in position 124 exhibited 13-fold higher activity than the native enzyme. However, introduction of noncoded amino acids allowed for further improvement of activity, as the enzyme with (4-nitrophenyl)alanine in position 124 had 30 times higher activity when compared to the native form.

It is not only the activity that can be altered by incorporation of noncoded amino acids. Introduction of structures possessing certain chemical functions leads to the possibility of highly regioselective modification of enzymes. For example, selective enzymatic modification of cystein residues with compounds containing azide groups has led to the preparation of enzymes that could be selectively immobilized using *click chemistry* methods [99].

Most of the aforementioned efforts are based on the *nonsense suppression methodology* as the method for peptide altering [100]. In this approach, a nonsense codon is introduced into the enzyme-coding mRNA in the site that has to be altered. Simultaneously, the tRNA-noncoded amino acid hybrids are prepared with nonsense anticodons. Finally, the translation of modified mRNA is performed *in vivo* [101] or *in vitro* [102].

5.3.4
Enzyme Catalytic Promiscuity – Old Enzymes for New Challenges

Although enzymes are known for their high reaction specificity, there are numerous examples in which they prove to be capable of catalyzing reactions that are different from the ones they evolved for [103–105]. Such a feature, called *enzyme promiscuity*, is believed to result from the ability of a single active site of an enzyme to catalyze more than one distinct chemical transformation. These transformations may differ in the type of bond formed or cleaved, or may differ in the catalytic mechanism of bond making and breaking. In most cases, enzyme promiscuity includes both changes. It is believed that the promiscuity of enzymes is evolutionarily developed: when a need for new enzymatic functions arises, Nature uses existing enzymes that promiscuously bind a new substrate or catalyze a new reaction, and then "tinkers" with their active sites to fit the new substrates or reaction [104].

Enzyme promiscuity is clearly advantageous to chemists since it broadens the applicability of enzymes in chemical synthesis. New catalytic activities in existing enzymes can be enhanced by protein engineering – appropriate mutagenesis of the enzymes [106]. Some of the most illustrative examples of this unusual activity of common enzymes are presented below.

Proteases, which originally catalyze the amidic carbon–nitrogen bond breaking, also catalyze ester hydrolysis. However, in this case, the catalytic mechanism is likely very similar and consists in the preliminary attack of the active site serine on the carbonyl carbon atom [103].

Lipases are the enzymes for which a number of examples of a promiscuous activity have been reported. Thus, in addition to their original activity comprising hydrolysis of lipids and, generally, catalysis of the hydrolysis or formation of carboxylic esters [107], lipases have been found to catalyze not only the carbon–nitrogen bond hydrolysis/formation (in this case, acting as proteases) but also the carbon–carbon bond-forming reactions. The first example of a lipase-catalyzed Michael addition to 2-(trifluoromethyl)propenoic acid was described as early as in 1986 [108]. Michael addition of secondary amines to acrylonitrile is up to 100-fold faster in the presence of various preparations of the lipase from *Candida antartica* (CAL-B) than in the absence of a biocatalyst (Scheme 5.20) [109].

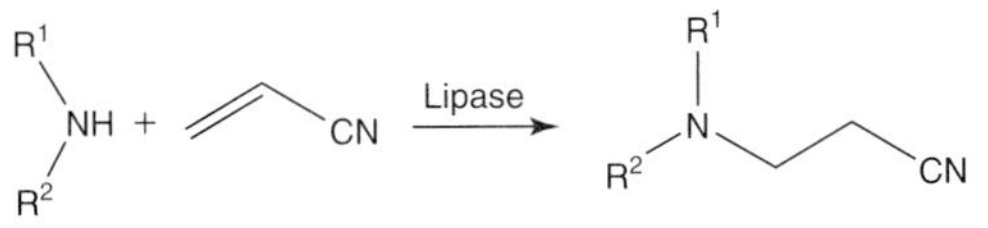

Scheme 5.20 Lipase-catalyzed Michael addition of amines to acrylonitrile.

In a similar way, lipases catalyze Michael addition of amines, thiols [110], and even 1,3-dicarbonyl derivatives [111, 112] to α,β-unsaturated carbonyl compounds (Scheme 5.21).

On the basis of the quantum-chemical studies, it has been established that it is the so called "oxyanion hole" of the enzyme that binds the carbonyl oxygen or nitrile nitrogen, enhancing the attack of a nucleophile, which is, in turn, activated by histidine (Scheme 5.22) [110].

This model clearly shows that the "catalytic machinery" involves a dyad of histidine and aspartate together with the oxyanion hole. Hence, it does not involve serine, which is the key amino acid in the hydrolytic activity of lipases, and, together with aspartate and histidine, constitutes the active site catalytic triad. This has been confirmed by constructing a mutant in which serine was replaced with alanine (Ser105Ala), and finding that it catalyzes the Michael additions even more efficiently than the wild-type enzyme (an example of "induced catalytic promiscuity") [105].

Lipase from *C.antarctica* also catalyzes carbon–carbon bond formation through aldol condensation of hexanal. The reaction is believed to proceed according to the same mechanism as the Michael additions [113]. Lipase from *Pseudomonas* sp.

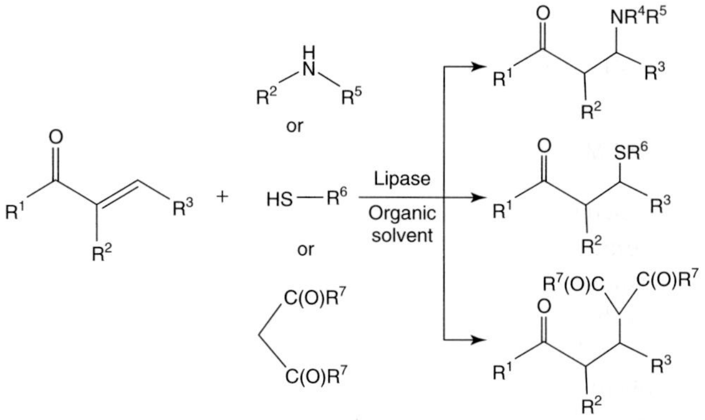

Scheme 5.21 Lipase-catalyzed Michael additions to α,β-unsaturated carbonyl compounds.

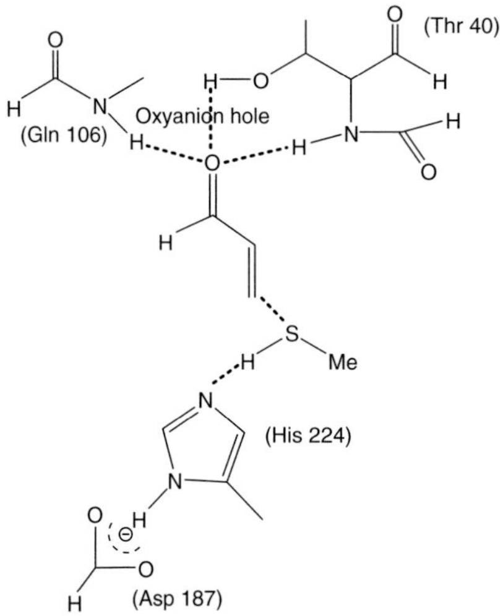

Scheme 5.22 Quantum-chemical model system of the lipase-catalyzed Michael addition of methanethiol to acrolein [110].

(PSL, AMANO) has been found to catalyze the Markovnikov-type addition of thiols to vinyl esters (Scheme 5.23) [114] and acylases: aminoacylase [115], penicillin G acylase [116], and acylase from *Aspergillus oryzae* [117], the same kind of addition of various nitrogen nucleophiles. Unfortunately, most additions described above are nonstereoselective.

Certain proteases, for example alkaline protease from *Bacillus subtilis*, have also proven to be good catalysts for the Michael addition of nitrogen nucleophiles to α,β-ethylenic compounds [118, 119].

Another intriguing behavior of some proteases is their ability to hydrolyze sulfur–nitrogen bond in *N*-acyl sulfinamides. Thus, when some *N*-acyl

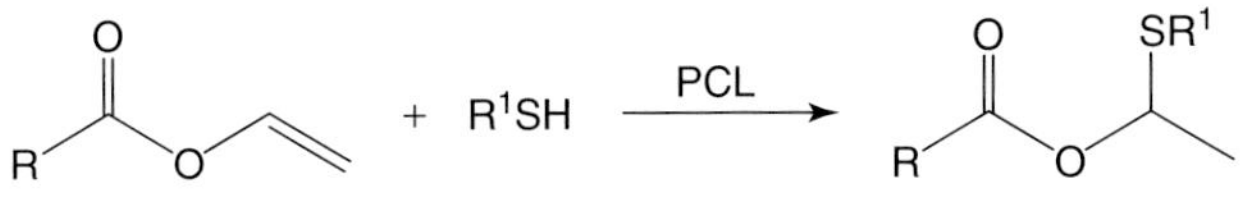

Scheme 5.23 Lipase-catalyzed Markovnikov addition of thiols to vinyl esters.

arenesulfinamides are treated with a buffer in the presence of subtilisin Carlsberg, hydrolysis of the S–N bond unexpectedly becomes the main hydrolytic process to give, under kinetic resolution conditions, the corresponding sulfinic acids and carboxamides, together with the enantiomerically enriched unreacted substrates (Scheme 5.24) [120]. The proof for the direct involvement of the enzyme active site in the process has been obtained from the electrospray MS, which clearly indicated the formation of an O-sulfinyl enzyme (sulfinylated most probably on the active site serine).

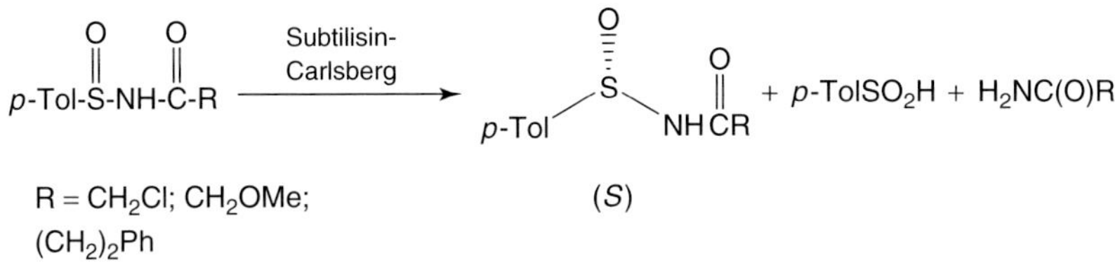

$R = CH_2Cl$; CH_2OMe; $(CH_2)_2Ph$

Scheme 5.24 Subtilisin-Carlsberg-catalyzed hydrolysis of sulfinamides.

Other interesting examples of proteases that exhibit promiscuous behavior are proline dipeptidase from *Alteromonas* sp. JD6.5, whose original activity is to cleave a dipeptide bond with a prolyl residue at the carboxy terminus [121, 122] and aminopeptidase P (AMPP) from *E. coli*, which is a proline-specific peptidase that catalyzes the hydrolysis of N-terminal peptide bonds containing a proline residue [123, 124]. Both enzymes exhibit phosphotriesterase activity. This means that they are capable of catalyzing the reaction that does not exist in nature. It is of particular importance, since they can hydrolyze unnatural substrates – triesters of phosphoric acid and diesters of phosphonic acids – such as organophosphorus pesticides or organophosphorus warfare agents (Scheme 5.25) [125].

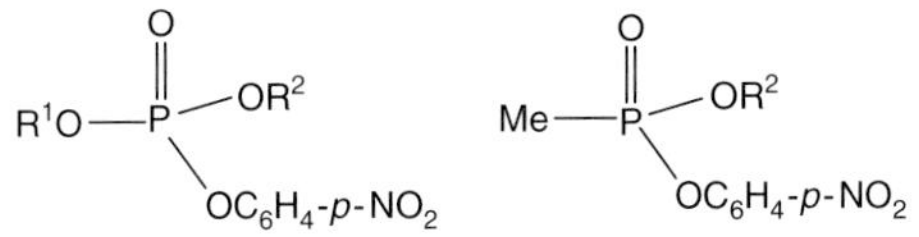

Scheme 5.25 Phosphorus esters as alternative substrates for peptidases.

Another family of enzymes that exhibit the same properties are mammalian paraoxonases, whose original activity is hydrolysis of five- and six-membered lactones [104].

Finally, it should be mentioned that recently a new type of enzyme catalytic promiscuity reaction has been reported, which does not involve any of the catalytic amino acids of the natural enzymatic process, nor does it use the natural active site. This phenomenon has been called *alternate-site enzyme promiscuity* [126].

5.4 Concluding Remarks

The aim of this short review was to signal the new possibilities offered by the fast-expanding repertoire of biocatalytic methods and approaches. These recent achievements make it possible to optimize enzymatic processes on levels reserved previously only for synthetic catalysts, and also to reach far beyond, employing the special features and origin of proteins.

The use of molecular biology methods, described in Section 5.3 seems to be especially worthwhile as it offers novel possibilities of optimization on process adjustment. Directed evolution leads to the formation of new biocatalysts with improved characteristics (selectivity, activity, stability, etc.). Incorporation of non-proteinogenic amino acids makes it possible to reach beyond the repertoire of building blocks used by nature. The prospect of bioconjugate preparation offers the possibility to form functional clusters of enzymes and to perform multiple synthetic steps in one pot.

These perspectives allow us to conclude that the constant rise of importance of biocatalysis in industrial processes is secured for years to come, and is limited only by further recognition from industrial society.

References

1. Faber, K. (2004) *Biotransformations in Organic Chemistry*, 5th edn, Springer-Verlag, Berlin.
2. Kiełbasiński, P. and Mikołajczyk, M. (2007) Chiral Heteroatom-containing Compounds, in *Future Directions in Biocatalysis* (ed. T.Matsuda), Elsevier, pp. 159–203.
3. Patel, R.N. (ed.) (2007) *Biocatalysis in the Pharmaceutical and Biotechnology Industries*, CRC Press, Boca Raton.
4. Aehle, W. (ed.) (2007) *Enzymes in Industry*, 3rd edn, Wiley-VCH GmbH & Co., KGaA., Weinheim.
5. Sugihara, J.M. (1953) *Adv. Carbohydr. Chem*, **8**, 1–44.
6. Kovac, P., Sokoloski, E.A., and Glaudemans, C.P. (1984) *J. Carbohydr. Res.*, **128**, 101–109.
7. Sweers, H.M. and Wong, C.H. (1986) *J. Am. Chem. Soc.*, **108**, 6421–6422.
8. Faber, K. (2000) Bio- and Chemo-catalytic Deracemisation Techniques, in *Enzymes in Action: Green Solutions for Chemical Problems* (eds B. Zwanenburg, M. Mikotajczyk, and P. Kiełbasiński), Kluwer Academic Publishers, Dordrecht, pp. 1–23.
9. Carrea, G. and Riva, S. (2008) *Organic Synthesis with Enzymes in Non-Aqueous Media*, Wiley-VCH GmbH & Co., KGaA., Weinheim.
10. (a) Yu, M.S., Lantos, I., Peng, Z.Q., Lu, J., and Cacchio, T. (2000) *Tetrahedron Lett.*, **41**, 5647–5651; (b) Liu, L.T., Hong, P.C., Huang, H.L., Chen, S.F., Wang, C.L.J., and Wen, Y.S. (2001) *Tetrahedron: Asymmetry*, **12**, 419–426.
11. Pifferi, G., Nizzola, R., and Cristoni, A. (1994) *Il Farmacol*, **49**, 453–455.
12. (a) Chenevert, R. and Desjardins, M. (1991) *Tetrahedron Lett.*, **32**, 4249–4253; (b) Chenevert, R. and Desjardins, M. (1994) *Can. J. Chem.*, **72**, 2312–2315; (c) Lam, L.K.P. and Jones, J.B. (1988) *J. Org. Chem.*, **53**, 2637–2344.

13. Chen, L.Y., Zaks, A., Chackalamannil, S., and Dugar, S. (1996) *J. Org. Chem.*, **61**, 8341–8343.
14. Homann, M.J., Vail, R., Morgan, B., Sabesan, V., Levy, C., Dodds, D.R., and Zaks, A.C. (2001) *Adv. Synth. Catal.*, **343**, 744–749.
15. (a) Fryszkowska, A., Komar, M., Koszelewski, D., and Ostaszewski, R. (2005) *Tetrahedron: Asymmetry*, **16**, 2475–2485; (b) Fryszkowska, A., Komar, M., Koszelewski, D., and Ostaszewski, R. (2006) *Tetrahedron: Asymmetry* **17**, 961–966.
16. Cao, L. (2005) *Curr. Opin. Chem Biol*, **9**, 217–226.
17. Szymanski, W. and Ostaszewski, R. (2006) *Tetrahedron: Asymmetry*, **17**, 2667–2671.
18. Szymanski, W., Zwolińska, M., and Ostaszewski, R. (2007) *Tetrahedron*, **63**, 7647–7653.
19. Pratesi, P., La Manna, A., Campiglio, A., and Ghislandi, V. (1959) *J. Chem. Soc.*, 4062–4065.
20. Pattenden, G., Gonzalez, M.A., Little, P.B., Millan, D.S., and Plowright, A.T. (2003) *Org. Biomol. Chem.*, **1**, 4173–4208.
21. Nicolas, P., Francis, D., Jean-Luc, V., Sopkova-de Olivera, S.J., Vincent, L., and Georges, D. (2003) *Tetrahedron*, **59**, 8049–8056.
22. Philipe, N., Deniret, F., Vasse, J.-L., and Dupas, G. (2003) *Tetrahedron*, **59**, 8049–8056.
23. Koszelewski, D., Redzej, A., and Ostaszewski, R. (2007) *J. Mol. Catal. B: Enzym.*, **47**, 51–57.
24. Houde, A., Kademi, A., and Leblanc, D. (2004) *Appl. Biochem. Biotechnol.*, **118**, 155–170.
25. Gotor-Fernandez, V., Brieva, R., and Gotor, V. (2006) *J. Mol. Catal. B: Enzym.*, **40**, 111–120.
26. Simeo, Y., Kroutil, W., and Faber, K. (2007) Biocatalytic Deracemization: Dynamic Resolution, Stereoinversion, Enantioconvergent Processes and Cyclic Deracemization, in *Biocatalysts in the Pharmaceutical and Biotechnology industries,*, (ed. R.N. Patel), CRC Press, Boca Raton, pp. 27–51.
27. Xue, C.-B. and Mousa, S.A. (1998) *Drugs Future*, **23**, 707–711.
28. Xue, C.-B., Wityak, J., Sielecki, T.M., Pinto, D.J., Batt, D.G., Cain, G.A., Sworin, M., Rockwell, A.L., Roderick, J.J., Wang, S., Orwat, M.J., Frietze, W.E., Bostrom, L.L., Liu, J., Higley, C.A., Rankin, F.W., Tobin, A.E., Emmett, G., Lalka, G.K., Sze, J.Y., Di Meo, S.V., Mousa, S.A., Thoolen, M.J., Racanelli, A.L., Hausner, E.A., Reilly, T.M., DeGrado, W.F., Wexler, R.R., and Olson, R.E. (1997) *J. Med. Chem.*, **40**, 2064–2084.
29. Wityak, J., Sielecki, T.M., Pinto, D.J., Emmett, G., Sze, J.Y., Liu, J., Tobin, E., Wang, S., Jiang, B., Ma, P., Mousa, S.A., Wexler, R.A., and Olson, R.E. (1997) *J. Med. Chem.*, **40**, 50–60.
30. Pesti, J.A., Yin, J., Zhang, L.-H., Anzalone, L., Waltermire, R.E., Ma, P., Gorko, E., Confalone, P.N., Fortunak, J., Silverman, C., Blackwell, J., Chung, J.C., Hrytsak, M.D., Cooke, M., Powell, L., and Ray, C. (2004) *Org. Proc. Res. Dev.*, **8**, 22–27.
31. Bommarius, A.S., Schwarm, M., and Drauz, K. (1996) *Chim. Oggi*, **14**, 61–64.
32. Henze, H. and Craig, W. (1945) *J. Org Chem.*, **10**, 2–9.
33. Pietzsch, M., Syldatk, C., and Wagner, F. (1990) *DECHEMA Biotechnol. Conf.*, **4**, 259–262.
34. Gross, C., Syldatk, C., and Wagner, F. (1987) *Biotechnol. Tech.*, **1**, 85–90.
35. Nishida, Y., Nakamichi, K., Nabe, K., and Tosa, T. (1987) *Enzyme Microb. Technol.*, **9**, 721–725.
36. May, O., Nguyen, P.T., and Arnold, F.H. (2000) *Nat. Biotechnol.*, **18**, 317–320.
37. May, O., Verseck, S., Bommarius, A., and Drauz, K. (2002) *Org. Proc. Res. Dev.*, **6**, 452–457.
38. (a) Larsson, A.L.E., Persson, B.A., and Bäckvall, J.E. (1997) *Angew. Chem., Int. Ed. Engl.*, **36**, 1211–1214; (b) Persson, B.A., Larsson, A.L.E., Le Ray, M., and Bäckvall, J.E. (1999) *J. Am. Chem. Soc.*, **121**, 1645–1650.
39. (a) Persson, B.A. and Bäckvall, J.E. (1999) *J. Org. Chem.*, **64**, 5237–5240;

(b) Edin, M. and Bäckvall, J.E. (2003) *J. Org. Chem.*, **68**, 2216–2222.

40. Pamies, O. and Bäckvall, J.E. (2003) *J. Org. Chem.*, **68**, 4815–4818.

41. Kiełbasiński, P., Rachwalski, M., Mikołajczyk, M., Moelands, M.A.H., Zwanenburg, B., and Rutjes, F.P.J.T. (2005) *Tetrahedron: Asymmetry*, **16**, 2157–2160.

42. Roengpithya, C., Patterson, D.A., Gibbins, E.J., Taylor, P.C., and Livingston, A.G. (2006) *Ind. Eng. Chem. Res.*, **45**, 7101–7109.

43. Pamies, O. and Bäckvall, J.E. (2002) *J. Org. Chem.*, **67**, 9006–9010.

44. Traff, A., Bogar, K., Warner, M., and Bäckvall, J.E. (2008) *Org. Lett.*, **10**, 4807–4810.

45. Fransson, A.-L., Boren, L., Pamies, O., and Bäckvall, J.E. (2005) *J. Org. Chem.*, **70**, 2582–2587.

46. Kim, M.-J., Choi, Y.K., Kim, S., Kim, D., Han, K., Ko, S.-B., and Park, J. (2008) *Org. Lett.*, **10**, 1295–1298.

47. Paetzold, J. and Bäckvall, J.E. (2005) *J. Am. Chem. Soc.*, **127**, 17620–17621.

48. Kim, N., Ko, S.-B., Kwon, M.S., Kim, M.-J., and Park, J. (2005) *Org. Lett.*, **7**, 4523–4526.

49. Gastaldi, S., Escoubet, S., Vanthuyne, N., Gil, G., and Bertrand, M.P. (2007) *Org. Lett.*, **9**, 837–839.

50. (a) Danda, H., Nagatomi, T., Maehara, A., and Umemura, T. (1991) *Tetrahedron*, **47**, 8701–8716. (b) Nakamura, K., Fuji, M., and Ida, Y. (2001) *Tetrahedron: Asymmetry*, **12**, 711–715.

51. Pedragosa-Moreau, S., Archelas, A., and Furstoss, R. (1993) *J. Org. Chem.*, **58**, 5533–5536.

52. Manoj, K.M., Archelas, A., Baratti, J., and Furstoss, R. (2001) *Tetrahedron*, **57**, 695–701.

53. (a) Pogorevc, M., Kroutil, W., Wallner, S.M., and Faber, K. (2002) *Angew. Chem., Int. Ed. Engl.*, **41**, 4052–4054; (b) Pogorevc, M., Strauss, U.T., Riermeier, T., and Faber, K. (2002) *Tetrahedron: Asymmetry*, **13**, 1443–1447.

54. (a) Wallner, S.M., Nestl, B., and Faber, K. (2005) *Tetrahedron*, **61**, 1517–1521; (b) Wallner, S.M., Nestl, B., and Faber, K. (2004) *Org. Lett.*, **6**, 5009–5010.

55. (a) Cull, S.G., Holbrey, J.D., Vargas-Mora, V., Seddon, K.R., and Lye, G.J. (2000) *Biotechnol. Bioeng.*, **69**, 227–233; (b) Erbeldinger, M., Mesiano, A.J., and Russell, A.J. (2000) *Biotechnol. Prog.*, **16**, 1129–1131; (c) Madeira Lau, R., van Rantwijk, F., Seddon, K.R., and Sheldon, R.A. (2000) *Org. Lett.*, **2**, 4189–4191.

56. For recent reviews, see: (a) van Rantwijk, F. and Sheldon, R.A. (2007) *Chem. Rev.*, **107**, 2757–2785; (b) Moon, Y.H., Lee, S.M., Ha, S.H., and Koo, Y.-M. (2006) *Korean J. Chem. Eng.*, **23**, 247–263; (c) Yang, Z. and Pan, W. (2005) *Enzyme Miocrob. Technol.*, **37**, 19–28; (d) Park, S. and Kazlauskas, R.J. (2003) *Curr. Opin. Biotechnol.*, **14**, 432–437; (e) van Rantwijk, F., Madeira Lau, R., and Sheldon, R.A. (2003) *Trends. Biotechnol.*, **21**, 131–138; (f) Kragl, U., Eckstein, M., and Kaftzik, N. (2002) *Curr. Opin. Biotechnol.*, **13**, 565–571.

57. Earle, M.J., Esperança, J.M.S.S., Gilea, M.A., Canongia Lopes, J.N., Rebelo, L.P.N., Magee, J.W., Seddon, K.R., and Widegren, J.A. (2006) *Nature*, **439**, 831–834.

58. Kosmulski, M., Gustafsson, J., and Rosenholm, J.B. (2004) *Thermochim. Acta*, **412**, 47–53.

59. Abraham, M.H. and Acree, W.E. (2006) *Green Chem.*, **8**, 906–915.

60. Wells, A.S. and Coombe, V.T. (2006) *Org. Process Res. Dev.*, **10**, 794–798.

61. Gathergood, N., Garcia, M.T., and Scammels, P.J. (2004) *Green Chem.*, **6**, 166–175.

62. de María, P.D. (2008) *Angew. Chem. Int. Ed.*, **47**, 6960–6968.

63. Yu, H., Wu, J., and Ching, C.B. (2005) *Chirality*, **17**, 16–21.

64. Lundell, K., Kurki, T., Lindroos, M., and Kanerva, L.T. (2005) *Adv. Synth. Catal.*, **347**, 1110–1118.

65. (a) Kaftzik, N., Wasserscheid, P., and Kragl, U. (2002) *Org. Process Res. Dev.*, **6**, 553; (b) Kaftzik, N., Neumann, S., Kula, M.R., and Kragl, U. (2003) Enzymatic Condensation Reactions in ionic liquids, in *Ionic Liquids as Green Solvents*, ACS Symposium Series, Vol. 856 (eds R.D. Rogers and

K.R. Seddon), American Chemical Society, Washington, DC, p. 206; (c) Kiełbasiński, P., Albrycht, M., Łuczak, J., and Mikołajczyk, M. (2002) *Tetrahedron: Asymmetry*, **13**, 735–738.
66. (a) Howarth, J., James, P., and Dai, J. (2001) *Tetrahedron Lett.*, **42**, 7517–7519; (b) Roberts, N.J., Seago, A., and Lye, G.J. (2002) Biocatalytic routes to the Efficient Synthesis of Pharmaceuticals in ionic liquids, in *Book of Abstracts, International Congress on Biocatalysis*, Hamburg, p. 117.
67. Gaisberger, R.P., Fechter, M.H., and Griengl, H. (2004) *Tetrahedron: Asymmetry*, **15**, 2959–2963.
68. Rumbau, V., Marcilla, R., Ochoteco, E., Pomposo, J.A., and Mecerreyes, D. (2006) *Macromolecules*, **35**, 8547–8549.
69. Lourenço, N.M.T. and Afonso, C.A.M. (2007) *Angew. Chem. Int. Ed.*, **46**, 8178–8181.
70. For recent review, see: Hobbs, H.R. and Thomas, N.R. (2007) *Chem. Rev.*, **107**, 2786–2820.
71. Celebi, N., Yildiz, N., Demir, A.S., and Calimli, A. (2007) *J. Supercrit. Fluids*, **41**, 386–390.
72. Vermue, M. and Tramper, J. (1995) *Appl. Chem.*, **67**, 345–373.
73. Gumí, T., Paolucci-Jeanjean, D., Belleville, M.-P., and Rios, G.M. (2007) *J. Memb. Sci.*, **297**, 98–103.
74. Matsuda, T., Watanabe, K., Harada, T., and Nakamura, K. (2004) *Catal. Today*, **96**, 103–111.
75. For recent review, see: Rezaei, K., Temelli, F., and Jenab, E. (2007) *Biotechnol. Adv.*, **25**, 272–280.
76. Matsuda, T., Harada, T., Nakamura, K., and Ikariya, T. (2005) *Tetrahedron: Asymmetry*, **16**, 909–915.
77. Overmeyer, A., Schrader-Lippelt, S., Kasche, V., and Brunner, G. (1999) *Biotechnol. Lett.*, **21**, 65–69.
78. Turner, C., Persson, M., Mathiasson, L., Adlercreutz, P., and King, J.W. (2001) *Enzyme Microbiol. Technol.*, **29**, 111–121.
79. Almeida, M.C., Ruivo, R., Maia, C., Freire, L., Cornea de Sampaio, T., and Barreiros, S. (1998) *Enzyme Microbiol. Technol.*, **22**, 494–499.
80. Matsuda, T., Watanabe, K., Harada, T., Nakamura, K., Arita, Y., Misumi, Y., Ichikawa, S., and Ikariya, T. (2004) *Chem. Commun.*, 2286–2287.
81. Mori, T., Li, M., Kobayashi, A., and Okahata, Y. (2002) *J. Am. Chem. Soc.*, **124**, 1188–1189.
82. Mishima, K., Matsuyama, K., Baba, M., and Chidori, M. (2003) *Biotechnol. Prog.*, **19**, 281–284.
83. Dijkstra, Z.J., Weyten, H., Willems, L., and Keurentjes, J.T.F. (2006) *J. Mol. Catal. B: Enzym.*, **39**, 112–116.
84. Matsuda, T., Tsuji, K., Kamitanaka, T., Harada, T., Nakamura, K., and Ikariya, T. (2005) *Chem. Lett.*, **34**, 1102–1103.
85. Gumí, T., Paolucci-Jeanjean, D., Belleville, M.-P., and Rios, G.M. (2007) *J. Memb. Sci.*, **297**, 98–103.
86. Chen, X., Liu, Z., Wang, J., Fang, J., Fan, H., and Wang, P.G. (2000) *J. Biol. Chem.*, **275**, 31594–31600.
87. Ljungcrantz, P., Carlsson, H., Mansson, M.A., Buckel, P., Mosbach, K., and Buelow, L. (1989) *Biochemistry*, **28**, 8786–8792.
88. Zhang, Y., Li, S.-Z., Li, J., Pan, X., Cahoon, R.E., Jaworski, J.G., Wang, X., Jez, J.M., Chen, F., and Yu, O. (2006) *J. Am. Chem. Soc.*, **128**, 13030–13031.
89. Torres Pazmino, D.E., Snajdrova, R., Baas, B.-J., Ghobrial, M., Mihovilovic, M.D., and Fraaije, M.W. (2008) *Angew. Chem. Int. Ed.*, **47**, 2275–2278.
90. For reviews, see: (a) Reetz, M.T. (2001) *Angew. Chem. Int. Ed.*, **40**, 284–310; (b) Wahler, D. and Reymond, J.-L. (2001) *Curr. Opin. Biotechnol.*, **12**, 535–544.
91. Reetz, M.T., Becker, M.H., Klein, H.-W., and Stöckigt, D. (1999) *Angew. Chem. Int. Ed.*, **38**, 1758–1761.
92. Drummond, D.A., Iverson, B.L., Georgiou, G., and Arnold, F.H. (2005) *J. Mol. Biol.*, **350**, 806–816.
93. Powell, K.A., Ramer, S.W., del Cardayré, S.B., Stemmer, W.P.C., Tobin, M.B., Longchamp, P.F., and Huisman, G.W. (2001) *Angew. Chem. Int. Ed.*, **40**, 3948–3959.
94. Funke, S.A., Eipper, A., Reetz, M.T., Otte, N., Thiel, W., van Pouderoyen, G., Dijkstra, B.W., Jaeger, K.-E., and

Eggert, T. (2003) *Biocatal. Biotransform.*, **21**, 67–73.
95. Reetz, M.T., Wilensek, S., Zha, D., and Jaeger, K.-E. (2001) *Angew. Chem. Int. Ed.*, **40**, 3589–3591.
96. Reetz, M.T., Bocola, M., Carballeira, J.D., Zha, D., and Vogel, A. (2005) *Angew. Chem. Int. Ed.*, **44**, 4192–4196.
97. Reetz, M.T., Wang, L.-W., and Bocola, M. (2006) *Angew. Chem. Int. Ed.*, **45**, 1236–1241.
98. Jackson, J.C., Duffy, S.P., Hess, K.R., and Mehl, R.A. (2006) *J. Am. Chem. Soc.*, **128**, 11124–11127.
99. Gauchet, C., Labadie, G.R., and Poulter, C.D. (2006) *J. Am. Chem. Soc.*, **128**, 9274–9275.
100. Dougherty, D.A. (2000) *Curr. Opin. Chem. Biol.*, **4**, 645–652.
101. Saks, M.E., Sampson, J.R., Nowak, M.W., Kearney, P.C., Du, F., Abelson, J.N., Lester, H.A., and Dougherty, D.A. (1996) *J. Biol. Chem.*, **271**, 23169–23175.
102. Noren, C.J., Anthony-Cahill, S.J., Griffith, M.C., and Schultz, P.G. (1989) *Science*, **244**, 182–188.
103. Bornscheuer, U.T. and Kazlauskas, R.J. (2004) *Angew. Chem. Int. Ed.*, **43**, 6032–6040.
104. Khershonsky, O., Roodveldt, C., and Tawfik, D.S. (2006) *Curr. Opin. Chem. Biol.*, **10**, 498–508.
105. Hult, K. and Berglund, P. (2007) *Trends Biotechnol.*, **25**, 231–238.
106. Kazlauskas, R.J. (2005) *Curr. Opin. Chem. Biol.*, **9**, 195–201.
107. Bornscheuer, U.T. and Kazlauskas, R.J. (2006) *Hydrolases in Organic Synthesis. Regio- and Stereoselective Biotransformations*, 2nd edn, Wiley-VCH GmbH & Co.KGaA, pp. 61–183.
108. Kitazume, T., Ikeya, T., and Murata, K. (1986) *J. Chem. Soc., Chem. Commun.*, 1331–1333.
109. Torre, O., Alfonso, I., and Gotor, V. (2004) *Chem. Commun.*, 1724–1725.
110. Carlqvist, P., Svedendahl, M., Branneby, C., Hult, K., Brinck, T., and Berglund, P. (2005) *Chembiochem*, **6**, 331–336.
111. Svedendahl, M., Hult, K., and Berglund, P. (2005) *J. Am. Chem. Soc.*, **127**, 17988–17989.
112. Xu, J.-M., Zhang, F., Wu, Q., Zhang, Q.-Y., and Lin, X.-F. (2007) *J. Mol. Catal. B: Enzym.*, **49**, 50–54.
113. Branneby, C., Carlqvist, P., Magnusson, A., Hult, K., Brinck, T., and Berglund, P. (2003) *J. Am. Chem. Soc.*, **125**, 874–875.
114. Lee, R.-S. (1997) *J. Chin. Chem. Soc.*, **44**, 77–80.
115. Wu, W.-B., Xu, J.M., Wu, Q., Lu, D.S., and Lin, X.-F. (2005) *Synlett*, 2433–2436.
116. Wu, W.-B., Wang, N., Xu, J.-M., Wu, Q., and Lin, X.-F. (2005) *Chem. Commun.*, 2348–2350.
117. Wu, W.-B., Xu, J.M., Wu, Q., Lu, D.S., and Lin, X.-F. (2006) *Adv. Synth. Catal.*, **348**, 487–492.
118. Cai, Y., Yao, S.-P., Wu, Q., and Lin, X.-F. (2004) *Biotechnol. Lett.*, **26**, 525–528.
119. Cai, Y., Sun, X.-F., Wang, N., and Lin, X.-F. (2004) *Synthesis*, 671–674.
120. Mugford, P.F., Magloire, V.P., and Kazlauskas, R.J. (2005) *J. Am. Chem. Soc.*, **127**, 6536–6537.
121. Hill, C.M., Wu, F., Cheng, T.-C., DeFrank, J.J., and Raushel, F.M. (2000) *Bioorg. Med. Chem. Lett.*, **10**, 1285–1288.
122. Hill, C.M., Wu, F., Cheng, T.-C., DeFrank, J.J., and Raushel, F.M. (2001) *Bioorg. Chem.*, **29**, 27–35.
123. Jao, S.-C., Huang, L.-F., Tao, Y.-S., and Li, W.-S. (2004) *J. Mol. Catal. B Enzym.*, **27**, 7–12.
124. Huang, L.-F., Su, B., Jao, S.C., Liu, K.-T., and Li, W.-S. (2006) *Chembiochem*, **7**, 506–514.
125. Cheng, T.-C. and DeFrank, J.J. (2000) Hydrolysis of Organophosphorus Compounds by Bacterial Prolidases, in *Enzymes in Action: Green Solutions for Chemical Problems* (eds B. Zwanenburg, M. Mikotajczyk, and P. Kiełbasiński), Kluwer Academic Publishers, Dordrecht, pp. 243–261.
126. Taglieber, A., Hobenreich, H., Carballeira, J.D., Mondiere, R.J.G., and Reetz, M.T. (2007) *Angew. Chem. Int. Ed.*, **46**, 8597–8600.

6
Oxidation Tools in the Synthesis of Catalysts and Related Functional Materials

Ignacio Melián-Cabrera and Jacob A. Moulijn

6.1
Introduction

The classical design of a heterogeneous catalyst is changing from the use of conventional supports to more sophisticated high-tech structures, such as carbon nanofibers (CNFs) and carbon nanotubes (CNTs), periodic organosilicas, and open-framework molecular sieves containing micro- and mesoporosity. These high-tech materials are in general not active themselves and require functionalization of the surface or bulk properties. Looking at recent reported work, we realized that oxidation reactions are increasingly more useful tools for functionalization, opening up opportunities to tune the catalytic performance. This chapter aims at compiling the state of the art on oxidation tools in the synthesis of catalysts and related functional materials.

6.2
Preparation Strategies Involving Chemical Oxidative Approaches

6.2.1
Calcination in Oxidative Atmospheres

Thermal treatment of a material in a gas oxidizing atmosphere is the simplest concept. This can be done in air, air diluted in N_2, dry air, or in ultrahigh purity O_2. In the laboratory practice, calcination is done in flowthrough beds, aided by fluidization, or in static box furnaces. Important aspects are the bed geometry, the removal of the generated gases, and temperature gradients.

Thermal treatments can be applied to modify the properties of a material, for example, dealumination and optimization of crystalline phases. These techniques do not require oxidants. Oxidative thermal treatments are generally employed to activate molecular sieves, by removing the organic templates employed during synthesis. This is one of the key steps when preparing porous catalysts or adsorbents. In air–atmosphere calcination, the templates are typically combusted between 400

Novel Concepts in Catalysis and Chemical Reactors: Improving the Efficiency for the Future.
Edited by Andrzej Cybulski, Jacob A. Moulijn, and Andrzej Stankiewicz

ISBN: 978-3-527-32469-9

and 700 °C. Occasionally the composite mesophases can contain up to 50 wt% of organics. The removal procedure is thus of the utmost importance.

Many studies on template thermal degradation have been reported on zeolites of industrial interest including ZSM5 [1–5], silicalite [1], and beta [6–8], as well as surfactant-templated mesostructured materials [9–13]. The latter are structurally more sensitive than molecular sieves. Their structure usually shrinks upon thermal treatment. The general practice is slow heating at 1 °C min^{-1} (N_2/air) up to 550 °C, followed by a temperature plateau.

It is evident that dedicated studies are required for each structure to optimize the template oxidation protocol. Many structures, in particular nonsiliceous, are thermally very sensitive [14, 15]. Calcination can result in a complete breakdown due to hydrolysis, redox processes, and phase transformations. The removal of templates in those systems is critical, making the development of mild detemplation techniques necessary [16].

6.2.2 Aerosol Flame Technologies

Aerosol flame synthesis is a mature technology. A solid phase is generated by dispersing the metal precursors in a flame. The first reports are dated from the 1970s to the 1980s [17–19]. Reviews can be found in [20, 21]. Three different approaches are identified, depending on the state of precursor:

1) **Vapor fed aerosol flame synthesis (VAFS)**: the precursor is in gas phase by using volatile metal precursors such as chlorides.
2) **Flame spray pyrolysis (FSP)**: a liquid precursor solution is sprayed into the flame and ignites; its combustion drives the flame process.
3) **Flame-assisted spray pyrolysis (FASP)**: this is similar to FSP, but the flame is sustained by a fuel.

These processes are very rapid and allow the preparation of inorganic supports in one step. This technique allows large-scale manufacturing of supports such as titania, fumed silica, and aluminas. Sometimes the properties of the material differ from the conventional preparation routes and make this approach unique. Multicomponent systems can be also prepared, either by multimetallic solutions or by using a two-nozzle system fed with monometallic solutions [22]. The as-prepared powder can be directly deposited onto substrates, and the process is termed *combustion chemical vapor deposition* [23].

Active heterogeneous catalysts have been obtained. Examples include titania-, vanadia-, silica-, and ceria-based catalysts. A survey of catalytic materials prepared in flames can be found in [20]. Recent advances include nanocrystalline TiO_2 [24], one-step synthesis of noble metal TiO_2 [25], Ru-doped cobalt–zirconia [26], vanadia–titania [27], Rh-Al_2O_3 for chemoselective hydrogenations [28], and alumina-supported noble metal particles via high-throughput experimentation [29].

6.2.3
Solution Combustion Synthesis

This methodology uses the heat produced from the exothermic reaction between an oxidizer, usually a metal nitrate, and a fuel possessing amino groups [30, 31]. The reactants are dissolved in water in a suitable ratio and heated until the mixture is ignited and self-propagation takes place. The main advantages are the high speed, simplicity of equipment, and easy scale-up. However, it requires ignition, relatively high temperatures, and use of specific metal precursors. Typical fuels used are urea and hydrazides due to their highly exothermic combustion [32].

Solution combustion belongs to a wider family entitled *combustion synthesis* [33], covering also solid-state processes. This approach leads to technologically interesting materials but not to many catalysts. Conversely, combustion using liquid solutions allows to prepare heterogeneous catalysts supports and multimetallic counterparts. Many applications have been reported: metal-supported αAl_2O_3 [34], Cu/CeO_2 [35], Ag/CeO_2 [36], Pt/Pd CeO_2 [37], $Ce_{1-x}Pt_xO_{2-\delta}$ [38], HDS based CoMo [39], *in situ* methods [40], perovskite structures [41–45], $Ce_{1-x}Ni_xO_2$ for propane oxidative steam reforming [46], $CuO/ZnO/ZrO_2/Pd$ for oxidative hydrogen production from methanol [47], and mesoporosity induction in MgO [48] and Al_2O_3 [49].

6.2.4
Sulfonic Acid Functionalization of Ordered Mesoporous Materials and Periodic Organosilicas

Mesoporous molecular sieves have drawn considerable attention as an alternative to zeolites. It was soon realized that additional functionalization is required. The Al-substituted formulations showed relatively low acid strengths. Research focused on the covalent attachment of HSO_3 groups to silica surfaces. This can be done by postsynthesis grafting, coating, or co-condensation of mercaptopropyl groups that are oxidized into -HSO_3. Another area of interest developed with the discovery of periodic organosilicas that required acidity incorporation too. In both cases, oxidation chemistry played an important role.

Organic modifications of the mesoporous materials were achieved by grafting on surface hydroxyl groups [50, 51] or co-condensation of siloxane and organosiloxane precursors [52–56]. Co-condensation allowed the use of alkylthiol precursors, in particular 3-mercaptopropyltrimethoxysilane (MPTMS). Stein's group reported the introduction of organosulfur groups on MCM-41 [57] that yields sulfonic acids by successive HNO_3 oxidations. The material showed acid-catalyzed behavior.

Jacob's group reported a protocol that used mild H_2O_2 oxidations and MPTMS for three approaches, namely, grafted, coated, and co-condensed MCM-41 and HMS materials [58] (Figure 6.1a). The propyl spacer was stable to the oxidation conditions while sulfonic groups were generated as evidenced by NMR spectra. These materials

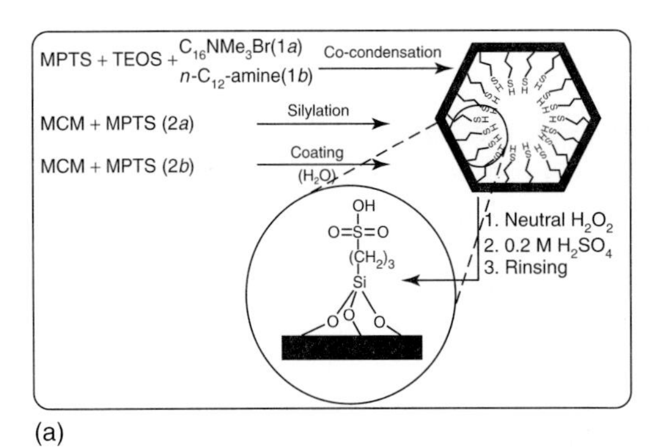

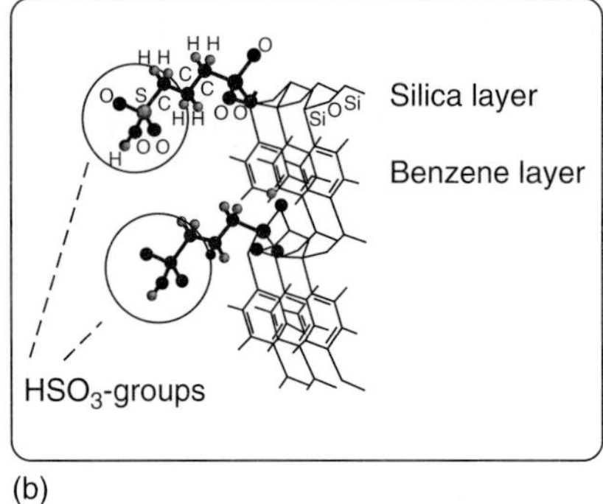

Figure 6.1 (a) HSO_3-functionalization on silica MCM/HMS materials via thiol incorporation (co-condensation, grafting, or coating) and mild H_2O_2 oxidation. (Adapted with permission from [58], Copyright 1998. Royal Society of Chemistry.) (b) Model for a phenylene periodic organosilica where the silica layers are functionalized with propylsulfonic-acid groups via HNO_3 oxidation of the thiol function. (Reprinted with permission from [68], Copyright 2002. American Chemical Society.)

proved to be active for the synthesis of 2,2-bis(5-methylfuryl)propane [58], esterification of sorbitol [58] and glycerol [59] with lauric acid, Bisphenol-A synthesis [60, 61], dehydration of xylose into furfural [62], and the esterification of high-free fatty acid soybean oil [63]. Pérez-Pariente *et al.* reported a modified MCM-41 by combining alkyl and mercaptopropyl groups [64]. H_2O_2 oxidation yielded the corresponding HSO_3 groups while the alkyl groups remained unchanged. The improved hydrophobicity of the channels benefits the esterification of fatty acids [65, 66]. Feng *et al.* [67] also applied the H_2O_2 oxidation of thiol-JLU-20 materials, obtaining a high -HSO_3 density.

In situ thiol oxidation can also be done just after the tetraethoxysilane hydrolysis. Both MPTMS and H_2O_2 are added simultaneously before aging. This was demonstrated by Stucky *et al.* [69] on SBA-15 containing HSO_3 and additional alkyl groups, followed by Mbaraka *et al.* [63], and Yang *et al.* [70], the latter including template removal. *In situ* oxidation avoids residual unreacted thiols, typically observed by postsynthetic routes.

Oxidation chemistry has also been useful for periodic organosilicas. Inagaki *et al.* reported the direct sulfonation of the phenylene group of mesoporous benzene–silica and its use as solid acid catalyst [71]. In a second protocol [68], the silica layers were functionalized while keeping the benzene sites intact (Figure 6.1b). This was done by co-condensation of 1,4-bis(triethoxy-silyl)benzene and MPTMS. Oxidation of the –SH groups was done in HNO_3 [57]. The crystal-like pore walls survived the oxidation but a decrease in BET area was observed. For the highest acidity only about 50% oxidation took place. H_2O_2 did not improve the oxidation yields [72]. Alkyl hydroperoxides were covalently functionalized onto SBA-15 by autoxidation using molecular oxygen and initiator or H_2O_2 [73]. Nonoxidative approaches have been reported as well [74–84].

6.2.5
Surface Oxidation of Carbon Nanofibers Prior to Functionalization

Oxidation chemistry has been practiced widely for carbon catalyst supports. Treatments with oxidants generate oxidic groups, which can be acidic or basic. It is not only performed with gas-phase reactants, for example, air, O_2, O_3, and CO_2, at 200–500 °C but also with aqueous solutions of HNO_3, H_2SO_4, HCl, $KMnO_4$, and H_2O_2 usually activated by temperature. Oxidation makes the carbon surface more hydrophilic and reactive. This is important in order to improve the wetting ability, dispersing metal (nano)particles, and grafting complexes. This topic has been extensively discussed for conventional supports [85–90] but less for CNTs and CNFs. These systems are more inert and oxidation is a prerequisite before functionalization. Most of the work reported concerns CNTs and is discussed separately in Section 6.2.6.

The first studies on CNFs oxidation discussed the impact of the surface treatments on bulk ordering [91]. Investigations for catalytic purposes came later with extensive contributions by the groups in Utrecht, Geus, and de Jong. For an optimal use of CNFs as catalyst supports, their surface has to be modified. One way is by introducing oxygen-containing groups. Protocols employ oxidations with concentrated HNO_3 or a mixture with H_2SO_4 under reflux [92–94]. The graphitic layers remain unchanged while the pore volume increases [93, 94]. Toebes *et al.* showed that the most predominant effect is the opening of the inner tubes of the fibers [93]. The oxidation did not only occur at the surface but also developed 2–3 nm into the subsurface [93]. Carbonyl groups are formed, which are subsequently converted into carboxyls and carboxylic anhydrides [92].

Successful applications of the oxygen-modified CNFs are reported on immobilization of metal complexes [95], incorporation of small Rh particles [96], supported Pt and Ru CNFs by adsorption and homogeneous deposition precipitation [97, 98], Co CNFs for Fischer–Tropsch synthesis [99], and Pt CNFs for PEM fuel cells [100].

6.2.6
Purification, Opening, and Size Reduction of Carbon Nanotubes by Oxidative Treatments

Oxidation reaction is essential for processing CNTs [101–104]. It enables purification by removing metals from synthesis and amorphous carbon domains, opening the long tubes, and cutting into shorter lengths that improve the wetting and filling. Such oxidative processes involve liquid-phase treatments under reflux with concentrated HNO_3 [105], H_2SO_4 [106], a mixture of both [107], or a two-step process involving HNO_3/H_2SO_4 and H_2SO_4/H_2O_2 [108]. These treatments introduce oxygen-containing surface groups, predominantly phenolic, carboxylic, and lactonic types. These groups stabilize dispersions of CNTs at higher concentrations than are possible for pristine CNTs [109]. Application of ultrasound prior to the acid treatment increases the number of O-groups [110] and completes the removal of wall-entrapped metallic impurities [111].

These approaches are crucial to measure the properties of the individual tubes, which are of large interest in theoretical studies [105, 112–115]. Confinement effects of metal nanoparticles within CNTs were also shown to be a promising concept in heterogeneous catalysts [116, 117]. However, several technical challenges were not solved until recently. Liquid treatments in HNO_3 mixtures remove metal catalyst impurities and amorphous carbons but at the expense of significant mass reduction of CNTs [118]. Alternative approaches include physical and gas-phase oxidations. For instance, multiwalled materials can be cut by ball milling [119, 120]. Electrical cutting in scanning tunneling [121] and atomic force microscopes [122] or electron beams [123] have been reported. Ozone treatment yields oxidation and etching of single-wall CNTs [124] too.

Recent investigations done on precise cutting of single-wall CNTs focus on (i) controlled liquid-phase oxidations after side wall damage and (ii) selective gas-phase oxidations, using metal cluster catalysts to promote an easier decomposition of the amorphous carbon domains. For instance, selective oxidation using gold particles has been reported. Clusters of about 20 nm after dispersion in carbonaceous soot containing single-wall carbon catalyze the oxidation of carbonaceous impurities at about 350 °C [125] leaving the CNTs undamaged. These results were later confirmed [126].

The first approach is carried out after purification by an uncatalyzed gas-phase oxidation. Chiang *et al.* proposed a multistep oxidation process at increasing temperatures followed by acid wash between steps [127, 128]. It is a method similar to that proposed by Dillon *et al.* [129] with the difference being the multistep procedure. At low-temperature amorphous carbon domains are removed, while the extraction of the metal residue embedded in the walls is carried out during the interstate acid washing. Such a protocol does not cut the tubes but induces sidewall damage. A second step cuts the CNTs through the damaged sites. This was achieved only by treating the purified nanotubes with 4 : 1 vol ratio of 96% H_2SO_4–30% H_2O_2 at room temperature [130]. Low mass loss, slow etch rates, and no extra sidewall damage were found. Tran *et al.* [131] proposed a cutting method of multiwalled CNTs based on repeated brief exposure to thermally oxidative conditions.

A novel process to control the cutting of long multiwalled tubes via catalytic oxidation with reduced losses was reported by Bao *et al.* [117, 132], Figure 6.2. The material was impregnated with $AgNO_3$ followed by decomposition at 300 °C in Ar, creating Ag particles of 10–15 nm on the exterior surface of the tubes. Subsequent catalytic oxidation was carried out and small pits were formed around the positions where the catalyst particles were located, due to oxidative etching of carbon. Ultrasound treatment in diluted nitric acid solution offered two functions. It completes the breaking of nanotubes around the pits and removes the silver clusters at the outer surface. Controlling the oxidation conditions was fundamental to avoid severe losses. At optimal conditions weight losses are < 20 wt%.

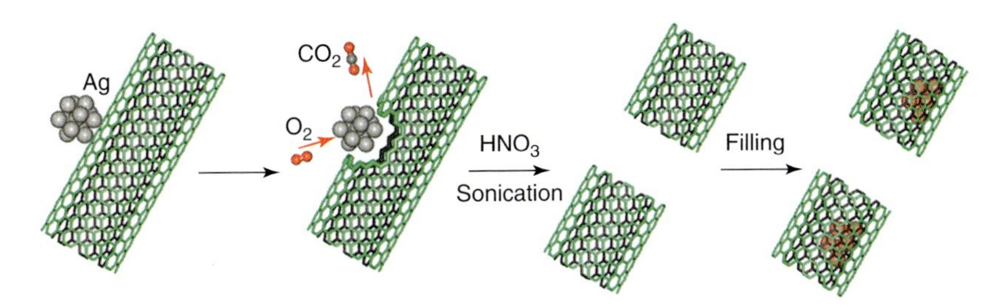

Figure 6.2 Two-step cutting of CNTs by hydrocarbon-assisted oxidation on silver clusters, followed by HNO_3 treatment mediated by ultrasound. The length of the CNTs is controlled through the Ag loading. (Reprinted with permission from [132], Copyright 2008, The Royal Society of Chemistry.)

6.2.7
Metal-Free Catalysis by Oxygen-Containing Carbon Nanotubes

Oxidized CNTs have shown unprecedented performance in oxidative dehydrogenation of alkanes. Su and coworkers [133] showed that multiwalled CNTs containing ketonic C=O groups catalyze the oxidative dehydrogenation of *n*-butane to butanes, especially butadiene. For conventional multimetal catalysts, the relatively fast alkenes oxidation limits the selectivity at higher conversion. The observed increase in selectivity comes from an inhibition of the secondary combustion. The presence of two neighboring C=O pairs is required to subtract two hydrogen atoms from butane. The catalytic effect of oxygen species on carbon materials has been already reported [134–137], but only for ethylbenzene, which is easier to activate than alkanes. Su *et al.* claim three effects: (i) higher selectivity, (ii) improved stability, and (iii) low partial pressures of oxygen. Addition of phosphorus improved the selectivity further by suppressing the combustion.

The active species are generated after refluxing the pristine CNT in HNO_3 [137, 138]. Other oxidation strategies can be implemented for tuning the type and density of the oxidized catalytic functions. Resasco [139] pointed out that these results open up an avenue for tuning the density and distribution of C=O pairs, in particular with controlled chiralities.

6.3
A Catalytic Oxidation Tool. Fenton Chemistry in Solid Catalyst Synthesis

6.3.1
What is the Fenton Reaction?

Fenton chemistry comprises reactions of H_2O_2 in the presence of iron species to generate reactive species such as the hydroxyl radical $OH^{\bullet}$. These radicals ($E^0 = 2.73$ V) lead to a more efficient oxidation chemistry than H_2O_2 itself ($E^0 = 1.80$ V).

The reaction was reported by Henry J. Fenton [140]. This reaction is applied in the treatment of hazardous organic wastes. A search for "Fenton reaction" in the website ISI Web of Knowledge® [141] throws up thousands of scientific papers due to the exponential growth in its use over the years. It has been reviewed in various papers [142–145]. Below, the reaction pathway in the absence of an organic compound is given:

$$Fe^{II} + H_2O_2 \rightarrow Fe^{III} + OH^- + OH^\bullet \quad (6.1)$$

$$Fe^{III} + H_2O_2 \rightarrow Fe^{II} + HO_2{}^\bullet + H^+ \quad (6.2)$$

$$OH^\bullet + H_2O_2 \rightarrow HO_2{}^\bullet + H_2O \quad (6.3)$$

$$OH^\bullet + Fe^{(II)} \rightarrow Fe^{III} + OH^- \quad (6.4)$$

$$Fe^{III} + HO_2{}^\bullet \rightarrow Fe^{II} + O_2H^+ \quad (6.5)$$

$$Fe^{II} + HO_2{}^\bullet + H^+ \rightarrow Fe^{III} + H_2O_2 \quad (6.6)$$

$$HO_2{}^\bullet + HO_2{}^\bullet \rightarrow H_2O_2 + O_2 \quad (6.7)$$

In the absence of any oxidizable compound, the net reaction is the Fe-catalyzed decomposition of H_2O_2 (Reaction 6.8). This reaction also occurs when a target contaminant is present.

$$2H_2O_2 \rightarrow 2H_2O + O_2 \quad (6.8)$$

This chemistry has been investigated and implemented for wastewater mineralization by oxidizing the organic pollutants. The process is very efficient, not selective and, as a consequence, almost all carbon matter can be removed. Topical areas also include soil and aquifer treatments, sometimes in combination with a secondary biotic process [145].

6.3.2 Can We Use Fenton Chemistry in Solid Catalyst Synthesis?

Although Fenton chemistry has been solely researched in water and soil purification, it can also be an added-value tool for catalyst preparation. The next sections review this concept including the following:

- shifting the complexation equilibria for ion-exchange by oxidation of the organic chelates [146];
- one-pot synthesis of metal-exchanged zeolites [147];
- mild detemplation of micro- and mesoporous materials [148–150].

The properties of the materials were evaluated by textural and structural techniques, while the final catalysts are compared on the basis of their N_2O-decomposition under simulated industrial conditions for nitric acid plants. This reaction is known to be activated by Fe-species [151, 152].

6.3.3
Kinetics of Fenton Chemistry

Fenton kinetics is quite complex. Different oxidizing species are present but hydroxyl radicals are regarded as the unique reactive species. Hydroxyl can be produced by thermal and photochemical reactions, and destroyed by the target pollutant, reaction intermediates, and by scavenging undesired reactions. Rate law formulae involve a large number of steps. Optimal conditions can be easily obtained by semiempirical modeling by accounting for the effect of the dominating parameters. Critical parameters are iron and hydrogen peroxide concentrations, temperature, light, and pH. Experimental design and response surface methodologies have been applied [145, 153, 154].

Haber and Weiss proposed that $OH^\bullet$ is the active oxidant [155] by combining an Fe^{II} salt with H_2O_2 (Reaction 6.1). Reaction 6.2 reduces the Fe^{III} into Fe^{II}, which will generate more $OH^\bullet$. The cycle continues until H_2O_2 is consumed. Reactions 6.1 and 6.2 show that the reaction is catalytically in Fe. Reaction 6.2 is several orders of magnitude slower compared to Reaction 6.1, being the rate-determining step of the overall process [145, 156–159]. This has consequences for the oxidation state of the Fe salt. The large majority of the iron ions are present as Fe^{III}. The system will behave independently of the initial oxidation state of iron. The Fe valence is relevant when the reaction is assisted by light. $OH^\bullet$ radicals can also be produced via Reaction 6.1 by supplying stoichiometric Fe^{II}. This however generates a large Fe^{III} residue.

The discussion above refers to the classical "dark" conditions where the chemical activation is achieved thermally. Fenton requires a moderate thermal activation, resulting in a reaction temperature ranging from 25 to 90 °C. The oxidizing capacity of the Fenton reaction can be increased by UV or UV–vis light irradiation [160, 161]. The increase is interpreted by means of the photoreduction ability of Fe^{III}:

$$Fe^{III}(L)_n + h\nu \rightarrow Fe^{II}(L)_{n-1} + L^\bullet_{ox} \tag{6.9}$$

Fe^{III} complexes display ligand to metal charge transfer excitation, dissociating to give Fe^{II} and oxidized ligand [162]. The photons introduce an alternative pathway to Reaction 6.2, being the rate-determining step under dark conditions, leading to an increased $OH^\bullet$ rate. $Fe(OH)^{2+}$ is the most important species (Reaction 6.9) because of its relatively high absorption coefficient and high concentration relative to other Fe^{III} species. Excitation of $Fe(OH)^{2+}$ alone can be used to generate radicals without oxidant, but stoichiometric amounts of Fe^{3+} are required. Electro-Fenton methods cover electrochemical reactions that are used to generate one or both of the reagents. The most promising approach is discussed elsewhere [163].

The pH effect in Fenton reactions is due to the Fe^{III} speciation. In highly acidic solutions containing noncoordinating species, Fe^{III} exists as $Fe(H_2O)^{6}_{3+}$. The composition as function of the pH is represented in Figure 6.3a. At increasing pH, Fe^{3+} undergoes hydrolysis forming $FeOH^{2+}$, $Fe(OH)_2{}^{+}$, and finally FeO_x via binuclear, $Fe_2(OH)^{2}_{4+}$, and polynuclear species. The aim of optimization is to avoid Fe precipitation, either bi-, poly-, or precipitated Fe oxides that are all inactive in

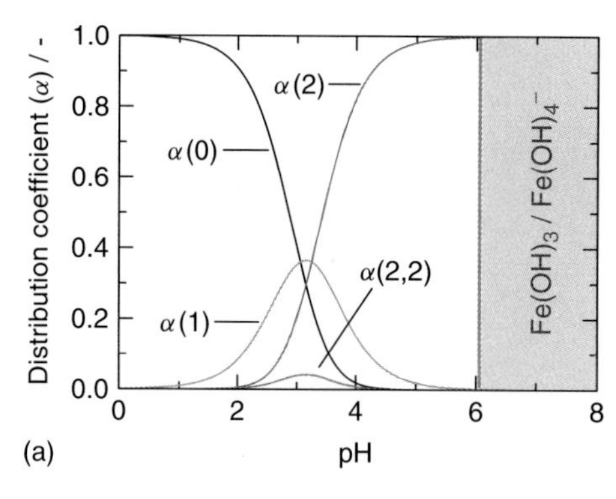

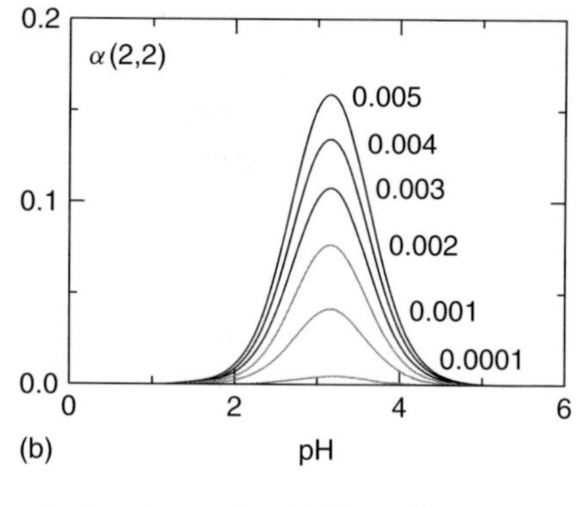

Figure 6.3 Calculated concentrations for the Fe^{III} speciation as a function of the pH in aqueous solutions taking into account the hydrolysis reactions: $Fe^{3+} + H_2O = Fe(OH)^{2+} + H^+ (\log K_1 = -3.05)$; $Fe^{3+} + 2H_2O = Fe(OH)_2{}^+ + 2H^+ (\log \beta_2 = -6.31)$; $2Fe^{3+} + 2H_2O = Fe_2(OH)_2{}^{4+} + 2H^+ (\log \beta_{22} = -2.91)$, where $\alpha(0) = [Fe^{3+}]/[Fe^{3+}]_{TOTAL}$; $\alpha(1) = [Fe(OH)^{2+}]/[Fe^{3+}]_{TOTAL}$; $\alpha(2) = [Fe(OH)_2{}^+]/[Fe^{3+}]_{TOTAL}$ and $\alpha(2,2) = [Fe_2(OH)_2{}^{4+}]/[Fe^{3+}]_{TOTAL}$. (a) Calculation for a $[Fe^{3+}]_{TOTAL}$ of 10^{-3} M. (b) $\alpha(2,2)$ for different $[Fe^{3+}]_{TOTAL}$ ranging from 10^{-4} to 5×10^{-3} M. (Equilibria constants were taken from [164–166].)

Fenton. The maximum performance is usually observed at pH slightly below 3 because of two reasons: (i) colloids that begin to precipitate at pH above 3 via the binuclear species ($\alpha_{2,2}$) are suppressed and (ii) the concentration of $Fe(OH)^{2+}$ (α_1) is close to its maximum. $Fe(OH)^{2+}$ species possess a high absorption coefficient under irradiation and maximize the oxidation yield. This holds for diluted systems. When the total Fe concentration is increased, the binuclear species become dominant and precipitation is favored. Figure 6.3b clarifies this aspect by showing the $Fe_2(OH)_2{}^{4+}$ concentration profiles for increasing Fe concentrations. In those cases, lowering the pH to about 2 is favorable.

6.4
First Concept in Catalyst Design. Shifting Complexation Equilibria for Ion-Exchange by Oxidation of the Organic Chelates

Conventional metal ion-exchange of zeolites involves aqueous solutions of soluble metal salts, usually nitrates. A disadvantage using nitrates is eutrophication of the wastewater. Metal–organic salts can be employed, but many of them form stable complexes, hindering the exchange reaction both kinetically and sterically. In some cases complexation is desired. Geus and coworkers benefited from chelation for impregnation [167, 168].

Our first example discusses the use of an oxidant to remove organic ligands in solution during ion-exchange [146]. The oxidant must be clean, strongly oxidizing, and should not generate residues on the catalyst. These properties are met by H_2O_2. It shifts complexation equilibria efficiently and the metal cations can be exchanged within the zeolite. The method was proven for the preparation of Fe-FER using Ferric citrate. The preparation temperature is a critical parameter. At low temperature, Fenton's kinetics is not fast enough to break down the ligands. At the highest titration temperature, the catalyst performance in N_2O decomposition dropped. This was explained by a simultaneous competitive Fe-hydrolysis process, leading to inactive FeO_x species. The optimal catalyst proved to be stable under tail gas conditions.

It was first assumed that the oxidant was H_2O_2. However, since traces of Fe-cations are present, a Fenton's type oxidation pathway, based on OH radicals, is more likely taking place.

6.5 Second Concept in Catalyst Design. One-Pot Synthesis of Fe Zeolite Catalysts

Iron zeolites are important in environmental catalysis, covering the abatement of nitrous oxide, NO_x control, and selective oxidation of organics [169]. The preparation of Fe-exchanged zeolite catalysts consists of various steps: (i) hydrothermal zeolite synthesis using organic templates; (ii) detemplation by thermal treatment; (iii) accommodation to the desired form, usually containing NH_4-groups; (iv) metal incorporation by wet chemistry; (v) drying; and (vi) calcination. Reducing the number of steps is challenging and can contribute to an improved and faster process.

We discuss here a combined process including detemplation and Fe incorporation by ion-exchange in the zeolite framework [147]. To achieve this, oxidants to decompose the organic template and Fe-cations for exchange are needed. Both requirements are in harmony with Fenton chemistry. The $OH^\bullet$ radicals can oxidize the template and the Fe-cations be exchanged simultaneously.

The preparation was performed on a commercial microcrystalline beta zeolite. The zeolite was treated with the Fenton's reagent and less than 0.3 wt% of carbon remained after the treatment. The porosity was fully developed as revealed by the pore-size distribution. Elemental analysis combined with TPR did confirm the high degree of Fe-exchange (98%) on the Brønsted sites.

Contacting the synthesized zeolite with the reagent leads to a ready-to-use catalyst. No further pretreatment before reaction is needed. The performance of such a novel one-pot catalyst was tested in N_2O decomposition, and was found to be even superior to the conventionally prepared counterpart. This was ascribed to the minimization of FeO_x formation.

It is concluded that zeolite beta can be simultaneously detemplated and Fe-exchanged without FeO_x formation by treating the parent zeolite with a Fenton reagent. The catalyst shows good performance on N_2O decomposition. This one-pot process simplifies its preparation protocol and can be extended to other systems. Indeed, our approach was followed by Liu *et al.* [170], for preparing Fe-SBA-15 for benzylation of benzene with interesting results.

6.6
Third Concept in Catalyst Design. Fenton Detemplation. Mild Organic Template Removal in Micro- and Mesoporous Molecular Sieves

By minimizing the Fe concentration (i.e., avoiding extensive Fe-exchange), zeolites, or mesoporous compounds can be detemplated at low temperatures without the need for high-temperature calcination. This third concept refers to the low-temperature Fenton detemplation. Strictly speaking, Fenton requires thermal activation but always below 100 °C. We refer here to "quasi room temperature" as compared to the high temperatures usually applied for calcination.

6.6.1
Role of Organic Templates and Drawbacks Associated to Calcination

Most of the microporous and mesoporous compounds require the use of structure-directing molecules under hydro(solvo)thermal conditions [14, 15, 171, 172]. A serious handicap is the application of high-temperature calcination to develop their porosity. It usually results in inferior textural and acidic properties, and even full structural collapse occurs in the case of open frameworks, (proto)zeolites containing small-crystalline domains, and mesostructures. These materials can show very interesting properties if their structure could be fully maintained. A principal question is, is there any alternative to calcination? There is a manifested interest to find alternatives to calcination to show the potential of new structures.

The organic templating approach was first introduced by Barrer and Denny [173]. Since then, organic amines, quaternary-ammonium bases, metal complexes, and other compounds have been extensively used in zeolite synthesis, acting as "space fillers" with low specificity, "structure-directing agent," or "true templates" [174]. Because such guest molecules usually interact with the frameworks through H-bonds, van der Waals's forces, or sometimes coordination bonds, it is crucial to remove the templates properly to form structurally stable, free-pore molecular sieves.

Microcrystalline zeolites such as beta zeolite suffer from calcination. The crystallinity is decreased and the framework can be notably dealuminated by the steam generated [175]. Potential Brønsted catalytic sites are lost and heteroatoms migrate to extra-framework positions, leading to a decrease in catalytic performance. Nanocrystals and ultrafine zeolite particles display aggregation issues, difficulties in regeneration, and low thermal and hydrothermal stabilities. Therefore, calcination is sometimes not the optimal protocol to activate such systems. Application of zeolites for coatings, patterned thin-films, and membranes usually is associated with defects and cracks upon template removal.

It must be noted that sometimes calcination is beneficial to create active species. Notable examples are the Sn-beta speciation [176] and generation of extra-framework Al-Lewis sites in beta zeolite for organic transformations

[175, 177]. However, even in these cases, it might be attractive to apply a mild detemplation followed by controlled heating for optimization.

For microporous compounds with special compositions, calcination effects are even more severe. As compared with zeolites, these compounds have lower thermal stability. Strictly speaking, most of them are nonporous since removal of the occluded guest molecules by calcination usually results in collapse. This is due to strong H-bonds with the framework, coordination bonds, and sometimes the templating molecule is shared with the inorganic polyhedra. Relevant examples of low-stability microporous compounds with interesting structural features are zeolitic open-framework phosphates made of Ga [178], In [179], Zn [180], Fe [181], Ni [182], V [183], and Al [184]. SU-M [185] is a mesoporous germanium oxide with crystalline pore walls, possessing one of the largest primitive cells and the lowest framework density of any inorganic material. The channels are defined by 30-rings. Structural and thermal information show that there exists a mismatch between framework stability and template decomposition. The latter requires temperatures higher than 450 °C, while the structure is preserved only until 300 °C.

Hence, for many promising materials milder template removal strategies are needed.

6.6.2
Alternative Approaches

The first option is to improve the calcination process, usually by a two-step calcination. Other different mild template-destructive methods have been proposed. They can be divided into thermal and chemical protocols. In addition, more ambitious protocols aimed at template recovery have been proposed. The major improvements are compiled in Table 6.1.

Solvent extraction is the most important technique for recovering surfactants from mesoporous materials. However, it is not very effective when applied to microporous compounds. Davis *et al.* [186] successfully extracted borosilicate and silicate BEA structures with acetic acid while a small template fraction could be removed for the aluminosilicate.

Recent reports describe more sophisticated detemplation methods. However, they are limited to mesoporous materials for the reasons described before. We show how Fenton chemistry can fulfill various missing challenges: (i) it provides a powerful oxidation capacity at low(er) temperatures and (ii) it can work for microporous compounds as well.

6.6.3
Fenton Detemplation. Concept and Proof-of-Principle

The approach consists of a liquid-phase oxidation using $OH^{\bullet}$ Fenton radicals from H_2O_2 for detemplation [148–150]. The radicals oxidize the organic template into CO_2 and H_2O while the porosity of the material is developed. The proof-of-principle of this concept is discussed for two case studies.

Table 6.1 Detemplation approaches of micro- and mesoporous materials.

Strategies aiming at template recovery
Solvothermal approaches [186, 187]
Use of withdrawable organic templates [188]
Supercritical extraction in the presence of co-solvents [189]
Destructive protocols I: thermal methods
High-temperature calcination
Two-step calcination [187e–f, 190]
Ammonolysis at moderate temperatures [191]
Platinum-catalyzed template removal [192]
Low-temperature plasma [193] and ozone treatment [194]
Destructive protocols II: chemical detemplation under relatively mild conditions
Detemplation during the crystallization [70]
Ozone in liquid phase [195] and UV–vis-assisted room temperature ozone [196]
UV–vis irradiation [197]
Digestion under microwaves [198]
$KMnO_4$ combined with H_2SO_4 [199]
Diluted hydrogen peroxide oxidation [200]
HNO_3/H_2O_2 treatments [201]
Fenton chemistry [148–150]

The first case study. Beta zeolite is the basis for many industrial applications due to its specific pore structure. It was successfully detemplated by controlled Fenton-oxidation. Remarkably, the template oxidation took place in the limited space of the microchannels. It was not obvious that the tiny pores in a zeolite would provide sufficient space for the process to work. The small size of the $OH^•$ compared to other oxidizing agents and its unique highly oxidizing potential must be the explanation. Characterization by low-pressure high-resolution Ar-physisorption shows an excellent Saito–Foley pore-size distribution (Figure 6.4a), comparable to reference zeolites with a pore-size development at 6.0 Å. No pore blockage due to possible impurities of remaining template was observed.

The aluminum coordination in the structure is another important aspect. The Al incorporation in a zeolite provides acidity as Brønsted sites, and these are applied as catalytic centers. When the material is air calcined, the steam generated from the template burning-off leads to a hydrolytic attack and migration of Al out of the structure. Calcination (Figure 6.4b) shows Al migration-dislodgement from the lattice as witnessed by the appearance of broad lines due to distorted tetra-, penta-, and octahedrally coordinated aluminum species around 50, 25, and 0 ppm (on the F2 axis) as well as a broadening of the line at about 55 ppm assigned to the tetrahedral framework aluminum, indicative of structural disorder. Such detrimental effects are avoided when applying $OH^•$ radicals. Fenton detemplation shows no damage at all, preserving the complete Brønsted capacity of the zeolite. Figure 6.4b

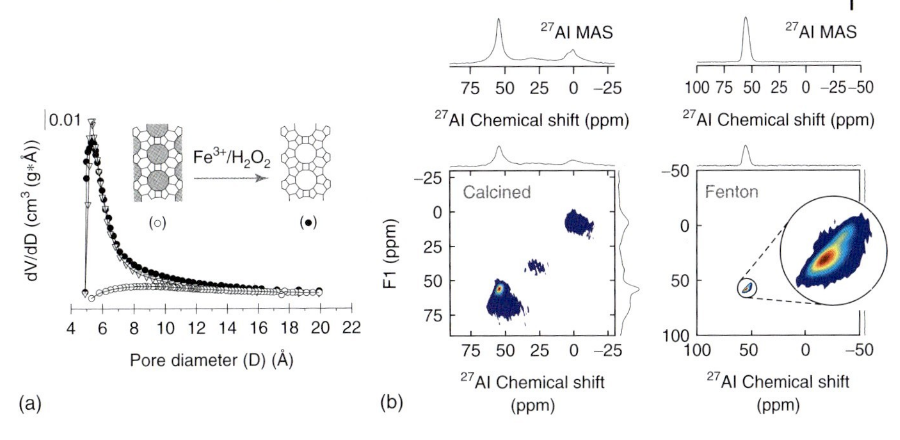

Figure 6.4 Features of beta zeolite after Fenton treatment. (a) Saito–Foley adsorption pore-size distribution from Ar-physisorption for: (O) parent zeolite containing the template (no porosity); (•) Fenton-detemplated; and (∇) commercial NH_4-form BEA. (b) Sheared ^{27}Al MQMAS NMR spectra obtained at 14.1 T showing the aluminum coordination for the calcined and Fenton-detemplated samples. (Data adapted from [149, 150].)

gives evidence for a total preservation of Al in the framework, showing T-site resolution (F1 projection), and no migration to extra-framework Al.

These results are promising. The challenge is to extend this methodology to other interesting structures for which calcination has proven detrimental.

The second case study. This involves all silica micro- and mesoporous SBA-15 materials. SBA-15 materials are prepared using triblock copolymers as structure-directing templates. Typically, calcined SBA-15 displays pore sizes between 50 and 90 Å and specific surface areas of 600–700 $m^2\,g^{-1}$ with pore volumes of 0.8–1.2 $cm^3\,g^{-1}$. Application of the Fenton concept to mesoporous materials looks simpler since mass transfer would be much less limited. However, it is not straightforward because hydrolysis can take place in the aqueous phase.

Characterization of the samples by TGA and CHN analysis shows that the template was effectively removed ($C < 0.2$ wt%). Small-angle X-ray scattering data of the calcined solid shows a reduction in the unit cell due to thermal shrinkage, while the values for the Fenton samples coincide with the starting precursor. Our approach therefore completely preserves the unit cell corresponding to the diameter of the micelles contained in the mesophase.

The benefits of the method are appreciated when the textural parameters are compared. Data derived from N_2-physisorption isotherms show that Fenton detemplation leads to improved textural parameters, with BET areas around 945 $m^2\,g^{-1}$ for a pore volume of 1.33 $cm^3\,g^{-1}$, while calcination leads to reduced textural parameters (667 $m^2\,g^{-1}$, 0.96 $cm^3\,g^{-1}$). T-plot analysis, strictly speaking, is not applicable for these bi-modal materials but it gives a good estimate. It shows that the micropore volume is doubled, which corresponds to an increase in the "calculated micropore area" from about

164 to 280 m^2 g^{-1}. Thus, the improved porosity features come from both micro- and mesoporosity due to the absence of shrinkage in mesopores and walls. The OH density was studied by ^{29}Si MAS NMR as Q_3/Q_4 ratio, indicating an increase in the concentration of OH-bonded silanols. In addition, the walls are thicker due to the absence of wall contraction. We are pleased that our approach has been followed by several groups. Xia and Mokaya applied it for ordered mesoporous aluminosilicates [202], Xing *et al.* on MCM-56 [203], Alam and Mokaya over mesoporous silicates from layered precursors [204], and Kecht and Bein on mesoporous silica nanoparticles [205] and vanadia supported SBA-15 [206].

6.7 Concluding Remarks

Besides current development of new catalysts and related functional materials, oxidation tools have always played an important role in their synthesis, activation, and functionalization. After a separate discussion per technique we have rationalized our literature findings (Table 6.2) as five principal oxidative functions with many proven applications.

Table 6.2 Rationalization of the oxidation chemistry applied in the synthesis of heterogeneous catalysts and related functional materials.

Oxidative function	Applications
1. Controlled and selective combustion of components via thermal or chemical routes	Calcination. Thermal detemplation of organic templates in micro- and mesoporous materials. Chemical detemplation protocols. Solution combustion synthesis
2. Use of flames as external heat source to induce quick drying and calcination	One-step aerosol flame synthesis
3. Removal of amorphous domains	Purification of CNT and CNF
4. Surface oxidation generating HSO_3, carboxylic, ketonic, and lactone groups	Oxidation of grafted or co-condensated alkylthiol groups in ordered mesoporous materials and periodic organosilicas Functionalization of hydrophobic carbons, CNFs, and CNTs Generation of catalytically active O-species
5. Controlled cutting and opening of closed carbon systems	Direct applications of CNTs (requires 20–100 nm in length) Inner filling and impregnation of CNTs with metal nanoparticles and complexes

References

1. Parker, L.M., Bibby, D.M., and Patterson, J.E. (1984) *Zeolites*, **4**, 168.
2. Gabelica, Z., Nagy, J.B., Bodart, P. *et al.* (1985) *Thermochim. Acta*, **93**, 749.
3. Tallon, J.L. and Buckley, R.G. (1987) *J. Phys. Chem.*, **91**, 1469.
4. Testa, F., Crea, F., Nastro, A. *et al.* (1991) *Zeolites*, **11**, 705.
5. Gao, X.T., Yeh, C.Y., and Angevine, P. (2004) *Microporous Mesoporous Mater.*, **70**, 27.
6. Párez Pariente, J., Martens, J.A., and Jacobs, P.A. (1988) *Zeolites*, **8**, 46.
7. Bourgeat-Lami, E., Di Renzo, F., Fajula, F., Hubert Mutin, P., and Des Courieres, T. (1992) *J. Phys. Chem.*, **96**, 3807.
8. Kanazirev1, V. and Price, G.L. (1996) *J. Catal.*, **161**, 156.
9. Keene, M.T.J., Gougeon, R.D.M., Denoyel, R., Harris, R.K., Rouquerol, J., and Llewellyn, P.L. (1999) *J. Mater. Chem.*, **9**, 2843.
10. Kleitz, F., Schmidt, W., and Schüth, F. (2001) *Microporous Mesoporous Mater.*, **44–45**, 95.
11. Kleitz, F., Schmidt, W., and Schüth, F. (2003) *Microporous Mesoporous Mater.*, **65**, 1.
12. Berube, F. and Kaliaguine, S. (2008) *Microporous Mesoporous Mater.*, **115**, 469.
13. Bagshaw, S.A. and Bruce, I.J. (2008) *Microporous Mesoporous Mater.*, **109**, 199.
14. Schüth, F., Kenneth, S.W.S., and Weitkamp, J. (2002) *Handbook of Porous Solids*, John Wiley & Sons, Inc., New York.
15. Xu, R., Pang, W., Yu, J., Huo, Q., and Chen, J. (2007) *Chemistry of Zeolites and Related Porous Materials, Synthesis and Structure*, John Wiley & Sons (Asia) Pte Ltd.
16. Patarin, J. (2004) *Angew. Chem. Int. Ed.*, **43**, 3878.
17. Formenti, M., Juillet, F., Meriaudeau, P., Teichner, S.J., and Vergnon, P. (1972) *J. Colloid Interface Sci.*, **39**, 79.
18. Ulrich, G.D. (1984) *Chem. Eng. News*, **62**, 22.
19. Kriegel, R., Töpfer, J., Preuss, N., Grimm, S., and Böer, J. (1994) *J. Mater. Sci. Lett.*, **13**, 1111.
20. Strobel, R., Baiker, A., and Pratsinis, S.E. (2006) *Adv. Powder Technol.*, **17**, 457.
21. Strobel, R. and Pratsinis, S.E. (2007) *J. Mater. Chem.*, **17**, 4743.
22. Strobel, R., Piacentini, M., Mädler, L., Maciejewski, M., Baiker, A., and Pratsinis, S.E. (2006) *Chem. Mater.*, **18**, 2532.
23. Hunt, A.T., Carter, W.B., and Cochran, J.K. Jr. (1993) *Appl. Phys. Lett.*, **63**, 266.
24. Balazs, N., Mogyorosi, K., Sranko, D.F. *et al.* (2008) *Appl. Catal., B Environ.*, **84**, 356.
25. Tiwari, V., Jiang, J., Sethi, V., and Biswas, P. (2008) *Appl. Catal., A - Gen.*, **345**, 241.
26. Teoh, W.Y., Setiawan, R., Madler, L. *et al.* (2008) *Chem. Mater.*, **20**, 4069.
27. Schimmoeller, B., Schulz, H., Ritter, A. *et al.* (2008) *J. Catal.*, **256**, 74.
28. van Vegten, N., Ferri, D., Maciejewski, M. *et al.* (2007) *J. Catal.*, **249**, 269.
29. Hannemann, S., Grunwaldt, J.D., Lienemann, P. *et al.* (2007) *Appl. Catal. A - Gen.*, **316**, 226.
30. Ravindranathan, P. and Patil, K.C. (1987) *Am. Ceram. Soc. Bull.*, **66**, 688.
31. Chick, L.A., Pederson, L.R., Maupin, G.D. *et al.* (1990) *Mater. Lett.*, **10**, 6.
32. Patil, K.C., Aruna, S.T., and Mimani, T. (2002) *Curr. Opin. Solid State Mater. Sci.*, **6**, 507.
33. Patil, K.C., Aruna, S.T., and Ekambaram, S. (1997) *Curr. Opin. Solid State Mater. Sci.*, **2**, 158.
34. Bera, P., Patil, K.C., Jayaram, V. *et al.* (1999) *J. Mater. Chem.*, **9**, 1801.
35. Bera, P., Aruna, S.T., Patil, K.C. *et al.* (1999) *J. Catal.*, **186**, 36.
36. Bera, P., Patil, K.C., and Hegde, M.S. (2000) *Phys. Chem. Chem. Phys.*, **2**, 3715.
37. Bera, P., Patil, K.C., Jayaram, V. *et al.* (2000) *J. Catal.*, **196**, 293.
38. Bera, P., Malwadkar, S., Gayen, A., Satyanarayana, C.V.V., Rao, B.S., and Hegde, M.S. (2004) *Catal. Lett.*, **96**, 213.

39. Gonzalez-Cortes, S.L., Xiao, T.C., Costa, P.M.F.J., Fontal, B., and Green, M.L.H. (2004) *Appl. Catal., A - Gen.*, **270**, 209.
40. Dinka, P. and Mukasyan, A.S. (2005) *J. Phys. Chem. B*, **109**, 21627.
41. Civera, A., Pavese, M., Saracco, G. *et al.* (2003) *Catal. Today*, **83**, 199.
42. Russo, N., Fino, D., Saracco, G., and Specchia, V. (2005) *J. Catal.*, **229**, 459.
43. Biamino, S., Fino, P., Fino, D., Russo, N., and Badini, C. (2005) *Appl. Catal., B - Environ.*, **61**, 297.
44. Fino, D., Russo, N., Saracco, G., and Specchia, V. (2006) *J. Catal.*, **242**, 38.
45. Russo, N., Furfori, S., Fino, D. *et al.* (2008) *Appl. Catal., B - Environ.*, **83**, 85.
46. Pino, L., Vita, A., Cipiti, F. *et al.* (2008) *Catal. Lett.*, **122**, 121.
47. Schuyten, S., Dinka, P., Mukasyan, A.S., and Wolf, E. (2008) *Catal. Lett.*, **121**, 189.
48. Nagappa, B. and Chandrappa, G.T. (2007) *Microporous Mesoporous Mater.*, **106**, 212.
49. Pavese, M. and Biamino, S. (2009) *J. Porous Mater.*, **16**, 59.
50. Mercier, L. and Pinnavaia, T.J. (1997) *Adv. Mater.*, **9**, 500.
51. Feng, X., Fryxell, G.E., Wang, L.Q. *et al.* (1997) *Science*, **276**, 923.
52. Burkett, S.L., Sims, S.D., and Mann, S. (1996) *Chem. Commun.*, 1367.
53. Huo, Q.S., Margolese, D.I., and Stucky, G.D. (1996) *Chem. Mater.*, **8**, 1147.
54. Macquarrie, D.J. (1996) *Chem. Commun.*, 1961.
55. Lim, M.H., Blanford, C.F., and Stein, A. (1997) *J. Am. Chem. Soc.*, **119**, 4090.
56. Fowler, C.E., Burkett, S.L., and Mann, S. (1997) *Chem. Commun.*, 1769.
57. Lim, M.H., Blanford, C.F., and Stein, A. (1998) *Chem. Mater.*, **10**, 467.
58. van Rhijn, W.M., de Vos, D.E., Sels, B.F. *et al.* (1998) *Chem. Commun.*, 317.
59. Bossaert, W.D., de Vos, D.E., van Rhijn, W.M. *et al.* (1999) *J. Catal.*, **182**, 156.
60. Das, D., Lee, J.F., and Cheng, S.F. (2001) *Chem. Commun.*, 2178.
61. Das, D., Lee, J.F., and Cheng, S.F. (2004) *J. Catal.*, **223**, 152.
62. Dias, A.S., Pillinger, M., and Valente, A.A. (2005) *J. Catal.*, **229**, 414.
63. Mbaraka, I.K., Radu, D.R., Lin, V.S.Y. *et al.* (2003) *J. Catal.*, **219**, 329.
64. Diaz, I., Marquez-Alvarez, C., Mohino, F. *et al.* (2000) *J. Catal.*, **193**, 283.
65. Diaz, I., Marquez-Alvarez, C., Mohino, F. *et al.* (2000) *J. Catal.*, **193**, 295.
66. Diaz, I., Mohino, F., Perez-Pariente, J. *et al.* (2004) *Thermochim. Acta*, **413**, 201.
67. Feng, Y.F., Yang, X.Y., Di, Y. *et al.* (2006) *J. Phys. Chem. B*, **110**, 14142.
68. Yang, Q.H., Kapoor, M.P., and Inagaki, S. (2002) *J. Am. Chem. Soc.*, **124**, 9694.
69. Margolese, D., Melero, J.A., Christiansen, S.C. *et al.* (2000) *Chem. Mater.*, **12**, 2448.
70. Yang, L.M., Wang, Y.J., Luo, G.S. *et al.* (2005) *Microporous Mesoporous Mater.*, **84**, 275.
71. Inagaki, S., Guan, S., Ohsuna, T. *et al.* (2002) *Nature*, **416**, 304.
72. Hamoudi, S. and Kaliaguine, S. (2003) *Microporous Mesoporous Mater.*, **59**, 195.
73. Sasidharan, M., Kiyozumi, Y., Mal, N.K. *et al.* (2006) *Adv. Funct. Mater.*, **16**, 1853.
74. Melero, J.A., Stucky, G.D., van Grieken, R. *et al.* (2002) *J. Mater. Chem.*, **12**, 1664.
75. Melero, J.A., van Grieken, R., Morales, G. *et al.* (2004) *Catal. Commun.*, **5**, 131.
76. Melero, J.A., van Grieken, R., and Morales, G. (2006) *Chem. Rev.*, **106**, 3790.
77. Alvaro, M., Corma, A., Das, D. *et al.* (2005) *J. Catal.*, **231**, 48.
78. Alvaro, M., Corma, A., Das, D. *et al.* (2004) *Chem. Commun.*, 956.
79. Macquarrie, D.J., Tavener, S.J., and Harmer, M.A. (2005) *Chem. Commun.*, 2363.
80. Nakajima, K., Tomita, I., Hara, M. *et al.* (2005) *Adv. Mater.*, **17**, 1839.
81. Hara, M., Yoshida, T., Takagaki, A., Takata, T., Kondo, J.N., Hayashi, S., and Domen, K. (2004) *Angew. Chem. Int. Ed.*, **43**, 2955.
82. Toda, M., Takagaki, A., Okamura, M., Kondo, J.N. *et al.* (2005) *Nature*, **438**, 178.

83. Takagaki, A., Toda, M., Okamura, M., Kondo, J.N., Hayashi, S., Domen, K., and Hara, M. (2006) *Catal. Today*, **116**, 157.
84. Okamura, M., Takagaki, A., Toda, M., Kondo, J.N., Domen, K., Tatsumi, T., Hara, M., and Hayashi, S. (2006) *Chem. Mater.*, **18**, 3039.
85. (a) Boehm, H.P. (1966) *Adv. Catal.*, **16**, 179; (b) Boehm, H.P. (1994) *Carbon*, **32**, 759.
86. Vinke, P., Vandereijk, M., Verbree, M. *et al.* (1994) *Carbon*, **32**, 675.
87. Barton, S.S., Evans, M.J.B., Halliop, E. *et al.* (1997) *Carbon*, **35**, 1361.
88. Rodriguez-Reinoso, F. (1998) *Carbon*, **36**, 159.
89. Figueiredo, J.L., Pereira, M.F.R., Freitas, M.M.A. *et al.* (1999) *Carbon*, **37**, 1379.
90. Boehm, H.P. (2002) *Carbon*, **40**, 145.
91. Darmstadt, H., Summchen, L., Ting, J.M. *et al.* (1997) *Carbon*, **35**, 1581.
92. Ros, T.G., van Dillen, A.J., Geus, J.W. *et al.* (2002) *Chem. Eur. J.*, **8**, 1151.
93. Toebes, M.L., van Heeswijk, E.M.P., Bitter, J.H. *et al.* (2004) *Carbon*, **42**, 307.
94. Lakshminarayanan, P.V., Toghiani, H., and Pittman, C.U. (2004) *Carbon*, **42**, 2433.
95. Ros, T.G., van Dillen, A.J., Geus, J.W. *et al.* (2002) *Chem. Eur. J.*, **8**, 2868.
96. Ros, T.G., Keller, D.E., van Dillen, A.J. *et al.* (2002) *J. Catal.*, **211**, 85.
97. Toebes, M.L., van der Lee, M.K., Tang, L.M. *et al.* (2004) *J. Phys. Chem. B.*, **108**, 11611.
98. Plomp, A.J., Vuori, H., Krause, A.O.I. *et al.* (2008) *Appl. Catal., A - Gen.*, **351**, 9.
99. Bezemer, G.L., Bitter, J.H., Kuipers, H.P.C.E., Oosterbeek, H., Holewijn, J.E., Xu, X.D., Kapteijn, F., van Dillen, A.J., and de Jong, K.P. (2006) *J. Am. Chem. Soc.*, **128**, 3956.
100. Guha, A., Lu, W.J., Zawodzinski, T.A., and Schiraldi, D.A. (2007) *Carbon*, **45**, 1506.
101. Iijima, S. (1991) *Nature*, **354**, 56.
102. Iijima, S. and Ichihashi, T. (1993) *Nature*, **363**, 603.
103. Bethune, D.S., Kiang, C.H., Devries, M.S. *et al.* (1993) *Nature*, **363**, 605.
104. Thess, A., Lee, R., Nikolaev, P. *et al.* (1996) *Science*, **273**, 483.
105. Tsang, S.C., Chen, Y.K., Harris, P.J.F. *et al.* (1994) *Nature*, **372**, 159.
106. Hiura, H., Ebbesen, T.W., and Tanigaki, K. (1995) *Adv. Mater.*, **7**, 275.
107. Esumi, K., Ishigami, M., Nakajima, A. *et al.* (1996) *Carbon*, **34**, 279.
108. Liu, J., Rinzler, A.G., Dai, H.J. *et al.* (1998) *Science*, **280**, 1253.
109. Shaffer, M.S.P., Fan, X., and Windle, A.H. (1998) *Carbon*, **36**, 1603.
110. Lago, R.M., Tsang, S.C., Lu, K.L. *et al.* (1995) *J. Chem. Soc., Chem. Commun.*, **13**, 1355.
111. Chattopadhyay, D., Galeska, I., and Papadimitrakopoulos, F. (2002) *Carbon*, **40**, 985.
112. Laasonen, K., Andreoni, W., and Parrinello, M. (1992) *Science*, **258**, 1916.
113. Mintmire, J.W., Dunlap, B.I., and White, C.T. (1992) *Phys. Rev. Lett.*, **68**, 631.
114. Saito, R., Fujita, M., Dresselhaus, G. *et al.* (1993) *Mat. Sci. Eng., B - Solid*, **19**, 185.
115. Hamada, N., Sawada, S., and Oshiyama, A. (1992) *Phys. Rev. Lett.*, **68**, 1579.
116. Chu, A., Cook, J., Heesom, R.J.R. *et al.* (1996) *Chem. Mater.*, **8**, 2751.
117. Pan, X.L. and Bao, X.H. (2008) *Chem. Commun.*, **47**, 6271.
118. Hu, H., Zhao, B., Itkis, M.E. *et al.* (2003) *J. Phys. Chem. B*, **107**, 13838.
119. Jia, Z.J., Wang, Z.Y., Liang, J. *et al.* (1999) *Carbon*, **37**, 903.
120. Pierard, N., Fonseca, A., Konya, Z. *et al.* (2001) *Chem. Phys. Lett.*, **335**, 1.
121. Rubio, A., Apell, S.P., Venema, L.C. *et al.* (2000) *Eur. Phys. J. B*, **17**, 301.
122. Park, J.Y., Yaish, Y., Brink, M. *et al.* (2002) *Appl. Phys. Lett.*, **80**, 4446.
123. Banhart, F., Li, J.X., and Terrones, M. (2005) *Small*, **1**, 953.
124. Mawhinney, D.B., Naumenko, V., Kuznetsova, A. *et al.* (2000) *J. Am. Chem. Soc.*, **122**, 2383.
125. Mizoguti, E., Nihey, F., Yudasaka, M. *et al.* (2000) *Chem. Phys. Lett.*, **321**, 297.

126. Zhang, M., Yudasaka, M., Nihey, F. *et al.* (2000) *Chem. Phys. Lett.*, **328**, 350.
127. Chiang, I.W., Brinson, B.E., Smalley, R.E. *et al.* (2001) *J. Phys. Chem. B.*, **105**, 1157.
128. Chiang, I.W., Brinson, B.E., Huang, A.Y. *et al.* (2001) *J. Phys. Chem. B*, **105**, 8297.
129. Dillon, A.C., Jones, K.M., Bekkedahl, T.A. *et al.* (1997) *Nature*, **386**, 377.
130. Ziegler, K.J., Gu, Z.N., Peng, H.Q. *et al.* (2005) *J. Am. Chem. Soc.*, **127**, 1541.
131. Tran, M.Q., Tridech, C., Alfrey, A. *et al.* (2007) *Carbon*, **45**, 2341.
132. Wang, C.F., Guo, S.J., Pan, X.L. *et al.* (2008) *J. Mater. Chem.*, **18**, 5782.
133. Zhang, J., Liu, X., Blume, R., Zhang, A.H., Schlögl, R., and Su, D.S. (2008) *Science*, **322**, 73.
134. Pereira, M.F.R., Orfao, J.J.M., and Figueiredo, J.L. (1999) *Appl. Catal., A - Gen.*, **184**, 153.
135. Fiedorow, R., Przystajko, W., and Sopa, M. (1981) *J. Catal.*, **68**, 33.
136. Mestl, G., Maksimova, N.I., Keller, N. *et al.* (2001) *Angew. Chem. Int. Ed.*, **40**, 2066.
137. Zhang, J., Su, D.S., Zhang, A.H. *et al.* (2007) *Angew. Chem. Int. Ed.*, **46**, 7319.
138. Supporting material available on Science Online ref. 133.
139. Resasco, D.E. (2008) *Nat. Nanotechnol.*, **3**, 708.
140. Fenton, H.J.H. (1894) *J. Chem. Soc.*, **65**, 899.
141. ISI Web of Knowledge®. Thomson Reuters. URL: *www.isiknowledge.com/*
142. (a) Tarr, M.A. (2003) *Chemical Degradation Methods for Wastes and Pollutants*, In Environmental, Science and Pollution Control Series, vol. 26, Marcel Dekker, Inc., p. 165; (b) Burkitt, M.J. (2003) *Prog. React. Kinet. Mechanism*, **28**, 75.
143. Neyens, E. and Baeyens, J. (2003) *J. Hazard. Mater.*, **98**, 33.
144. Parsons, S. (2004) *Advanced Oxidation Processes for Water and Wastewater Treatment*, IWA Publishing, London.
145. Pignatello, J.J., Oliveros, E., and MacKay, A. (2006) *Crit. Rev. Env. Sci. Technol.*, **36**, 1.
146. Melián-Cabrera, I., Kapteijn, F., and Moulijn, J.A. (2005) *Chem. Commun.*, (12), 1525.
147. Melián-Cabrera, I., Kapteijn, F., and Moulijn, J.A. (2005) *Chem. Commun.*, (16), 2178.
148. Melián-Cabrera, I., Kapteijn, F., and Moulijn, J.A. (2006) EP1690831 A1.
149. Melián-Cabrera, I., Kapteijn, F., and Moulijn, J.A. (2005) *Chem. Commun.*, (16), 2744.
150. Melián-Cabrera, I., Osman, A.H., van Eck, E.R.H., Kentgens, A.P.M., Polushkin, E., Kapteijn, F., and Moulijn, J.A. (2007) *Stud. Surf. Sci. Catal.*, **170**, 648.
151. Pérez-Ramírez, J. (2002) "Catalyzed N_2O Activation, Promising (new) Catalyst for Abatement and utilization", PhD Thesis dissertation, TU-Delft, The Netherlands.
152. Pérez-Ramírez, J., Kapteijn, F., Schoffel, K. *et al.* (2003) *Appl. Catal., B-Environ.*, **44**, 117.
153. Oliveros, E., Goeb, S., Bossmann, S.H., Braun, A.M., Nascimento, C.A.O., and Guardani, R. (2000) Waste water treatment by the photochemical enhanced Fenton reaction: Modeling and optimization using experimental design and artificial neural networks, in *Sustainable Energy and Environmental Technology, Proceedings of the Third Asia Pacific Conference* (eds X. Hu and P.L. Yue), World Scientific, Singapore, p. 577.
154. Oliveros, E., Legrini, O., Braun, A.M. *et al.* (1997) *Water Sci. Technol.*, **35**, 223.
155. Haber, F. and Weiss, J. (1934) *Proc. R. Soc. A.*, **134**, 332.
156. Buxton, G.V., Greenstock, C.L. *et al.* (1988) *J. Phys. Chem. Ref. Data*, **17**, 513.
157. Walling, C. and Goosen, A. (1973) *J. Am. Chem. Soc.*, **95**, 2987.
158. De Laat, J. and Gallard, H. (1999) *Environ. Sci. Technol.*, **33**, 2726.
159. Rossetti, G.H., Albizzati, E.D., and Alfano, O.M. (2002) *Ind. Eng. Chem. Res.*, **41**, 1436.
160. Pignatello, J.J. (1992) *Environ. Sci. Technol.*, **26**, 944.

161. Sun, Y. and Pignatello, J.J. (1993) *Environ. Sci. Technol.*, **27**, 304.
162. Sima, J. and Makanova, J. (1997) *Coord. Chem. Rev.*, **160**, 161.
163. Qiang, Z., Chang, J.H., and Huang, C.P. (2003) *Water Res.*, **37**, 1308.
164. Baes, C.F. and Mesmer, R.E. (1976) *The Hydrolysis of Cations*, Wiley Interscience, New York.
165. Flynn, C.M. Jr. (1984) *Chem. Rev.*, **84**, 31.
166. Gallard, H., De Laat, J., and Legube, B. (1999) *Water Res.*, **33**, 2929.
167. van den Brink, P.J. *et al.* (1991) *Stud. Surf. Sci. Catal.*, **63**, 527.
168. van Dillen, A.J. *et al.* (2003) *J. Catal.*, **216**, 257.
169. Weckhuysen, B.M. (2005) Proceedings of the International Workshop on Microporous and Mesoporous Materials as Catalytic Hosts for Fe, Cu and Co, Scheveningen, The Netherlands.
170. Liu, Y.M. *et al.* (2008) *J. Phys. Chem. C*, **112**, 16575.
171. Cejka, J., van Bekkum, H., Corma, A., and Schueth, F. (eds) (2007) *Introduction to Zeolite Science and Practice*, 3rd revised edn, Studies in Surface Science and Catalysis, vol. 168, Elsevier, Amsterdam.
172. Reviews: (a) Corma, A. (1995) *Chem. Rev.*, **95**, 559; (b) Corma, A. (2003) *J. Catal.*, **216**, 298; (c) Corma, A. (1997) *Chem. Rev.*, **97**, 2373; (d) Davis, M.E. (2002) *Nature*, **417**, 813; (e) Marcilly, C. (2001) *Stud. Surf. Sci. Catal.*, **135**, 37; (f) Davis, M.E. (1993) *Acc. Chem. Res.*, **26**, 111; (g) Yu, J.H. and Xu, R.R. (2003) *Acc. Chem. Res.*, **36**, 481; (h) van Donk, S., Janssen, A.H., Bitter, J.H., and de Jong, K.P. (2003) *Catal Rev.-Sci. Technol.*, **45**, 297; (i) Cundy, C.S. and Cox, P.A. (2005) *Microporous Mesoporous Mater.*, **82**, 1.
173. Barrer, R.M. and Denny, P.J. (1961) *J. Chem. Soc.*, 971.
174. (a) Davis, M.E. and Lobo, R.F. (1992) *Chem. Mater.*, **4**, 389; (b) Lobo, R.F., Zones, S.I., and Davis, M.E. (1995) *J. Inclusion Phenom. Mol.*, **21**, 47.
175. (a) Camblor, M.A. and Pérez-Pariente, J. (1991) *Zeolites*, **11**, 202; (b) Camblor, M., Corma, A., and Valencia, S. (1998) *Microporous Mesoporous Mater.*, **25**, 59; (c) Müller, M., Harvey, G., and Prins, R. (2000) *Microporous Mesoporous Mater.*, **34**, 135; (d) van Bokhoven, J.A., Koningsberger, D.C., Kunkeler, P., van Bekkum, H., and Kentgens, A.P.M. (2000) *J. Am. Chem. Soc.*, **122**, 12842; (e) Beers, A.E.W., van Bokhoven, J.A., de Lathouder, K.M. *et al.* (2003) *J. Catal.*, **218**, 239.
176. (a) Corma, A., Nemeth, L.T., Renz, M., and Valencia, S. (2001) *Nature*, **412**, 423; (b) Boronat, M., Concepción, P., Corma, A., Renz, M., and Valencia, S. (2005) *J. Catal.*, **234**, 111.
177. (a) Corma, A. and García, H. (1997) *Catal. Today*, **38**, 257; (b) Sheldon, R.A. and van Bekkum, H. (2001) *Fine Chemicals Through Heterogeneous Catalysis*, Wiley-VCH Verlag GembH, Weinheim; (c) Corma, A. and García, H. (2003) *Chem. Rev.*, **103**, 4307.
178. (a) Sassoye, C., Loiseau, T., Taulelle, F., and Férey, G. (2000) *Chem. Commun.*, 943; (b) Sassoye, C., Marrot, J., Loiseau, T., and Ferey, G. (2002) *Chem. Mater.*, **14**, 1340; (c) Beitone, L., Marrot, J., Loiseau, T., Férey, G., Henry, M., Huguenard, C., Gansmuller, A., and Taulelle, F. (2003) *J. Am. Chem. Soc.*, **125**, 1912; (d) Walton, R.I., Millange, F., Loiseau, T., O'Hare, D., and Férey, G. (2000) *Angew. Chem. Int. Ed.*, **39**, 4552; (e) Chippindale, A.M., Peacock, K.H., and Cowley, A.R. (1999) *J. Solid State Chem.*, **145**, 379.
179. (a) Dhingra, S.S. and Haushalter, R.C. (1993) *J. Chem. Soc., Chem. Commun.*, 1665; (b) Williams, I.D., Yu, J., Du, H., Chen, J., and Pang, W. (1998) *Chem. Mater.*, **10**, 773; (c) Thirumurugan, A. and Natarajan, S. (2003) *Dalton Trans.*, 3387.
180. (a) Yang, G. and Sevov, S.C. (1999) *J. Am. Chem. Soc.*, **121**, 8389; (b) Rodgers, J.A. and Harrison, W.T.A. (2000) *J. Mater. Chem.*, **10**, 2853; (c) Neeraj, S., Natarajan, S., and Rao, C.N.R. (1999) *Chem. Commun.*, 165.
181. (a) Moore, P.B. and Shen, J. (1983) *Nature*, **306**, 356; (b) Choudhury, A., Natarajan, S., and Rao, C.N.R. (1999) *Chem. Commun.*, 1305.

182. Guillou, N., Gao, Q., Forster, P.M., Chang, J., Nogués, M., Park, S.E., Férey, G., and Cheetham, A.K. (2001) *Angew. Chem. Int. Ed.*, **40**, 2831.
183. Soghomonian, V., Chen, Q., Haushalter, R.C. *et al.* (1993) *Science*, **259**, 1596.
184. Davis, M.E., Saldarriag, C., Montes, C., Garces, J., and Zhao, X. (1998) *Nature*, **331**, 698.
185. Zou, X., Conradsson, T., Klingstedt, M. *et al.* (2005) *Nature*, **437**, 716.
186. Jones, C.W., Tsuji, K., Takewaki, T., Beck, L.W., and Davis, M.E. (2001) *Microporous Mesoporous Mater.*, **48**, 57.
187. (a) Hitz, S. and Prins, R. (1997) *J. Catal.*, **168**, 194; (b) Jones, C.W., Tsuji, K., and Davis, M.E. (1998) *Nature*, **393**, 52; (c) Takewaki, T., Beck, L.W., and Davis, M.E. (1999) *Top. Catal.*, **9**, 35; (d) Jones, C.W., Tsuji, K., Takewaki, T., Beck, L.W., and Davis, M.E. (2001) *Microporous Mesoporous Mater.*, **48**, 57; (e) Yang, C.M., Zibrowius, B., Schmidt, W., and Schüth, F. (2003) *Chem. Mater.*, **15**, 3739; (f) Yang, C.M., Zibrowius, B., Schmidt, W., and Schüth, F. (2004) *Chem. Mater.*, **16**, 2918; (g) Whitehurst, D.D. (1992) US Patent 5,143,879.
188. (a) Lee, H., Zones, S.I., and Davis, M.E. (2003) *Nature*, **425**, 385; (b) Lee, H., Zones, S.I., and Davis, M.E. (2004) *Recent Advances in the Science and Technology of Zeolites and Related Materials, Parts A-C*, Elsevier Science BV, Amsterdam.
189. (a) Kawi, S. and Lai, M.W. (1998) *Chem. Commun.*, 1407; (b) van Grieken, R., Calleja, G., Stucky, G.D., Melero, J.A., Garcia, R.A., and Iglesias, J. (2003) *Langmuir*, **19**, 3966.
190. Aguado, J., Escola, J.M. *et al.* (2005) *Microporous Mesoporous Mater.*, **83**, 181.
191. Kresnawahjuesa, O., Olson, D.H. *et al.* (2002) *Microporous Mesoporous Mater.*, **51**, 175.
192. Krawiec, P., Kockrick, E., Simon, P. *et al.* (2006) *Chem. Mater.*, **18**, 2663.
193. (a) Maesen, T.L.M. *et al.* (1987) *J. Chem. Soc., Chem. Commun.*, **17**, 1284; (b) Maesen, T.L.M. *et al.* (1989) *Appl. Catal., A*, **48**, 373; (c) Maesen, T.L.M. *et al.* (1990) *J. Chem. Soc., Chem. Faraday Trans.*, **86**, 3967.
194. (a) Keene, M.T.J. *et al.* (1998) *Chem. Commun.*, 2203; (b) Heng, S., Pui Sze Lau, P. *et al.* (2004) *J. Membr. Sci.*, **243**, 69; (c) Motuzas, J. *et al.* (2007) *Microporous Mesoporous Mater.*, **99**, 197; (d) Kuhn, J. *et al.* (2009) *Microporous Mesoporous Mater.*, **120**, 12; (e) Kuhn, J. *et al.* (2009) *Microporous Mesoporous Mater.*, **120**, 35.
195. Buchel, G., Denoyel, R., and Llewellyn, P.L. (2001) *J. Rouquerol, J. Mater. Chem.*, **11**, 589.
196. Li, Q., Amweg, M.L., Yee, C.K., Navrotsky, A., and Parikh, A.N. (2005) *Microporous Mesoporous Mater.*, **87**, 45.
197. Parikh, A.N., Navrotsky, A., Li, Q., Yee, C.K. *et al.* (2004) *Microporous Mesoporous Mater.*, **76**, 17.
198. Tian, B., Liu, X., Yu, C., Gao, F., Luo, Q., Xie, S. *et al.* (2002) *Chem. Commun.*, 1186.
199. Lu, A.H., Li, W.C., Schmidt, W., and Schüth, F. (2006) *J. Mater. Chem.*, **16**, 3396.
200. Xiao, L., Li, J., Jin, H., and Xu, R. (2006) *Microporous Mesoporous Mater.*, **96**, 413.
201. Masuda, T. *et al.* (2003) *Sep. Purif. Technol.*, **32**, 181.
202. Xia, Y.D. and Mokaya, R. (2006) *J. Phys. Chem. B*, **110**, 9122.
203. Xing, H.J., Zhang, Y., Jia, M.J. *et al.* (2008) *Catal. Commun.*, **9**, 234.
204. Alam, N. and Mokaya, R. (2008) *J. Mater. Chem.*, **18**, 1383.
205. Kecht, J. and Bein, T. (2008) *Microporous Mesoporous Mater.*, **116**, 123.
206. Xu, J., Chen, M., Liu, Y.M. *et al.* (2009) *Microporous Mesoporous Mater.*, **118**, 354.

7
Challenges in Catalysis for Sustainability

Riitta L. Keiski, Tanja Kolli, Mika Huuhtanen, Satu Ojala, and Eva Pongrácz

7.1
Introduction

The rapid pace of development of our world over the last century has been largely based on easy access to fossil fuels. These resources are, however, limited, while their demand is growing rapidly. It is also becoming clear that the scale of carbon dioxide (CO_2) emissions following the use of fossil fuels is threatening the climate of the Earth. This makes the development of sustainable production and energy solutions in industry, transportation, and households the most important scientific and technical challenge of our time.

The term *sustainable development* was popularized by the report of the World Commission on Environment and Development [1] entitled Our Common Future, also known as the *Brundtland Report*. Our Common Future was written after 3 years of public hearings and over 500 written submissions. Commissioners from 21 countries analyzed this material, with the final report being submitted to the United Nations General Assembly in 1987 [2]. It is also to be noted that the by-now well-known definition of sustainable development as "development that meets the needs of the present without compromising the ability of future generations to meet their own needs" is essentially a value concept, and it requires the promotion of values that encourage consumption standards that are within the bounds of the ecologically possible and which all can reasonably aspire to. In essence, it also combines the concepts of intragenerational and intergenerational equity.

Our Common Future reported on many global realities and demanded urgent action on eight key issues [1]: (i) population and human resources, (ii) food security, (iii) species and ecosystems, (iv) energy, (v) industry, (vi) the urban challenge, (vii) managing the commons, and (viii) conflict and environmental degradation. Our Common Future was not intended to be "a prediction of ever increasing environmental decay, poverty, and hardship in an ever more polluted world among ever decreasing resources", yet that is exactly what has happened. After more than 20 years, the projected "new era of economic growth based on policies that sustain and expand the environmental resource base" has not materialized yet. Indeed the magnitude of environmental crisis has deepened [3]. In this chapter,

Novel Concepts in Catalysis and Chemical Reactors: Improving the Efficiency for the Future.
Edited by Andrzej Cybulski, Jacob A. Moulijn, and Andrzej Stankiewicz

ISBN: 978-3-527-32469-9

we would like to address some of the issues highlighted in the Brundtland report, and contemplate on the role of catalysis in addressing some of the key issues.

7.2 Population and Human Resources

Our global problem still is that we are consuming vital resources at rates we cannot sustain or at costs to the environment that we cannot or should not pay. This is happening because of both rising consumption by the rich and rising numbers of poor people who consume the bare minimum [4]. Catalysis has a great potential in efficiently using the resources we have.

Several consumables are produced via catalytic processes and reactions and, over the course of the last few decades, new and more efficient catalytic routes have been developed for the production of goods. However, novel and improved ways are still needed, for example, to fulfill the growing fuel and energy demands of developed and developing countries. Catalytic methods can have a significant role in fulfilling this demand, and thus provide more sustainable ways for the world's economical development [5].

7.3 Food Security

There has been an increasing debate concerning the use of foodstuff for fuel and energy production. In recent years, the demand to produce bioethanol has led to a rapid increase in the prices of corn and seeds. When aiming for the EU's target of 10% biofuels in transport fuels by 2020 (2003/30/EC), great efforts need to be put into research on producing biofuels from nonfood sources. New processes to utilize nonfood sources such as wastes, both wood and nonwood matter, are the subject of much research at the moment. These processes can be performed sustainably via catalytic routes to produce, for instance, energy and fuels from nonfood raw materials [6].

7.4 Species and Ecosystem

Tropical rainforests and biological diversity had been a great concern at the time of the Brundtland report. Back in 1987, 900 million ha were left of the 1.6 billion original cover. Since then, we have lost over 200 million ha of rainforests, practically one-fourth of what existed then. Now, we have less than half of the original cover: around 43% [3].

Eighteen percent of Greenhouse Gases (GHGs) stem from "land use change", a euphemism for deforestation. A crucial challenge is how to avoid the problem of energy crops fueling deforestation even further, and impacting on food and livelihood security [3].

Biofuels can be produced from nonfood, nonrainforest sources by catalysis. Catalysts are used for increased energy efficiency and fuel economy (lean-burn engines), increased production of chemicals from natural gas and biomass, improved efficiency for the use of raw-oil-based feedstocks in chemical processes, lower emissions in petroleum refineries, and reduced emissions of gases contributing to global warming. Catalysts are designed for better means to produce H_2 as a future fuel, development of cheaper and more efficient fuel cells, and development of solar cells for the production of chemicals, such as H_2. Catalysts are used to facilitate hydrolysis of cellulose, conversion of lignin to aromatic carboxylic acids, selective hydrogenation and deoxygenation of cellulose, and selective functionalization of carbohydrates [7].

7.5 Energy

The Brundtland report was key in bringing the expression "global warming". After 20 years, the prevalence of fossil fuels in the world energy matrix has not changed, but aggravated, and discussion on global warming is no longer "if", but "how much". To a large extent, the transition from unsustainable to sustainable energy expected by the Brundtland report has been a great failure. Notwithstanding, the report brought concepts of energy efficiency and renewables to forefront [3].

7.5.1 Sustainable Fuels through Catalysis

Crude oil is primarily formed of paraffins, naphthene, and aromatic hydrocarbon, the proportions of which vary considerably depending on the origin. Crude oil also contains other elements such as sulfur, nitrogen, oxygen, and, to a certain extent, metals. There are many different types of crude oil, and their constituents vary from one oil field or area to another. Reformulation means that the improvement in the gasoline formula has resulted in lower emissions from the vehicles. In reformulated gasoline, the amounts of components that are critical in view of emissions, such as sulfur, aromatic compounds, olefins, and benzene, are lower than in conventional gasoline. As fuel oxygenates bring additional oxygen to combustion, it is possible to reduce the hydrocarbon (HC), carbon monoxide (CO), and particulate emissions (PM) to a considerable extent. This is particularly useful in cars without catalytic converters [8].

The demand for environment-friendly fuels requires the removal of organosulfur compounds present in crude-oil fractions. SO_2 or SO_3 contribute to the formation of acid rain and have an effect on pollution control devices [9]. Very stringent environmental regulations will limit the sulfur levels in diesel fuels in EU to less than 10 ppm by the end of 2010 [10]. The conventional sulfur-compound removing method is catalytic hydrodesulfurization (HDS), which requires both high temperature and high pressure of hydrogen gas to produce light oil having low levels of sulfur compounds. This brings about a number of problems including high investments, high operation costs, reduction of catalyst cycle time, and increase in the hydrogen consumption due to hydrogenation of aromatics present in the diesel fuel. Nevertheless, the efficiency of HDS is limited to the removal of bentzothiophenes (BTs) and dibentzothiophenes (DBTs) [11, 12].

Desulfurization processes are absolutely necessary for producing clean fuels. Possible strategies to realize ultradeep sulfurization currently include adsorption, extraction, oxidation, and bioprocesses. Oxidative desulfurization (ODS) combined with extraction is considered one of the most promising of these processes [13]. Ultradeep desulfurization of diesel by selective oxidation with amphiphilic catalyst assembled in emulsion droplets has given results where the sulfur level of desulfurized diesel can be lowered from 500 ppm to about 0.1 ppm without changing the properties of the diesel [12].

Ultradeep desulfurization of fuel oils is used for producing not only clean fuels but also sulfur-free hydrogen used in fuel-cell systems, in which the hydrogen can be produced potentially through the reforming of fuel oils. Fuel-cell systems must be run with little-to-no sulfur content, because sulfur can irreversibly poison the precious metal catalysts and electrodes used [12].

7.5.2
Hydrogen Production for Fuel Cells

An alternative energy source to substitute fossil fuels is hydrogen. During the last decade, a lot of effort has been put into the production of hydrogen for fuel cells and also for industrial processes as a feedstock. The interest in developing fuel cells as an energy source for small-scale portable devices and distributed energy production is increasing rapidly. However, sustainable production of hydrogen is a challenge. Producing H_2 from water is very energy intensive and, therefore, expensive.

Photocatalysis on semiconductor nanoparticles (wires and nanotubes) such as Si, TiO_2, WO_3, and VO_2 offers a facile route for hydrogen production from water and aqueous mixtures of alcohols. A great advantage of the photocatalytic approach is its low temperature operation and possible exploitation of solar energy in hydrogen production. Photocatalytic reforming may be a good solution as the process is driven under ambient conditions and by sunlight [14].

Production of hydrogen and its derivatives from bio-based raw materials has also been a subject of much research. Several reports can be found on the photocatalytic reforming of biomass to hydrogen [14].

A viable way for hydrogen production is reforming HCs such as methane to produce synthesis gas. This can only be achieved by designing proper catalysts for the reforming and fuel-cell units. Bioethanol reforming over various metals such as Group VIII metals (e.g., Ni, Co, Pt, Rh) has been studied and found effective for the reforming reactions. Reforming of alcohols [15, 16], natural gas [17], and HCs [18] has been studied over various catalysts, supports, and their combinations.

Ethanol is a nontoxic substance with relatively high H_2 content, and its advantage is that it can be produced from renewable sources, for example, from various biomasses and wastes. In addition, purification of the produced reforming gas has been of interest to researchers. Hydrogen purification has been studied, for instance, with membranes [19] which can also have catalytic performances.

7.5.2.1 Case – Use of Carbon Nanotube-Based Catalysts in Hydrogen Production

Carbon nanotubes (CNTs) are a set of materials with different structures and properties. They are among the most important materials of modern nanoscience and nanotechnology field. They combine inorganic, organic, bio-organic, colloidal, and polymeric chemistry and are chemically inert. They are insoluble in any solvent and their chemistry is in a key position toward interdisciplinary applications, for example, use as supports for catalysts and catalytic membranes [20, 21].

The studies on isomerization, hydroformulation, hydrogenation, and oxidation as well as syngas conversion have been carried out over, for example, Rh-, Fe-, Pd-loaded CNTs and also CNT-only catalysts. The results give an evidence of suitability of the CNT-based catalysts for various reactions. The recent studies have been focused on multiwalled CNTs; however, the number of studies on single-walled CNTs have nowadays increased due to their wall and nanotube properties. According to Pan and Bao [22] CNTs can be an alternative for active carbons due to their unique properties, such as electrical conductivity, mechanical strength, thermal stability as well as hydrogen storage capacity. On introduction of active metal particles on and inside the nanotubes, the combination of these can produce interesting metal–support interaction which may be different depending on where the metal particle is located. CNTs have been studied with various metals, and reduction and oxidation characters are reported. The activation energies of oxidation over metal particles located outside and inside the CNT differ remarkably. Thus, it can be concluded that understanding the interactions between CNT and metal particles is crucial in CNT-based catalyst manufacturing [22]. In our studies, the CNT-based materials have been investigated both as reforming catalysts and as fuel-cell electrode membranes. Pt-, Rh-, Ni-, and Co-loaded CNT catalysts have been tested in steam reforming of bioethanol. The results indicate the suitability of CNT-based catalysts for reforming and fuel-cell electrodes [23, 24].

7.6 Industry

In the 20 years since the Brundtland report, great developments have taken place in industries toward sustainable practices. As a case in point, the problem of acid rain, an issue of concern in 1987, has improved to a large extent, thanks to catalytic pollution abatement both in stationary and automotive emissions. Catalysis for Green Chemistry and Engineering will continue to have a crucial role in improving the environmental performance of industry [25–27]. Nowadays, catalytic procedures are often implemented according to the green chemistry principles and good industrial examples are found that fulfill several of the 12 principles of green chemistry at the same time [28–31].

Catalysts have an enormous importance especially in the chemical industry via enabling reactions to take place, making processes more efficient, increasing the operating profit, and making processes environmentally friendly, economic, and safe. In addition to chemical productions, a major area of application of environmental catalysis is in agro-food production, pulp and paper, electronic, and metal finishing industry. A major potential for catalysis in improving the industry's environmental profile is also in clean air technologies such as in nitrogen oxides (NO_x) control, photocatalysis, VOC control as well as in catalytic CO_2 utilization technologies.

7.6.1 Catalysis for Sustainable Production

The principles of sustainable production (Table 7.1) highlight not only the significance of nonpolluting, energy-saving, and resource-efficient processes, but also the importance of economic profitability as well as the safety and well-being of workers. Already 10–15 years ago the challenges for catalysis in sustainable production were stated as, for example by Thomas and Thomas, development of "zero-waste" processes, minimization of hazardous products and GHGs, replacement of corrosive liquid acids by benign solid acid catalysts, evolution of sustainable systems, reduction in volume of by-products, elimination of voluminous by-products, and development of processes requiring less consumption of catalysts [32]. The value of catalysis in 2010 in terms of goods/services created by catalytic processes is assumed to be over $3 trillion. This is based on chemical processes employing catalysts

Table 7.1 The principles of sustainable production [33].

1.	Products and packaging are designed to be safe and ecologically sound throughout their life cycle; analogously, services are designed to be safe and ecologically sound.
2.	Wastes and ecologically incompatible byproducts are reduced, eliminated, or recycled.
3.	Energy and materials are conserved, and the forms of energy and materials used are most appropriate for the desired ends.
4.	Chemical substances or physical agents and conditions that present hazards to human health or the environment are eliminated.
5.	Work spaces are designed to minimize or eliminate physical, chemical, biological, and ergonomic hazards.
6.	Management is committed to an open, participatory process of continuous evaluation and improvement, focused on the long-term viability of the firm.
7.	Work is organized to conserve and enhance the efficiency and creativity of employees.
8.	The security and well-being of employees is a priority, as is the continuous development of their talents and capacities.
9.	The communities around workplaces are respected and enhanced economically, socially, culturally, and physically; equity and fairness are promoted.

valued at over $12 billion, for example, polymers, pharmaceuticals, petrochemicals, environmental protection. Improvements in catalyst activity and selectivity – 100% selectivity expected in the twenty-first century – offer tremendous potential benefits for the manufacturing processes. In heterogeneous catalysis, an increased attention will be paid to sustainable production, environmental protection, and alternative energy sources.

The main objectives in environmental protection by catalysis, which may be called *catalysis for sustainability*, are preventing pollution by developing nonpolluting and energy-saving processes, decreasing pollution by achieving higher selectivity in fuel and chemicals manufacture, decreasing pollution by developing new and improved catalytic processes for removal of pollutants, and developing competitive processes for production of chemicals as well as hydrogen and other energy sources, providing reduced CO_2 production.

7.6.2 Catalytic CO_2 Utilization Technologies

Humans in essence have been transferring carbon from the lithosphere to the atmosphere for two centuries, in the scale of 7 billion tons per year at present, and natural systems are incapable of closing this cycle [34]. Therefore, apart from the urgent need to decrease our dependence on fossil carbon, every opportunity for utilizing man-made CO_2 needs to be explored. Pure CO_2 is a valuable chemical feedstock, which conceptually competes with the CO-based C1 building blocks, that is, CO, phosgene, and methanol. It needs to be pointed out, however, that CO_2 is seldom available in a pure form. Most emissions are a mixture of a number of components such as nitrogen oxides, sulfur oxides, water vapor, and volatile organic compounds (VOCs).

The utilization of CO_2 as a feedstock for producing chemicals is an interesting challenge to explore new concepts and new opportunities for catalysis and industrial chemistry [35–38]. The conversion of CO_2 to organics is effective through either the transfer of the entire molecule or the cleavage of one C–O bond. Besides the nature of the products formed, the major difference resides in the energy input. Cleavage of C–O bond involves the reduction of CO_2 and a hydrogen source, being, therefore, an energy-intensive reaction. In contrast, the transfer of the entire molecule occurs through an acido-basic mechanism and is considered as a non-energy-intensive process [39]. Both scenarios emphasize the role of catalysts in promoting activity and selectivity [40].

Both new catalysts and new processes need to be developed for a complete exploitation of the potential of CO_2 use [41]. The key motivation to producing chemicals from CO_2 is that CO_2 can lead to totally new polymeric materials; and also new routes to existing chemical intermediates and products could be more efficient and economical than current methods. As a case in point, the conventional method for methanol production is based on fossil feedstock and the production of dimethyl carbonate (DMC) involves the use of toxic phosgene or CO. A proposed alternative production process involves the use of CO_2 as a raw material (Figure 7.1)

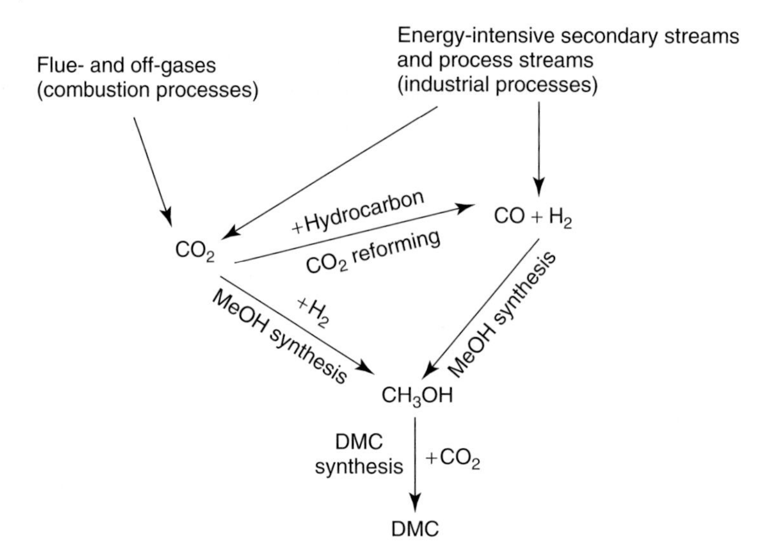

Figure 7.1 CO_2 reforming and methanol and DMC synthesis from CO_2.

[42, 43]. Production of carbonic esters can also take place via reactions between higher alcohols and CO_2. These compounds find use in a variety of sectors, for example, organic synthesis, perfumes, pharmaceuticals, polymers, solvents, and lubricants [44].

7.6.3
NO_x Emissions Reduction

In modern lean-burn engines, the traditionally used three-way catalysts (TWCs) do not remove NO_x emission. To solve this problem, the so-called lean NO_x catalysts (LNCs), NO_x storage and reduction (NSR) catalysts, and plasma-assisted catalysis (PAC) are employed. Lean NO_x traps are essentially TWCs that contain precious metals as well as high concentrations of NO_x storage compounds, for example, alkaline earth metals and/or alkali metals [45]. The LNCs are mainly designed to reduce NO_x with unburned HCs existing in the exhaust streams in the presence of O_2 but result in deNO_x efficiencies of lower than 50% and significant amounts of N_2O formation during the course of the reaction. The NSR catalysts were developed to remove NO_x from lean-burn gasoline engines operating in cyclic fuel-lean and - rich modes. The lean NO_x trap concept is based on switches between lean and sort fuel-rich phases. During the lean operation phase NO is first oxidized to NO_2, which reacts with the catalyst storage compound, for example, barium and alkaline oxides, to form nitrates. For the conversion of stored NO_x, the engine is periodically shifted to fuel-rich operation phase ($\lambda = 0.8$). Stored NO_x reacts on noble metals with hydrogen (H_2), carbon monoxide (CO), and HCs to form nitrogen. The system is controlled by the lambda-, temperature-, and optionally NO_x-sensors [46].

The PAC system utilizes plasma to oxidize NO to NO_2, which then reacts with a suitable reductant over a catalyst. LNC, NSR, and PAC systems have still several challenging tasks to be solved. Consequently, all these technologies are not yet appropriate for commercial applications to diesel and lean-burn engine exhausts [47].

Emission control from heavy duty diesel engines in vehicles and stationary sources involves the use of ammonium to selectively reduce NO_x from the exhaust gas. This NO_x removal system is called selective catalytic reduction by ammonium (NH_3-SCR) and it is additionally used for the catalytic oxidation of CO and HCs. The ammonia primarily reacts in the SCR catalytic converter with NO_2 to form nitrogen and water. Excess ammonia is converted to nitrogen and water on reaction with residual oxygen. As ammonia is a toxic substance, the actual reducing agent used in motor vehicle applications is urea. Urea is manufactured commercially and is both ground water compatible and chemically stable under ambient conditions [46].

Selective catalytic reduction of NO by hydrocarbons (HC-SCR) is one of the promising technologies for the removal of NO in the exhaust from diesel and lean-burn gasoline engines. Catalysts explored for the HC-SCR control include platinum (Pt), copper (Cu), and iridium (Ir), and recently there has been considerable interest in the behavior of silver (Ag) catalysts [48, 49]. The first advantage in HC-SCR is that the system no longer requires an extra reductant, since raw fuels appropriate for reducing NO_x can be readily supplied into the engine exhaust stream. Another benefit is in its on-board application without any installation of additional infrastructure, unlike in urea-SCR processes for diesel NO_x emission control. Finally, this HC-SCR technology has the possibility to simultaneously remove both NO_x and unburned and/or partially burned HCs from engine emissions [47].

The plasma-catalyst system utilizes plasma to oxidize NO to NO_2 which then reacts with a suitable reductant over a catalyst; however, this plasma-assisted catalytic technology still comprises challenging tasks to resolve the formation of toxic by-products and the catalyst deactivation due to the deposition of organic products during the course of the reaction as well as to prepare cost effective and durable on-board plasma devices [47].

7.6.4 Photocatalysis

Photocatalysis is a method that can be used for water purification, decomposition of gas-phase impurities as well as for fouling prevention on different surfaces. In photocatalysis, TiO_2-based materials have been reported to have good activity. The removal of very low amounts of organic matters such as VOCs as well as solvents from water via traditional methods can be difficult and in many cases quite energy consuming. Hence, there is a need to find and develop new and efficient catalytic materials for this purpose; solar or ultraviolet light can be utilized to further decrease the pollutants in water streams by a sustainable and energy-efficient way [50].

The challenges to be faced in air-purification systems using photocatalysis involve the treatment of relatively large gas flows in devices with low pressure drops, good catalyst irradiation, and efficient reactant species as well as good photocatalyst contacting [51–53].

7.6.5
Catalytic Oxidation – a Way to Treat Volatile Organic Emissions

Catalytic incineration is already a quite mature technology, and it is used, for example, in low-emission power generation [54, 55]. This is not, however, the only application for the technology. It is widely used in the abatement of VOC emissions of industrial processes [56]. It has the general advantages of operation at lower temperature, smaller size, less demands for high-temperature construction materials, and so on, compared with thermal oxidation [57]. Owing to the variety of VOCs, several catalytic materials including metal oxides and noble metals are used in oxidation processes, and several research groups are searching for more active, selective, and durable materials [55]. This task is not straightforward, since VOC emissions are mixtures and often contain certain VOCs or some other gaseous compounds that can deactivate the catalyst. It has been shown that platinum and palladium are very active and durable in solvent emission abatement [58]. Indeed, Pt particularly maintains good activity for years [58], but in some cases it may be too active causing overoxidation of VOCs that further causes problems of corrosion as in the case of sulfur-containing VOCs [59]. Because of this, selection of appropriate construction material is an important issue in the field of catalytic incineration relating to the design of a reactor [60].

The selectivity of the catalyst is of major importance in the case of chlorinated VOCs: the oxidation products should not contain even more harmful compounds than the parent-molecule, for example, formation of dioxins should be avoided. In addition, the minimization of Cl_2 and maximization of HCl in a product gas should be achieved [61]. These are just a few examples of why researchers are continuing the search for VOC oxidation catalysts as well as new reactor concepts. The new possibilities include, for example, utilization of nanosized gold catalysts in the oxidation of sulfur-containing VOCs and microwave-assisted processes where combination of adsorption and oxidation is used in low-concentration VOC oxidation [62, 63].

7.6.5.1 Case – Utilization of Reverse Flow Reactor (RFR) in Industrial Cl-VOC Oxidation

The reverse flow reactor (RFR) concept was originally patented by Cottrell in 1938 [64] in the United States and further developed and applied to different purposes by several researchers, for example, Matros and coworkers [65]. This technology has found its application also in the field of catalytic oxidation [66]. The incinerator containing two sets of regenerative-type heat exchangers and at least two catalyst honeycombs uses flow reversal to recover the heat produced in exothermal oxidation reactions [67]. A regenerative heat exchanger can typically achieve a heat

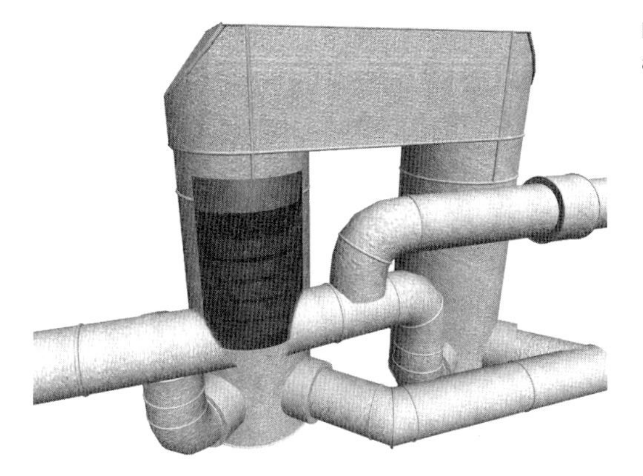

Figure 7.2 The reverse flow reactor (RFR).

recovery effectiveness of 85% or even higher [57]. The major advantage of heat recovery is that this type of reactor may run without external heat supply, when the so-called auto thermal point, which is about 0.7 g m^{-3} of VOC, is achieved. Extra dilution air is used when the temperature of the system increases over the upper temperature limit value. The dilution is used to maintain the reaction mixture below explosion limits [59]. In addition to heat economy, it has been shown that the transient operation caused by flow reversal is also advantageous for catalyst performance (Figure 7.2) [68].

Catalytic incineration has been applied in the abatement of chlorinated VOC emissions in the pharmaceutical industry. The major compounds in the emission mixture are dichloromethane, perchloroethylene, dimethylformamide, oxitol, and toluene. The incinerator operates normally at 400–500 °C, but when emissions contain perchloroethylene the temperature is increased up to 500–600 °C. The emission mixture also contains water, which pushes the selectivity further toward HCl formation instead of formation of Cl_2. After oxidation, the product gases are washed with NaOH scrubbers. The purification level of over 99% can be achieved with the incinerator, the activity of which has been shown to be very stable after one year of continuous operation [69–71].

7.7 The Urban Challenge

In 2008, the world reached an invisible but momentous milestone: For the first time in history, more than half its human population, 3.3 billion people, lived in urban areas. By 2030, this is expected to swell to almost 5 billion [72]. This means that the number of people living in cities will be more than that was present on the whole planet at the time of writing the Brundtland report. The concentration of poverty, slum growth, and social disruption in cities that had been the subject of concern in 1987 is still valid. Cities embody the environmental damage done by modern civilization, yet experts and policy makers today increasingly recognize the potential value of cities to long-term sustainability [72]. The rapid development in urbanization is also changing the requirements of modern transport. The driving forces are climbing oil price, city emissions with impact on human health, and decoupling CO_2 from growth and global warming. For most people, public transportation with hybrid technology, high capacity buses, and new engine technology are the most cost-efficient solutions for the society to meet prioritized requirements [73]. The way in which catalysis can play a great role is by improving the quality of urban air through control of automotive exhaust emissions. One research line is the utilization of various catalytic materials such as metal-loaded oxides, zeolites, and carbon nanotubes in the reduction of NO_x from stationary sources [74, 75].

7.7.1 Challenges in Control of Emission from Automotive Exhaust

The pollutants that are limited by regulations in automotive exhaust gas control are NO_x, CO, HCs, and particulate matters (PMs) [46]. TWCs are typically used to treat exhaust gas of gasoline engines. The catalyst converts the pollutants NO_x, CO, and HCs into harmless compounds via oxidation and reduction reactions. To get the best possible conversion of the pollutants, Lambda-sensors are used to control the air-to-fuel ratios of the engine and catalyst performance [46]. Traditionally used TWC contains platinum group metals (PGMs), for example, Rh, Pt, and Pd, on a very porous washcoat material, such as alumina, mounted on ceramic or metallic monolith (Figure 7.3). The price of the PGMs is high and their availability is low. Therefore, there has been a lot of research finding new catalyst materials that also fulfill tightened regulatory demands.

Oxidative catalytic converters are used to reduce CO and HCs originating from imperfect combustion in engines. At certain temperatures, these converters may also oxidize NO to NO_2. Original equipment manufacturer (OEM) particle filters (PFs) employ NO_2 to oxidize trapped soot at lower temperatures. However, the excess NO_2 may escape from the system as tailpipe emissions. NO_2 is very toxic to humans, and it also has impacts on atmospheric ozone-forming chemistry. Alvaraz *et al.* have stated that the primary NO_2 emissions of modern diesel cars are increasing [76].

7.7.1.1 Case – Automotive Exhaust Gas Catalyst Research

Diesel and gasoline exhaust gas purification has drawn increasing attention. Catalyst materials containing different noble metals, such as Pd, Pt, and Rh,

Figure 7.3 Overview of the metallic monolith structure.

on different washcoats (Al_2O_3, La_2O_3, zeolites), and additive materials (oxygen storage materials, barium oxides) have been under intensive studies for many years [77–95]. The catalyst materials are observed to behave differently depending on the reaction conditions, which can be rich, stoichiometric, or lean [79]. The activity tests with different reactant gas mixtures (NO, HCs, CO, and oxygen) together with surface analysis methods such as *in situ* DRIFT [84–88] give important information to understand the oxidation and reduction reactions occurring on the catalyst's surface. Since the lifetime of a modern catalytic converter is high (over 100 000 km), the development of laboratory-scale accelerated aging methods is vital for finding out the reasons for the catalytic decline by thermal degradation [80, 81, 89] as well as by chemical poisoning via phosphorus [89–94], calcium [90], and sulfur [95–97]. These poisons cause many changes in the properties of a catalyst.

Understanding particulate emissions, their formation and control, is another key issue in meeting the new particulate emission limits set by the new EURO emission standards. The particulate emissions formed in diesel engines have a mass median diameter of 0.05–1.0 μm. Particle consists of carbon with various HCs adsorbed on it including polyaromatic hydrocarbon (PAH) as well as nitro-PAH compounds. The small particles are reported to be very harmful for human health [98]. To remove particulate emissions from diesel engines, diesel particulate filters (DPF) are used. Filter systems can be metallic and ceramic with a large number of parallel channels. In applications to passenger cars, only ceramic filters are used. The channels in the filter are alternatively open and closed. Consequently, the exhaust gas is forced to flow through the porous walls of the honeycomb structure. The solid particles are deposited in the pores. Depending on the porosity of the filter material, these filters can attain filtration efficiencies up to 97%. The soot deposits in the particulate filter induce a steady rise in flow resistance. For this reason, the particulate filter must be regenerated at certain intervals, which can be achieved in the passive or active process [46].

Studies on the particulate distributions from compressed natural gas (CNG) or diesel-fuelled engines with diesel oxidation catalyst (DOC) or partial diesel particle filter (pDPF) have also been performed. The results obtained are used as data for the model, to study the particle penetration into the human respiratory tracts. As a result, the number distribution of particles in different parts of lungs can be modeled [99–101]. Understanding the particle formation and their effects and finding the methods to eliminate the formed particulates from exhaust gas contribute to a cleaner urban environment and thus to a better quality of life.

7.8 Future Advances in Catalysis for Sustainability

The potential for the use of catalysis in support of sustainability is enormous [102, 103]. New heterogeneous and homogeneous catalysts for improved reaction selectivity, and catalyst activity and stability, are needed, for example, new catalytic materials with new carbon modifications for nanotubes, new polymers, ligands for selective and efficient catalysis, and catalysts [22]. The use of alternative reaction media and activation for catalytic reactions offer new ideas to enhance catalytic reactions. Good examples already exist in the use of supercritical fluids and ionic liquids as reaction media with improved conditions, for example, preventing carbon formation during reactions, in microwave activation for targeted energy transfer, in catalysis for the generation of new materials, and in the synthesis and biosynthesis of materials, fuels, and chemicals [104, 105]. New catalytic atom-economic reactions and processes are emerging. There are already real-life examples from industries of how catalysis has fulfilled the principles of green chemistry and technology [26–28]. Catalytic systems for the conversion and storage of energy have already been under extensive research for quite a long time. This concerns, for example, nanostructured materials for fuel cells, solar cells, ceramic batteries as well as for the storage of hydrogen from sustainable energy sources. Chemo-enzymatic transformations of renewable feedstocks to, for example, biohydrogen are underway [102]. On the process engineering side, microstructured catalytic reactors, including membrane reactors with phenomena integration, are emerging into the market. Catalysts for electrocatalysis and photo-activation have lately become one of the most sought-after materials for research in the field of catalysis [103]. New functionalized materials and biomass-based innovations are the key driving forces in this approach. The role of catalysis in process intensification has without doubt a key role. High-throughput methods and molecular modeling for catalyst discovery are the methods quite extensively used to make the catalyst development more systematic and economical [102, 106].

Catalysis will have a great impact on chemicals manufacturing in the future via new reaction routes (chemo-enzymatic) and design of catalysts through a combination of the best features of biocatalysis, and homogeneous and heterogeneous catalysis [102, 107]. The design of catalysts based on nanomaterials approach will offer new tools for chemical process integration at the nanoscale level [108–110]. This guarantees that the design of plants can be achieved with a better integration of units to conserve energy and raw materials. In sustainable production, catalysis by nanomaterials offers optimized direct chemical transformation of raw materials into desired products. This leads to minimized energy use, waste generation, and environmental hazards, and improved process safety, which are the key principles in Green Chemistry and Engineering approach. One of the key reactions, F–T synthesis, which is a mature technology, could be utilized to reduce the CO_2 emissions: this can be achieved by the efficient use of various sustainable raw material sources via synthesis gas production and production of clean fuels with low S and N contents [108, 111, 112]. Research and investments are increasing very rapidly, and new plants are being built. Biodiesel production leads to expansion of, for example, the glycerol market and new derivatives made out of glycerol are a challenge to environmental sustainability [113, 114]. CO_2 utilization, when being successful, has a positive impact on the global carbon balance. All these will determine the way in which chemicals and other products are going to be manufactured, distributed, and used in our society in the future.

The effort to combine the best features of homogeneous, heterogeneous, and biocatalysis into one pot for the benefit of sustainable production of energy and society at large, is considered to have a key role in tomorrow's sustainable approach [107, 115]. Scientists are doing a great work to achieve this goal. At the nanometer scale, for example, in catalysis by nanomaterials, better control of activity, selectivity, and deactivation of catalysts is attained [109, 110]. Recent advances have shown that design, synthesis, characterization, and manipulation of nanoscale catalysts are possible. Catalysis nanoscience includes research to find out the relationships between catalyst synthesis, active-site structure on the atomic- and nanoscale, reaction mechanisms, and catalyst activity, selectivity, and lifetime. Advances will lead to design of catalysts for novel catalytic reactions that take advantage of, for example, self-assembly of catalytic sites in predetermined two- and three-dimensional configurations. Recent attempts and desires to bridge different catalysis areas, from biocatalysis to homogeneous and heterogeneous catalysis, together by combining the high selectivity of homogeneous cluster catalysts with the stability and versatility of supported heterogeneous catalysts have in some cases already shown its potential. Research activities and innovations are needed and underway in this area of research [107].

7.9 Conclusions

In its communication toward stimulating technologies for sustainable development (COM (2004) 38 final), the European Commission defined environmentally sound technologies as those that protect the environment, are less polluting, use all resources in a more sustainable manner, recycle more of their wastes and products, and handle residual wastes in a more acceptable manner than the technologies for which they were substitutes [116]. Environmentally sound technologies in the context of pollution are process and product technologies that generate low or no waste, for the prevention of pollution. They also cover end-of-pipe technologies for treatment of pollution after it has been created.

It is evident that catalysis has a key role as a technology for sustainable development. The main objectives of catalysis for sustainability are prevention of pollution by developing nonpolluting and energy-saving processes, reduction in pollution by achieving higher selectivity in fuel and chemicals manufacture through developing new and improved catalytic processes, as well as removal of pollutants, and development of competitive processes for the production of chemicals, hydrogen, and other energy sources, providing reduced CO_2 production. Catalytic technologies also reduce emissions in agro-food production, pulp and paper, and electronic and metal finishing industries, remediate water and soil contamination, help recycle solid waste such as polymers, improve quality of indoor or outdoor air, and improve the use of natural resources and energy efficiency (including solar energy). Catalysts have an enormous impact on industry and everyday life by enabling reactions to take place, making processes more efficient, increasing the operating profit, and making processes environmentally friendly, economic, and safe.

Environmental catalysis has its potential in improving innovations in the field of catalysis and highlighting the new directions for research driven by market, social, and environmental needs. Therefore, it can be concluded that environmental catalysis plays a key role in demonstrating the role of catalysis as a driver of sustainability by improving the quality of life and protecting human health and the environment.

References

1. World Commission on Environment and Development (1987) Report of the World Commission on Environment and Development: Our Common Future. Transmitted to the General Assembly as an Annex to document A/42/427 – Development and International Co-operation: Environment. *http://www.un-documents.net/wced-ocf.htm.* (Accessed May 2010).
2. UNESCO (United Nations Educational, Scientific and Cultural Organisation) (2002) Teaching and Learning for a Sustainable Future.
3. Lacerda, L. (2007) Presented at the 18th Annual General Assembly (AGA) and Conference, June 1–3, 2007, Madrid. *http://www.efc.be/ftp/public/AGA_Presentations/2007/Sustainabilityplus_Lacerda.pdf.* (Accessed May 2010).
4. McKenzie, D. (2008) *New Sci.*, **200** (2683), 20.
5. Jin, Y., Wang, D., and Wei, F. (2004) *Chem. Eng. Sci.*, **59**, 1885–1895.
6. Wei, L., Pordesimo, L.O., Igathinathane, C., and Batchelor, W.D. (2009) *Biomass Bioenergy*, **33**, 255–266.
7. Armor, J.N. (2000) *Appl. Catal. A: Gen.*, **194-195**, 3–11.
8. Neste Oil Oyj. Bensiiniopas. *http://www.neste.fi/binary.aspx?path=2589;2655;2710;2821;2944;2708;3237&page=3379&field=File Attachment&version=2.* (Accessed May 2010).
9. García-Gutiérrez, J.L., Fuentes, G.A., Hernández-Terán, M.E., Murrieta, F., Navarrete, J., and Jiménez-Cruz, F. (2006) *Appl. Catal. A: Gen.*, **305**, 15–20.
10. Air Pollution Control Programme 2010 (2002) The Finnish National Programme for the Implementation of Directive 2001/81/EC, Approved by the Government on September 26, Ministry of Education, Helsinki, pp. 1–40.
11. Zhao, D., Liu, R., Wang, J., and Liu, B. (2008) *Energy Fuel*, **22**, 1100–1103.
12. Li, C., Jiang, Z., Gao, J., Yang, Y., Wang, S., Tian, F., Sun, F., Sun, X., Ying, P., and Han, C. (2004) *Chem. Eur. J.*, **10** (9), 2277–2280.
13. Lu, H., Gao, J., Jiang, Z., Jing, F., Yang, Y., Wang, G., and Li, C. (2006) *J. Catal.*, **239**, 369–375.
14. Fu, X., Long, J., Wang, X., Leung, D.Y.C., Ding, Z., Wu, L., Zhang, Z., Li, Z., and Fu, X. (2008) *Int. J. Hydrogen Energy*, **33**, 6484–6491.
15. Soyal-Baltacıoğlua, F., Erhan Aksoylua, A., and Ilsen Önsan, Z. (2008) *Catal. Today*, **183** (3-4), 183–189.
16. Le Valanta, A., Garrona, A., Biona, N., Epron, F., and Duprez, D. (2008) *Catal. Today*, **138** (3-4), 169–174.
17. Seo, J.G., Youn, M.H., Jung, J.C., Cho, K.M., Park, S., and Song, I.K. (2008) *Catal. Today*, **138** (3-4), 130–134.
18. Leea, S.Y., Ryua, B.H., Hana, G.Y., Leeb, T.J., and Yoona, K.J. (2008) *Carbon*, **46** (14), 1978–1986.
19. Li, H., Goldbach, A., Li, W., and Xu, H. (2008) *J. Membr. Sci.*, **324** (1-2), 95–101.
20. Gogotsi, Y. (ed.) (2006) *Nanomaterials Handbook*, CRC Taylor & Francis, Boca Raton.

21. O'Connell, M.J. (ed.) (2006) *Carbon Nanotubes, Properties and Applications*, CRC Taylor & Francis, Boca Raton.
22. Pan, X. and Bao, X. (2008) *Chem. Commun.*, 6271–6281.
23. Huuhtanen, M., Turpeinen, E., Mustonen, T., Baliñas Gavira, A., Seelam, P.K., Kordás, K., Tóth, G., and Keiski, R.L. (2008) 13th Nordic Symposium on Catalysis, October 5–7, 2008, Gothenburg.
24. Keiski, R.L., Huuhtanen, M., Kordás, K., Turpeinen, E., Tóth, G., Mustonen, T., and Leiviskä, K. (2008) CopenMind, September 1–3, Copenhagen.
25. Anastas, P. and Williamson, T.C. (eds) (1998) *Green Chemistry; Frontiers in Benign Chemical Syntheses and Processes*, Oxford University Press, Inc., New York, 364 p.
26. Anastas, P.T. and Warner, J.C. (1998) *Green Chemistry, Theory and Practice*, Oxford University Press, Inc., New York, 135 p.
27. Anastas, P.T. and Zimmerman, J.B. (2003) *Environ. Sci. Technol.*, **37**, 94A–101A.
28. Anastas, P.T., Bartlett, L.B., Kirchhoff, M.M., and Williamson, T.C. (2000) *Catal. Today*, **55**, 11–22.
29. Anastas, P.T., Kirchhoff, M.M., and Williamson, T.C. (2001) *Appl. Catal. A: Gen.*, **221**, 3–13.
30. Armor, J.N. (1999) *Appl. Catal. A: Gen.*, **189**, 153–162.
31. Centi, G. and Perathoner, S. (2003) *Catal. Today*, **77**, 287–297.
32. Thomas, J.M. and Thomas, W.J. (1997) *Principles and Practice of Heterogeneous Catalysis*, VCH Publishers, Weinheim.
33. Quinn, M., Kriebel, D., and Moure-Eraso, R. (1998) *Am. J. Ind. Med.*, **34**, 297–304.
34. Beckman, E.J. (2003) *Ind. Eng. Chem. Res.*, **42**, 1598–1602.
35. Centi, G. and Perathoner, S. (2004) *Stud. Surf. Sci. Catal.*, **153**, 1–8.
36. Gong, J., Ma, X., and Wang, S. (2007) *Appl. Catal. A: Gen.*, **316** (1), 1–21.
37. Ballivet-Tkatchenko, D. and Sorokina, S. (2003) Linear organic carbonates, in *Carbon Dioxide Recovery and Utilization* (ed. M. Aresta), Kluwer Publishers, pp. 261–277.
38. Sakakura, T., Choi, J.-C., and Yasuda, H. (2007) *Chem. Rev.*, **107**, 2365–2387.
39. Arakawa, H., Aresta, M., Armor, J.N., Barteau, M.A., Beckman, E.J., Bell, A.T., Bercaw, J.E., Creutz, C., Dinjus, E., Dixon, D.A., Domen, K., DuBois, D.L., Eckert, J., Fujita, E., Gibson, D.H., Goddard, W.A., Goodman, D.W., Keller, J., Kubas, G.J., Kung, H.H., Lyons, J.E., Manzer, L.E., Marks, T.J., Morokuma, K., Nicholas, K.M., Periana, R., Que, L., Rostrup-Nielson, J., Sachtler, W.M.H., Schmidt, L.D., Sen, A., Somorjai, G.A., Stair, P.C., Stults, B.R., and Tumas, W. (2001) *Chem. Rev.*, **101** (4), 953–996.
40. Dinjus, E., Fornika, R., Pitter, S., and Zevaco, T. (2002) Carbon dioxide as a C_1 building block, in *Applied Homogeneous Catalysis with Organometallic Compounds*, vol. 3 (eds B. Cornils and W.A. Herrmann), 2nd edn, Wiley-VCH Verlag GmbH, Weinheim, pp. 1189–1213.
41. Sunggyu, L. (1999) *Methanol Synthesis Technology*, CRC Press.
42. Raudaskoski, R., Turpeinen, E., Lenkkeri, R., Pongrácz, E., and Keiski, R.L. (2009) *Catal. Today*, **144**, 318–323.
43. Ballivet-Tkatchenko, D., Chermette, H., Plasseraud, L., and Walter, O. (2006) *Dalton Trans.*, 5167–5175.
44. Ballivet-Tkatchenko, D., Chambrey, S., Keiski, R., Ligabue, R., Plasseraud, L., Richard, P., and Turunen, H. (2006) *Catal. Today*, **115** (1-4), 80–87.
45. Theis, J.R. and Gulari, E. (2007) *Appl. Catal. B: Env.*, **74**, 40–52.
46. Alkemade, U.G. and Schumann, B. (2006) *Solid State Ions*, **177**, 2291–2296.
47. Hyeon Kim, M. and Nam, I.-S. (2004) New opportunity for HC-SCR technology to control NO_x emission from advanced internal combustion engines, in *Catalysis*, **18**, (ed. J.J. Spivey), Royal Society of Chemistry, pp. 116–185.
48. Twigg, M.V. (2007) *Appl. Catal. B: Env.*, **70**, 2–15.

49. Twigg, M.V. (2006) *Catal. Today*, **117**, 407–418.
50. De Lasa, H., Serrano, B., and Salaices, M. (2005) *Photocatalytic Reaction Engineering*, Springer Science +Business Media, New York, 187 p.
51. Ilisz, I., Dombi, A., Mogyorósi, K., Farkas, A., and Dékány, I. (2002) *Appl. Catal. B: Env.*, **39**, 247–256.
52. Ohko, Y., Tryk, D., Hashimoto, K., and Fujishima, A. (1998) *J. Phys. Chem. B.*, **102** (15), 2699–2704.
53. Robert, D. and Malato, S. (2002) *Sci. Total Env.*, **291**, 85–97.
54. Lee, J.H. and Trimm, D.L. (1995) *Fuel Proc. Tech.*, **42**, 339–359.
55. Hayes, R.E. and Kolaczkowski, S.T. (1997) *Introduction to Catalytic Combustion*, Gordon and Breach Science Publishers, Amsterdam.
56. Ojala, S. (2005) Catalytic oxidation of volatile organic compounds. Doctoral Thesis. Oulu University Press, Finland.
57. Jennings, M.S. (1985) *Catalytic Incineration for Control of Volatile Organic Compound Emissions*, Noyes Publishers.
58. Ojala, S., Lassi, U., Härkönen, M., Maunula, T., Silvonen, R., and Keiski, R.L. (2006) *Chem. Eng. J.*, **120**, 11–16.
59. Ojala, S., Lassi, U., Ylönen, R., Keiski, R.L., Laakso, I., Maunula, T., and Silvonen, R. (2005) *Tappi J.*, **4** (1), 9–14.
60. Ojala, S., Lassi, U., Ylönen, R., Karjalainen, H., and Keiski, R.L. (2005) *Catal. Today*, **100**, 367–372.
61. Gonzáles-Velasco, J.R., Aranzabal, A., López-Fonseca, R., Ferret, R., and Gonzáles-Marcos, J.A. (2000) *Appl. Catal. B: Env.*, **24**, 33–43.
62. Kucherov, A.V., Sinev, I.M., Slovetskaya, K.I., Ojala, S., Keiski, R.L., Golosman, E.Z., and Kustov, L.M. (2008) 13th Nordic Symposium on Catalysis, October 5–7, Gothenburg.
63. Ojala, S., Mikkola, J.P., and Keiski, R.L. (2008) 18th International Congress of Chemical and Process Engineering, August 27–31, Prague.
64. Cottrell, F.G. (1938) Purifying gas and apparatus. US Patent 2121733.
65. Matros, Y.S., Noskov, A.S., and Chumachenko, V.A. (1993) *Chem. Eng. Proc.*, **32**, 89–98.
66. Ojala, S., Lassi, U., and Keiski, R.L. (2005) *Chem. Eng. Trans.*, **6**, 569–574.
67. Kolios, G., Frauhammer, J., and Eigenberger, G. (2002) *Chem. Eng. Sci.*, **57**, 1505–1510.
68. Matros, Y.S. (1990) *Chem. Eng. Sci.*, **45** (8), 2097–2102.
69. Pitkäaho, S., Ojala, S., and Keiski, R.L. (2008) IWCC7, 27 September-1 October, Zürich.
70. Pitkäaho, S., Ojala, S., and Keiski, R.L. (2008) 13th Nordic Symposium on Catalysis, October 5–7, Gothenburg.
71. Keiski, R.L., Pitkäaho, S., and Ojala, S. (2008) CopenMind, September 1–3, Copenhagen.
72. UNFPA (United Nations Population Fund) (2007) State of the World Population 2007. Unleashing the Potential of Urban Growth, 108 p.
73. Jobson, E. (2008) oral presentation, 5th International Conference on Environmental Catalysis, 31 August–3 September, Belfast.
74. Huang, B., Huang, R., Jin, D., and Ye, D. (2007) *Catal. Today*, **126**, 279–283.
75. Valdés-Solis, T., Marbán, G., and Fuertes, A.B. (2003) *Appl. Catal. B: Env.*, **46**, 261–271.
76. Alvarez, R., Weilenmann, M., and Favez, J.-Y. (2008) *Atmos. Env.*, **42**, 4699–4707.
77. Shimizu, K.-I., Kawabata, H., Satsuma, A., and Hattori, T. (1998) *Appl. Catal. B: Env.*, **19**, L87–L92.
78. Burch, R., Breen, J.P., and Meunier, F.C. (2002) *Appl. Catal. B: Env.*, **39**, 283–303.
79. Kolli, T., Rahkamaa-Tolonen, K., Lassi, U., Savimäki, A., and Keiski, R.L. (2004) *Topics Catal.*, **30/31**, 341–346.
80. Kolli, T., Rahkamaa-Tolonen, K., Lassi, U., Savimäki, A., and Keiski, R.L. (2005) *Catal. Today*, **100**, 297–302.
81. Kolli, T., Rahkamaa-Tolonen, K., Lassi, U., Savimäki, A., and Keiski, R.L. (2005) *Catal. Today*, **100**, 303–307.

82. Kolli, T., Lassi, U., Rahkamaa-Tolonen, K., Kinnunen, T.-J.J., and Keiski, R.L. (2006) *Appl. Catal. A: Gen.*, **298**, 65–72.
83. Kolli, T., Kröger, V., and Keiski, R.L. (2007) *Topics Catal.*, **45**, 165–168.
84. Huuhtanen, M., Määttä, T., Rahkamaa-Tolonen, K., Maunula, T., and Keiski, R.L. (2004) *Topics Catal.*, **30-31**, 359–363.
85. Huuhtanen, M., Rahkamaa-Tolonen, K., Maunula, T., and Keiski, R.L. (2005) *Catal. Today*, **100**, 321–325.
86. Ahola, J., Huuhtanen, M., and Keiski, R.L. (2003) *Ind. Eng. Chem. Res.*, **42**, 2756–2766.
87. Huuhtanen, M., Kolli, T., Maunula, T., and Keiski, R.L. (2002) *Catal. Today*, **75**, 379–384.
88. Huuhtanen, M., Maunula, T., and Keiski, R.L. (2005) *Stud. Surf. Sci. Catal.*, **158**, 1867–1874.
89. Kanerva, T., Kröger, V., Rahkamaa-Tolonen, K., Lepistö, T., and Keiski, R.L. (2007) *Topics Catal.*, **45**, 137–142.
90. Kröger, V., Hietikko, M., Lassi, U., Ahola, J., Kallinen, K., Laitinen, R., and Keiski, R.L. (2004) *Topics Catal.*, **30-31**, 469–473.
91. Kröger, V., Lassi, U., Kynkäänniemi, K., Suopanki, A., and Keiski, R.L. (2006) *Chem. Eng. J.*, **120**, 113–118.
92. Kröger, V., Hietikko, M., Andove, D., French, D., Lassi, U., Suopanki, A., Laitinen, R., and Keiski, R.L. (2007) *Topics Catal.*, **42-43**, 409–413.
93. Kröger, V., Kanerva, T., Lassi, U., Rahkamaa-Tolonen, K., Lepistö, T., and Keiski, R.L. (2007) *Topics Catal.*, **42-43**, 433–436.
94. Kröger, V., Kanerva, T., Lassi, U., Rahkamaa-Tolonen, K., Lepistö, T., and Keiski, R.L. (2007) *Topics Catal.*, **45**, 153–157.
95. Kolli, T., Huuhtanen, M., Hallikainen, A., Kallinen, K., and Keiski, R.L. (2009) *Catal. Lett.*, **127**, 49–54.
96. Spencer, M.S. and Twigg, M.V. (2005) *Ann. Rev. Mater. Res.*, **35**, 427–464.
97. Rodriguez, J.A. and Hrbek, J. (1999) *Acc. Chem. Res.*, **32**, 719–728.
98. Cohen, J.T. and Nikula, K. (1999) The health effects of diesel exhaust: laboratory and Epidemiologic Studies, in *Air Pollution and Health* (eds S.T. Holgate, J.M. Samet, H.S. Koren, and R.L. Maynard), Academic Press, San Diego, pp. 707–745.
99. Oravisjärvi, K., Pietikäinen, M., and Keiski, R.L. (2006) Particulate Formation in Engines and their Health Effects (Review). Report 324, Department of Process and Environmental Engineering, University of Oulu, 83 p.
100. Oravisjärvi, K., Pietikäinen, M., Rautio, A., Haataja, M., Voutilainen, M., Ruuskanen, J., and Keiski, R.L. (2009) Exposure to particles originating from a diesel bus with catalyst after-treatment; estimated lung deposition for the school age children, in *CaPoL8: 8th International Congress on Catalysis and Automotive pollution Control, April 15–17, 2009*, **3**, (ed. N. Kruse, A. Fennet, J.-M. Bastin, and T. Visart de Bocarmé), Brussels, Belgium, pp. 435–442.
101. Pietikäinen, M., Oravisjärvi, K., Rautio, A., Voutilainen, A., Ruuskanen, J., and Keiski, R.L. (2009), *Sci. Total Environ.* **408**, 163–188.
102. Sheldon, R.A. (2008) *Chem. Commun.*, 3352–3365.
103. Centi, G. and Perathoner, S. (2008) *Catal. Today*, **138**, 69–76.
104. Welton, T. (2008) *Green Chem.*, **10**, 483.
105. Ballivet-Tkatchenko, D., Piquet, M., Solinas, M., Francio, G., Wasserscheid, P., and Leitner, W. (2003) *Green Chem.*, **5** (2), 232–235.
106. Centi, G. and van Santen, R.A. (eds) (2007) *Catalysis for Renewables, from Feedstock to Energy Production*, Wiley-VCH Verlag GmbH, 448 p.
107. Somorjai, G.A. (2004) *Nature*, **430**, 730.
108. Somorjai, G.A. and Rioux, R.M. (2005) *Catal. Today*, **100**, 201–215.
109. Somorjai, G.A. and Park, J.Y. (2007) *Phys. Today*, **60** (10), 48–53.

110. Somorjai, G.A. and Park, J.Y. (2007) *Catal. Lett.*, **115** (3-4), 87–98.

111. Morales, F. and Weckhyusen, B.M. (2006) *Catalysis*, **19**, 1–40.

112. Dry, M.E. (2002) *Catal. Today*, **71**, 227–241.

113. Behr, A., Eilting, J., Irawadi, K., Leschinski, J., and Linder, F. (2008) *Green Chem.*, **10**, 13–30.

114. Gu, Y., Azzouzi, A., Pouilloux, Y., Jerome, F., and Barrault, J. (2008) *Green Chem.*, **10**, 164–167.

115. Grunes, J., Zhu, J., and Somorjai, G.A. (2003) *Chem. Commun.*, 2257–2260.

116. Commission of the European Communities (COM) (2004) Communication from the Commission to the Council and the European Parliament Stimulating Technologies for Sustainable Development: An Environmental Technologies Action Plan for the European Union. 38 final. *http://ec.europa.eu/environment/etap/pdfs/com_2004_etap_en.pdf*. (Accessed May 2010).

8
Catalytic Engineering in the Processing of Biomass into Chemicals

Tapio Salmi, Dmitry Murzin, Päivi Mäki-Arvela, Johan Wärnå, Kari Eränen, Narendra Kumar, and Jyri-Pekka Mikkola

8.1
Introduction

Chemical reaction engineering has traditionally focused on the applications related to energy production from fossil fuels and production of bulk chemicals. In these fields, a lot of research and development effort has taken place both in academia and industry and, consequently, the industrial processes and catalytic operations have been fine-tuned to an incredible sophistication level during the course of nearly 100 years. The breakthrough in numerical methods and computing, along with lumping techniques, has enabled the quantitative treatment of very complex chemical systems in various reactors, such as catalytic fixed beds and fluidized beds [1]. Simultaneously, new concepts for catalytic reactors have emerged, particularly various structured solutions such as monoliths, foams, and catalyst fibers [2].

Now mankind is facing a new challenge: the era of easily accessible, environmentally sound, and affordable fossil raw materials is approaching the end and we should prepare for the consequences of the "peak oil." The resources of fossil raw materials diminish, particularly the access to crude oil. The most important change is, however, caused by the concern for the state of the atmosphere and the imminent threat of global warming. Because of the alarming climate change and the continuous increase in the carbon dioxide content as well as other green house gases in the atmosphere caused by the burning of fossil fuels, it is necessary to shift to a more sustainable technology such as the use of biomass as a source of fuels, chemicals, and materials [3, 4]. Catalytic conversion of platform molecules plays a key role in the treatment of biomass [4]. A promising perspective has been opened to a sustainable global order: in future, we will not be any more dependent on fossil resources produced in politically unstable areas and controlled by dictatorships.

Chemical engineering in general and chemical reaction engineering in particular are in key position to carry out this transformation. One of the most important sources of biomass is forests, not only the rain forests and eucalyptus trees growing in the tropical areas of the earth but also the forests in the Northern hemisphere, for instance in Canada, USA, Russia, and in the Fenno-Scandic region. The big

Novel Concepts in Catalysis and Chemical Reactors: Improving the Efficiency for the Future.
Edited by Andrzej Cybulski, Jacob A. Moulijn, and Andrzej Stankiewicz

ISBN: 978-3-527-32469-9

benefit of the use of biomass from wood is that it does not compete with the food chain. In the present chapter, we give some indication of which raw materials one can get from biomass and how to apply the concepts of catalysis and chemical reaction engineering in processing them.

8.2 Chemicals and Fuels from Biomass

An overview of the use of hardwood and softwood is given in Figure 8.1 (B. Holmbom, 2009, private communication). As the schematic picture reveals, many possibilities exist for the use of wood – and biomass in general – as a source of chemicals.

In line with the current industrial practices, wood is typically processed mechanically to chips, which are further treated chemically or mechanically, the ultimate aim being pulp and paper. Roughly speaking, Nordic coniferous trees consist of cellulose (around 50%), hemicelluloses (10%), lignin (<40%), and extractives (few percent), depending on the particular species in question. It should be emphasized that the main components from wood biomass, cellulose, and hemicelluloses exist as main components in agricultural residues, too [3]. Glucose can equally well be obtained from cellulose as from starch. An interesting starting molecule for chemicals is lactose, the milk sugar. The use of lactose as a raw material for synthesis does not compete with the food chain, because huge amounts of lactose

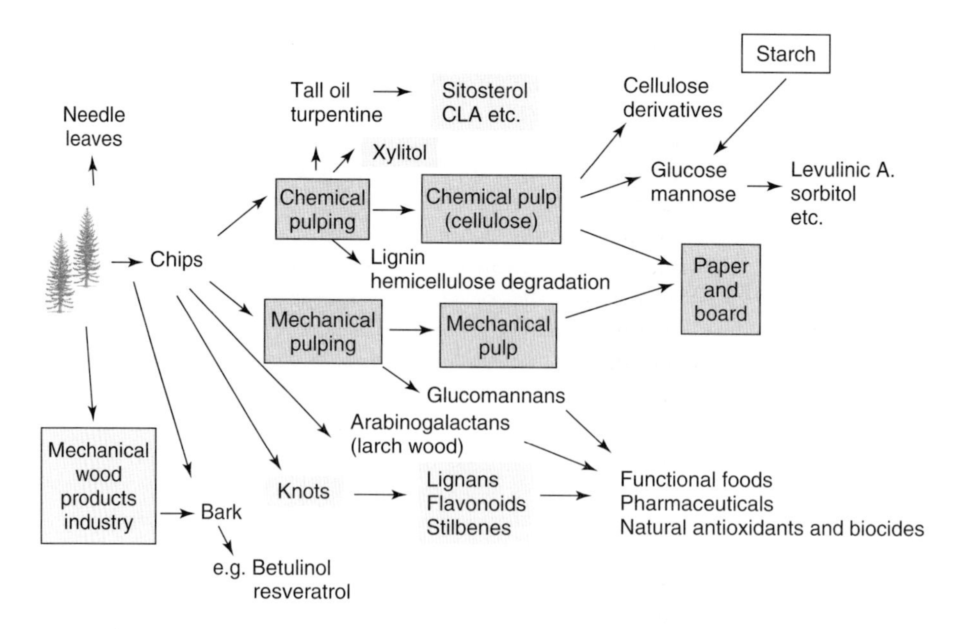

Figure 8.1 Overview of products available from wood biomass.

go waste in the form of whey as a result of the production of cheese and other milk products (see Section 8.3.6).

The aim of chemical pulping is to dissolve the lignin material and most part of the hemicelluloses into the cooking liquor. The black liquor of the dominating process, the sulfate process (i.e., Kraft pulping), contains NaOH and NaHS; lignin is dissolved in an alkaline process and cellulose fibers remain for papermaking. It is possible to recover some chemical fractions from chemical pulping, particularly tall oil and turpentine [5].

The turpentine fraction is gasified during the process, and valuable chemicals, such as α-pinene and carene can be separated by distillation. They can be further processed by catalytic treatment, such as isomerization and oxidation [6–8].

From the tall oil fraction, valuable chemicals such as triterpenoids, fatty acids, and resin acids, can be separated. Sitosterol, a triterpenoid, is the raw material for a cholesterol-suppressing agent, sitostanol [9, 10]. Conjugated fatty acids (CLA), which are important antioxidants, can be produced by isomerization of fatty acids and fatty acid esters with homogeneous [11–13] or heterogeneous catalysts [14]. Fatty acids and resin acids can also be saturated through catalytic hydrogenation and used as fuel components after decarboxylation over heterogeneous catalysts, such as supported Pd [15].

Some part of the cellulose fraction is redirected to make cellulose derivatives, such as cellulose acetate, methyl and ethyl cellulose, carboxymethyl cellulose, hydroxyethyl cellulose, and hydroxypropyl cellulose. These derivatives find multiple applications, for instance, as additives in current products (e.g., paints, lacquers) of chemical industry. Typically, the preparation of cellulose derivatives takes place as a two-phase reaction: cellulose is pretreated, for example, with alkali, and a reagent is added to get the substitution. Usually no catalyst is needed [5].

The use of the lignin fraction is much more cumbersome; currently the best-known chemical of a real commercial importance is vanillin, which is obtained by oxidation of the black liquor. Another example is a product called *spray-dried lignosulfonate* (as sodium salt) obtained from the older, acidic sulfite pulping process. It is sold as a commercial product primarily as a concrete additive for enhanced strength. Since the cement industry is one of the big contributors of carbon dioxide emissions (due to the production of calcium oxide from calcium carbonate), the use of this renewable, wood-derived product not only is fossil-carbon neutral in itself but also reduces carbon dioxide emission due to the diminished need for cement in large infrastructures made of concrete.

The black liquor from the sulfate process, on the other hand, contains a lot of hydroxyacids (in pinewood black liquor around 30 wt%) [5], which might be of potential use as a source of chemicals, but their separation from the black liquor mixture is a challenge. Nevertheless, the whole energy-economy as well as recycling of cooking chemicals for chemical pulping relies heavily on the use of recovery boilers. In these units, the aqueous mixture of primarily lignin residues and major part of the cooking chemicals are pumped into a large furnace that besides reduction of the cooking chemicals (that can be fed back to the pulping) provides an integrated process site with the much needed process steam, which, besides being used in the processes, is also regularly utilized for the coproduction of electricity.

Wood chips can also be utilized as such to produce bioethanol. The cellulose and hemicellulose material is hydrolyzed in the presence of acids (H_2SO_4, HCl, or HCOOH) or enzymes to yield glucose and other monosaccharides [16]. Lignin is separated by filtration as a solid residue and the monosaccharides are fermented to ethanol, which, in turn, is separated from water and catalyst by distillation. Ethanol can be used not only as energy source but also as a platform component to make various chemicals, such as ethene and polyethene. Today "green" acetaldehyde and acetic acid from wood-derived bioethanol is manufactured by SEKAB Ab, at the Örnsköldsvik "Biorefinery of the Future" industrial park.

To make chemicals, it is much more straightforward to start from the wood chips, knots, and bark directly rather than from the black liquor produced in the sulfate process. In mechanical pulping, some hemicelluloses, such as glucomannans, can be recovered and processed further by hydrolysis, fermentation, oxidation, and hydrogenation. An even more attractive way is to extract hemicelluloses from wood chips with water – as an example it should be mentioned that larch wood (*Larix sibirica*) contains around 15% and even more arabinogalactan by weight. In an analogous manner, galactoglucomannan and arabinoglucuronoxylan can be extracted from the wood material [17, 18] and hydrolyzed [19, 20] to yield monosaccharides which can be fermented or exposed to catalytic transformation.

The separation of various hemicelluloses and their catalytic treatment has a great perspective. A world-famous product is the natural sweetener, anticaries, and anti-inflammatory agent, xylitol, which is obtained from glucuronoxylan and arabinoglucuronoxylan [21]. The main source of xylanes is birch (*Betula verrucosa*). The hemicellulose is hydrolyzed to monosaccharides, which are separated and xylose is hydrogenated over a heterogeneous catalyst to xylitol. The classical catalyst for this purpose has been Raney nickel [22, 23], but supported Ru catalysts have given very promising results [24–26]. It has recently been shown [26] that arabinose, galactose, mannose, rhamnose, and maltose can be successfully hydrogenated to the corresponding sugar alcohols on Ru/active carbon catalysts. In an analogous manner, glucose can be hydrogenated to sorbitol [24]. Sugar alcohols are platform chemicals, which can be used for many purposes, such as the production of fuels via aqueous reforming and biolubricants through esterification.

From bark, valuable chemicals, such as betulinol, can be obtained [5]. Betulinol is used as a health-promoting agent. Recently Holmbom *et al.* [27, 28] discovered that hydroxymatairesinol (HMR) is concentrated in the stems and knots of Norway spruce (*Picea abies*). It can be extracted and transformed catalytically to matairesinol (MAT), which is an antioxidant and anticarcinogenic agent [29–31].

Besides the polysaccharides existing in the biomass originating from wood, two disaccharides appearing in large amounts in the nature should be mentioned: saccharose (mainly in sugar beats and sugar cane) and lactose (in the milk of mammals). Saccharose consists of glucose and fructose units, while lactose consists of glucose and galactose units. They can be transformed to various chemicals, either as such or after hydrolysis [32].

Table 8.1 Examples of molecules from biomass through catalysis.

Reaction	Reactant	Product	Catalyst
Hydrolysis of cellulose (−O−)	Cellulose	Glucose	Acid, enzyme (HO)
Hydrolysis of hemicelluloses (−O−)	Hemicellulose	Arabinose, galactose, glucose, mannose, xylose, etc.	Acid, enzyme (HO)
Hydrolysis of disaccharides	Saccharose, maltose, lactose	Corresponding monosaccharides	Acid, enzyme (HO)
Hydrogenation of sugars (−CHO)	Arabinose, galactose, glucose, mannose, xylose, lactose + hydrogen	Arabinitol, galactitol, sorbitol, mannitol, xylitol, lactitol, etc.	Trad. Raney nickel, Ru/C better (HE)
Oxidation of sugars (−CHO)	Sugars (see above) + oxygen	Sugar acids	Trad. Pd, Au, and Au/Pd (HE) catalysts promising
Fermentation of sugars	Mainly glucose, other hexoses	Ethanol	Yeasts (Y)
Hydrogenation of phytosterols (C=C)	Sterol (e.g., sitosterol)	Stanol (e.g., sitosterol)	Pd on various supports (C, zeolite, + hydrogen polymer) (HE)
Hydrogenation of fatty acids (C=C)	Unsaturated acid	Saturated acid	Ni, Pt, Pd, Rh, Ru (HE)
Hydrogenolysis of lignans (−OH)	For example, HMR + hydrogen	For example, MAT	Pd/C, Pd/C nanofibers, Pd-H-Beta-150 (HE)
Isomerization of fatty acids	Linoleic acid	CLA	Rh-phosphine complexes Cr-carbonyls (HO), Ru/C, Ru/Al_2O_3 (HE)

HE = heterogeneous catalyst, HO = homogeneous catalyst, E = enzyme, Y = yeast.

A brief summary of current and potential processes is given in Table 8.1. As shown in the table, most of the reactions are hydrolysis, hydrogenolysis, hydration, hydrogenation, oxidation, and isomerization reactions, where catalysis plays a key role. Particularly, the role of heterogeneous catalysts has increased in this connection in recent years; therefore, this chapter concerns mostly the application of heterogeneous solid catalysts in the transformation of biomass. An extensive review of various chemicals originating from nature is provided by Mäki-Arvela *et al.* [33].

When looking at the list of catalysts being successful, it is striking to notice that only few catalyst metals seem to be of importance. For the double bond hydrogenation and hydrogenolysis of hydroxyl groups, Pd is both active and selective [34]. For the hydrogenation of the carboxyl group (e.g., production of sugar alcohols) sponge nickel (Raney nickel) and Ru on active carbon work well, but Raney nickel will probably be replaced by Ru because of environmental concerns and health reasons. Ru is also an active and selective catalyst in isomerization in the production of conjugated acids. Oxidation of sugars to sugar acids [35], for instance oxidation of lactose to lactobionic acids, is traditionally carried out on Pd catalysts, but the risk is the oxidation of metallic Pd by molecular oxygen, which impairs the activity and selectivity. Recent studies have shown that gold-ceria catalysts could provide a promising alternative for this oxidation process [36].

Besides the catalyst activity, the selectivity is a central issue in the treatment of the molecules from biomass. A typical example is catalytic hydrogenation of sugar molecules to sugar alcohols. A product selectivity of about 90% is easily achieved on Ni and Ru catalysts, but approaching 100% requires a fine-tuning of the catalyst properties and optimization of the reaction conditions. The selectivity of the catalyst can sometimes be improved with promoters; for instance, Mo is used as a promoter in Raney nickel catalysts. A too high reaction temperature leads to decomposition products and lack of hydrogen on the catalyst surface promotes the isomerization reaction and enhances catalyst deactivation by fouling. Thus rather low temperatures (around 100 °C) and moderate hydrogen pressures (40–80 bar) have proven to be the optimal ones for sugar hydrogenations on Ni and Ru catalysts.

The control of pH is an issue for hydrogenation and oxidation processes, particularly for oxygenation of sugars to sugar acids. During the oxidation process, which can be carried out in continuous Pd monolith reactors [37], the pH of the solution drops because of the sugar acid formed. Increased acidity leads to suppressed catalyst activity. The remedy is a controlled neutralization of the solution during the progress of the reaction [36].

The role of support is of crucial importance for some of the transformations. A typical example is hydrogenolysis, where an acidity function is needed to accomplish the reaction. Active carbon has proven to be an excellent support material for cases where Pd and Ru catalysts are used (Table 8.1). The problem with active carbon is that it originates from different sources, with different functional groups attached to the surface. This implies that the catalyst does not have the best possible reproducibility. Another drawback with active carbon is its rather poor mechanical strength. This has led researchers to investigate alternative support materials, for example, instead of Pd/C catalysts Pd on zeolite supports can be used in the hydrogenation of the double bond. The recent development of synthetic carbon materials might help to solve the problem. Catalyst particles of synthetic carbon, for example, Sibunit, have a good mechanical strength and can be used in fixed beds. Furthermore, it is possible to impregnate the outer layer of the carbon particle only with the active metal and in this way create an egg-shell catalyst, which is suitable for the treatment of large organic molecules with a substantial diffusion resistance in the porous catalyst structure. Carbon nanofibers with a reproducible structure can be produced by pyrolysis of a gaseous aromatic component on a metal surface. This kind of nanofiber can be impregnated with metals, for example, to prepare a Pd catalyst for hydrogenation of double bonds or hydrogenolysis of hydroxyl groups or a Ru catalyst for hydrogenation of carbonyl groups.

Catalyst deactivation is a big problem in the treatment of molecules originating from biomass. Leaching, fouling, poisoning, and sintering are in general the typical reasons for catalyst deactivation. Catalysts, which exhibit leaching of metals are, of course, excluded from the very beginning, since many applications concern food, health-promoting chemicals, and pharmaceuticals. Sintering is not a typical phenomenon for catalysts used in biomass transformation, since the reaction temperatures are moderate. The organic molecules incorporated in the transformations have a tendency to deposit on the surface and initiate coking even at rather low reaction temperatures ($<100\,^{\circ}C$). The purity of the raw material is a crucial issue. If the reactant molecules originate from the black liqueur, they carry sulfur and sulfurous compounds as impurities, which have a deteriorating effect on the catalyst, poisoning it irreversibly. Thus a precleaning of the feed is very important. It is applied, for instance, to the hydrogenation of sitosterol to sitostanol. The sitosterol–sitostanol mixture is cleaned via adsorption on active carbon before entering the hydrogenation reactor.

An interesting way to retard catalyst deactivation is to expose the reaction mixture to ultrasound. Ultrasound treatment of the mixture creates local hot spots, which lead to the formation of cavitation bubbles. These cavitation bubbles bombard the solid, dirty surface leading to the removal of carbonaceous deposits [38]. The ultrasound source can be inside the reactor vessel (ultrasound stick) or ultrasound generators can be placed in contact with the wall of the reactor. Both designs work in practice, and the catalyst lifetime can be essentially prolonged, leading to process intensification. The effects of ultrasound are discussed in detail in a review article [39].

8.3 Chemical Reaction Engineering in Biomass Transformation

A rational scale-up of a production process should be based on the correct stoichiometric and kinetic model, as discussed by many authors, for example [40, 41]. Moreover, kinetics is exciting per se, since it yields valuable information about the underlying reaction mechanisms [42]. Thus it is very important that the kinetic information is not obscured by catalyst deactivation and mass transfer effects. To obtain correct stoichiometric information is a challenge in itself in the case of transformation of organic molecules from biomass, since by-products typically appear in low concentrations (around 1 wt%) in a concentrated environment (the concentrations of reactant and main product being typically 20–60 wt%). Thus very precise high-pressure liquid chromatography (HPLC) or sophisticated gas chromatography coupled to mass spectrometry (GC–MS) is typically used, and the analysis work is completed with NMR analysis. The use of GC methods for many chemical species of biological origin requires tedious and costly preprocessing of samples (silylation) in order to transfer them into a volatile form. Any attempts to boil and vaporize, for example carbohydrates (a prerequisite of GC analysis), inevitably leads to caramelization reactions irreversibly destroying the samples. Therefore,

HPLC has become a very dominant analysis method for monosaccharides and their derivatives.

Most of the actual reactions involve a three-phase process: gas, liquid, and solid catalysts are present. Internal and external mass transfer limitations in porous catalyst layers play a central role in three-phase processes. The governing phenomena are well known since the days of Thiele [43] and Frank-Kamenetskii [44], but transport phenomena coupled to chemical reactions are not frequently used for complex organic systems, but simple – often too simple – tests based on the use of first-order Thiele modulus and Biot number are used. Instead, complete numerical simulations are preferable to reveal the role of mass and heat transfer at the phase boundaries and inside the porous catalyst particles.

Thus it is important to obtain reliable models for catalyst deactivation and to investigate, whether it is possible to decouple the deactivation model from the kinetic model or if it is necessary to treat the catalyst deactivation as one of the surface reactions on the catalyst [45].

The reactions are still most often carried out in batch and semi-batch reactors, which implies that time-dependent, dynamic models are required to obtain a realistic description of the process. Diffusion and reaction in porous catalyst layers play a central role. The ultimate goal of the modeling based on the principles of chemical reaction engineering is the intensification of the process by maximizing the yields and selectivities of the desired products and optimizing the conditions for mass transfer.

8.3.1
Reaction and Mass Transfer in Porous Catalyst Structures

Reaction, diffusion, and catalyst deactivation in a porous catalyst layer are considered. A general model for mass transfer and reaction in a porous particle with an arbitrary geometry can be written as follows:

$$\frac{dc_i}{dt} = \varepsilon_p^{-1}\left(r_i\rho_p f(r) - r^s\frac{d(N_i r^s)}{dr}\right) \tag{8.1}$$

The component generation rate (r_i) is calculated from the reaction stoichiometry,

$$r_i = \sum_j \nu_{ij} R_j a_j \tag{8.2}$$

where R_j is the initial rate of reaction j and a_j is the corresponding activity factor; $f(r)$ is the radial distribution function of the active sites: $f(r) = 1.0$ for conventional catalyst pellets but an increasing function for egg-shell catalysts. The form of the function can be determined experimentally with inductive current plasma (ICP) analysis and various mathematical functions can be fitted to the data, for instance $f = f_0(r/R)^\alpha$.

For the catalyst activity factor (a_j), several models have been proposed, depending on the origin of catalyst deactivation, that is, sintering, fouling, or poisoning. The following differential equation can semiempirically represent different kinds of separable deactivation functions:

$$\frac{da_j}{dt} = -k_j'\left(a_j - a_j^*\right)^n \tag{8.3}$$

where k_j' is the deactivation parameter and a_j is the asymptotic value of the activity factor – for irreversible deactivation, $a_j^* = 0$. Depending on the value of the exponent (n), the solution of Eq. (8.3) becomes [46]

$$a_j = a_j^* + \left(a_{0j} - a_j^*\right)e^{-k_j t} \tag{8.4a}$$

and

$$a_j = a_j^* + \left((a_0 - a_*)^{1-n} + k_j(n-1)t\right)^{\frac{1}{1-n}}, n \neq 1 \tag{8.4b}$$

Some special cases of Equation (8.4) are of interest: for irreversible first ($n = 1$) and second ($n = 2$) order deactivation kinetics we get $a_j = a_{0j}e^{-k_j t}$ and $a_j = a_{0j}/\left(1 + a_{0j}k_j^t\right)$, respectively.

In the case that the effective diffusion coefficient approach is used for the molar flux, it is given by $N_i = -D_{ei}(dc_i/dr)$, where $D_{ei} = (\varepsilon_p/\tau_p)D_{mi}$ according to the random pore model. Standard boundary conditions are applied to solve the particle model Eq. (8.1).

The molecular diffusion coefficients in liquid phase can be estimated from the correlations of Wilke and Chang [47] for organic solutions and Hayduk and Minhas [48] for aqueous solutions, respectively. An extensive comparison of the available correlations is provided by Wild and Charpentier [49].

8.3.2 Rate Equations

The exact formulation of rate equations depends heavily on the particular chemical case of interest. Some common features in the biomass transformation reactions are of notable interest. Typically, the process involves a reaction between a large organic molecule and a much smaller reagent molecule (e.g., oxygen or hydrogen). In this case, the overall rate is often controlled by the surface reaction step, whereas the adsorption and desorption steps are assumed to be rapid. Further assumptions are needed concerning the details of the reactant adsorption. Both competitive and noncompetitive adsorption models have been proposed and used to describe the kinetics. A noncompetitive adsorption model can be justified by the size difference of the reacting molecules. A general overview of kinetic models is provided in [40, 42], where also a semi-competitive adsorption model is discussed. It has been successfully applied to sugar hydrogenation [22]. In most cases, rate equations of the type

$$R_j = \frac{k_j\left(\prod c_i^{\alpha i} - \prod c_i^{\beta i}/K_j\right)}{\left(\sum K_k c_k^{\gamma k} + 1\right)\left(\sum K_l c_l^{\delta ll} c_l + 1\right)} \tag{8.5}$$

are sufficient in kinetic modeling of biomass transformations. Equation (8.5) comprises both competitive and noncompetitive adsorption. The recent progress in quantum chemistry has enabled the calculation of adsorption states of complex organic molecules, which gives important mechanistic aspects to the derivation of rate equations.

8.3.3 Reactor Models

Most of the transformations of molecules originating from biomass are still carried out in (semi)batch tank reactors; therefore continuous reactors are discarded here. For a semi-batch stirred tank reactor, the model can be based on the following assumptions: the reactor is well agitated and therefore, no concentration differences appear in the bulk of the liquid; gas–liquid and liquid–solid mass transfer resistances can prevail; and finally, the liquid phase is in batch, while the gas-phase component (typically hydrogen or oxygen) is continuously fed into the reactor. The gas-phase pressure is maintained constant. The liquid and gas volumes inside the reactor vessel can in many cases be regarded as constant, since the changes of the fluid properties due to reaction are minor. The total pressure of the gas phase as well as the reactor temperature are continuously monitored and stored on a PC. The volatilities of the reactive organic components should be checked and their contribution terms should be included in the model equations, if the amounts are considerable. It is important to emphasize that species present in even rather moderate amounts can sometimes provide important insights into the actual reaction mechanisms, for example, due to the complex nature and still incompletely understood carbohydrate transformation chemistry in aqueous systems.

A mass balance for an arbitrary liquid-phase component in the stirred tank reactor is thus written as follows:

$$\frac{dc_i}{dt} = N_i a_p - N_{\mathrm{GL}i} a_{\mathrm{GL}} \tag{8.6}$$

The initial condition is $c_i = c_i(0)$ at $t = 0$. The flux at the particle surface is obtained from the solution of Eq. (8.1) and the gas–liquid flux can be estimated from the film theory.

8.3.4 Model Simulation and Parameter Estimation Techniques

The partial differential equations describing the catalyst particle are discretized with central finite difference formulae with respect to the spatial coordinate [50]. Typically, around 10–20 discretization points are enough for the particle. The ordinary differential equations (ODEs) created are solved with respect to time together with the ODEs of the bulk phase. Since the system is stiff, the computer code of Hindmarsh [51] is used as the ODE solver. In general, the simulations progressed without numerical problems. The final values of the rate constants, along with their temperature dependencies, can be obtained with nonlinear regression analysis. The differential equations were solved *in situ* with the backward difference method implemented in ModEst. The kinetic parameters were estimated by using a combined Simplex–Levenberg–Marquardt method [52, 53], which minimizes the residual sum of squares between the estimated and the experimental concentrations with nonlinear regression.

Here we consider three important categories of reactions that appear in the treatment of biomass: hydrolysis of polysaccharides (homogeneous catalysis) as well as hydrogenation of carbonyl groups (monosaccharides; heterogeneous catalysis) and hydrogenation of double bonds (triterpenoids; heterogeneous catalysis). Hydrolysis of polysaccharides is a crucially important step, whatever source is used (plants, trees, wastes of biomass), since it provides the basic sugars. Hydrogenation of the basic sugars gives polyalcohols, which are platform molecules for a biorefinery [4]. Treatment of double bonds is of importance in the production of fine chemicals and in the preparation of biofuels.

8.3.5
Hydrolysis of Polysaccharides – from Arabinogalactan to Monomers

In the acid hydrolysis of arabinogalactane (AG), the two main monosaccharides, arabinose and galactose, were obtained (Figure 8.2) at temperatures up to 100 °C, pH $= 1, 2$, and 3, and the initial AG concentration $c_{AG} = 0.5, 2.5$, and 5.0 mass %. Typical kinetic curves of the AG hydrolysis and the released monomers are shown in Figure 8.3. A complete conversion of AG was achieved after 1400 min. It can be noted from the results that arabinose was released totally within 120 min, whereas there was a continuous gradual increase of the galactose concentration during 1400 min. As can be seen, the monomers, arabinose and galactose, were stable at the above-mentioned conditions. There was a sharp decrease in AG molar weight within the first 120 min, which is explained by the presence of the two simultaneous reactions until complete release of arabinose followed by the gradual hydrolysis of galactose residue.

The results of AG hydrolysis at 90 °C and different pH values are shown in Figure 8.4. The residual AG concentrations were plotted against the reaction times for the different reaction conditions. In each of the experiments, the pH value of the solution was adjusted to 1, 2, and 3, respectively. The experimental results show a very strong dependence of the hydrolysis rates on the acid concentration (pH). A complete conversion of AG was achieved at pH 1 after 1400 min with

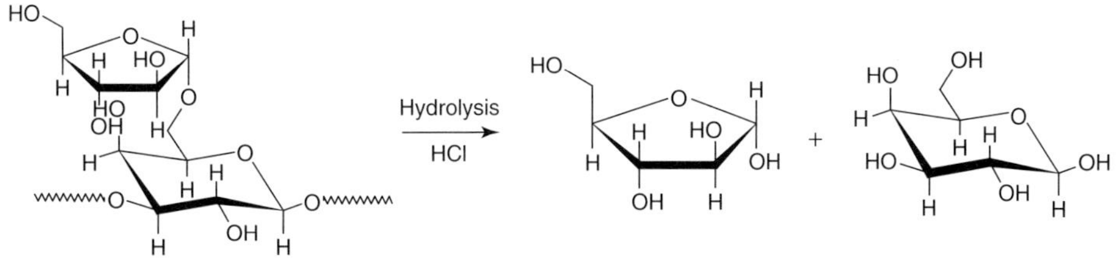

Figure 8.2 Acid-catalyzed hydrolysis of a hemicellulose (arabinogalactane).

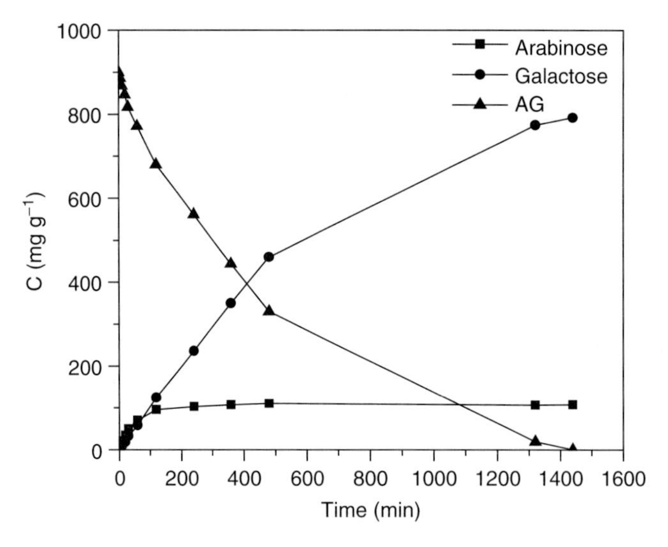

Figure 8.3 Typical kinetic curves of AG hydrolysis to arabinose and galactose at 90 °C and pH 1 [20].

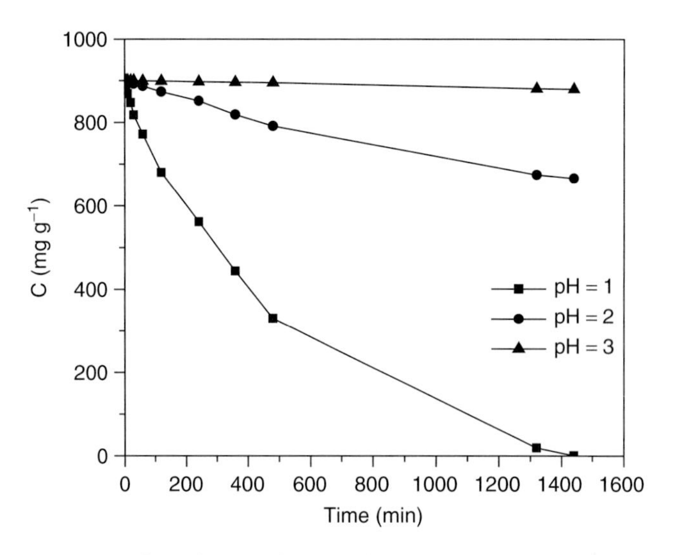

Figure 8.4 Effect of pH on the AG hydrolysis [20].

only 25% and less than 1% conversion at pH 2 and 3, respectively. At pH 1, AG was completely hydrolyzed to monomers achieving the maximum quantities of free sugar units from the polysaccharide. The slowest hydrolysis rates of AG with hardly any galactose sugar units being cleaved from the chain were observed at pH 3. Under these conditions, the obtained results suggest that the lowest acid concentration was not sufficient to hydrolyze the glucosidic bonds of AG. Thus we conclude here that it is easy to hydrolyze the arabinose unit from the side chains of AG followed by the release of galactose from the main chain.

The reaction rate increased with temperature. The hydrolysis rate of AG at temperatures lower than 70 °C was very slow. At 80 °C, only complete release of arabinose was achieved, but partially hydrolyzed galactose residue was left. The conversion of AG was 43%. A complete conversion of AG to monomers was achieved at 90 and 100 °C. After the hydrolysis at 100 °C, traces of degradation products such as furfural were observed. For this reason, the temperature for AG hydrolysis shall not exceed 100 °C.

A kinetic model based on the experimental results was developed. The hemicellulose structure was considered to be composed of two distinct fractions, one that is relatively easy to hydrolyze and the other more difficult [21]. The arabinose residue, because of its easier accessibility, might release faster, while the galactose units would be produced more slowly. The kinetic model was therefore set into the reaction scheme described. Applying the first-order kinetics with respect to the functional groups for both reactions, the following rate equations can be obtained: $r_1 = k_1 c_H c_{A,AG}$, $r_2 = k_2 c_H c_{G,AG}$, where r_1 and r_2 are the rates of AG hydrolysis to arabinose and galactose respectively, k_1 and k_2 are the kinetic rate constants, $c_H = [H^+]$, the acid concentration, and $c_{A,AG}$ and $c_{G,AG}$ denote the arabinose and galactose fractions in the polysaccharide, that is, easy to hydrolyze (arabinose) and more difficult to hydrolyze (galactose) hemicellulose fractions, respectively. The temperature dependence of the acid hydrolysis was modeled using the Arrhenius equation.

The constant-volume batch reactor model can be applied to the mass balances of the functional groups. The concentrations of arabinose (A) and galactose (G) units liberated in the hydrolysis are thus obtained from the differential equations $dc_A/dt = k_1 c_H c_{A,AG}$, $dc_G/dt = k_2 c_H c_{G,AG}$, where $c_{A,AG}$ and $c_{G,AG}$ give the concentrations of the arabinose and galactose units in AG. The following total balance is valid for the monomer units $c_A + c_{A,AG} = c_{0A,AG}$ and $c_G + c_{G,AG} = c_{0G,AG}$, which implies that the total amounts of the arabinose and galactose units remain constant. This assumption is justified, since the amounts of the degradation products were negligible. Finally the expressions $dc_A/dt = k_1 c_H (c_{0A,AG} - c_A)$, $dc_G/dt = k_2 c_H (c_{0G,AG} - c_G)$ are obtained.

The initial amounts correspond to the final amounts of arabinose and galactose in the case of complete hydrolysis, that is, $c_{0A,AG} = c_{\infty A}$ and $c_{0G,AG} = c_{\infty G}$. These relations are inserted in the mass balances, which are integrated with the initial condition $t = 0$, $c_A = 0$, and $c_G = 0$. The logarithmic functions are obtained: $-\ln(1 - c_A/c_{\infty A}) = k_1 c_H t$, and $-\ln(1 - c_G/c_{\infty G}) = k_2 c_H t$. The validity of the proposed model can be checked by plotting the right-hand sides of the equations versus the reaction time. Furthermore, division of the above equations gives a double logarithmic plot $\ln(1 - c_A/c_{\infty A})$ versus $\ln(1 - c_G/c_{\infty G})$, which should become a straight line, if the model is valid. The slope gives the ratio between the rate constants k_1/k_2 which is independent of the acid catalyst concentration (c_H). The plots obtained with different acid catalyst concentrations coincide as predicted by the model. The test plots gave preliminary values for the rate constants.

The final values of the rate constants along with their temperature dependencies were obtained with nonlinear regression analysis, which was applied to the differential equations. The model fits the experimental results well, having an explanation factor of 98%. Examples of the model fit are provided by Figures 8.3 and 8.4. An analogous treatment can be applied to other hemicelluloses.

8.3.6
Hydrogenation of a Carbonyl Group – from Sugars to Sugar Alcohols

Hydrogenation of the carbonyl group of a mono- or disaccharide gives the corresponding sugar alcohol. On paper, this reaction is simple, and it can schematically be written as

$$\mathrm{R{-}\overset{\overset{\displaystyle O}{\|}}{C}{-}H} + \mathrm{H_2} \longrightarrow \mathrm{R{-}\underset{\underset{\displaystyle H}{|}}{\overset{\overset{\displaystyle OH}{|}}{C}}{-}H} \tag{8.7}$$

However, the pattern is complicated by several factors. The sugar molecules to be hydrogenated mutarotate in aqueous solutions thus coexisting as acyclic aldehydes and ketoses and as cyclic pyranoses and furanoses and reaction kinetics are complicated and involve side reactions, such as isomerization, hydrolysis, and oxidative dehydrogenation reactions. Moreover, catalysts deactivate and external and internal mass transfer limitations interfere with the kinetics, particularly under industrial circumstances.

Catalytic hydrogenation is typically carried out in slurry reactors, where finely dispersed catalyst particles ($<100\,\mu m$) are immersed in a dispersion of gas and liquid. It has, however, been demonstrated that continuous operation is possible, either by using trickle bed [24] or monolith technologies [37]. Elevated pressures and temperatures are needed to have a high enough reaction rate. On the other hand, too high a temperature impairs the selectivity of the desired product, as has been demonstrated by Kuusisto *et al.* [23]. An overview of some feasible processes and catalysts is shown in Table 8.1.

Kinetic experiments were carried out isothermally in autoclave reactors of sizes 300 and 600 ml. The stirring rate was typically 1800 rpm. In most cases, the reactors were operated as slurry reactors with small catalyst particles (45–90 μm), but comparative experiments were carried out with a static basket using large catalyst pellets. HPLC analysis was applied for product analysis [22, 23].

Catalytic hydrogenation of various sugars, such as lactose, maltose, arabinose, galactose, xylose, and mannose were studied experimentally in a pressurized laboratory-scale batch reactor. Ruthenium on active carbon was used as a catalyst. The main product was always the corresponding sugar alcohol, the selectivity of which typically exceeded 95%. Examples of hydrogenation of arabinose and galactose are shown in Figures 8.5 and 8.6. On the basis of the experimental data, models were developed for simultaneous reaction, diffusion, and catalyst deactivation. Examples of the complete reaction schemes, which incorporate simultaneous, consecutive-competitive reactions, are displayed in Figure 8.7.

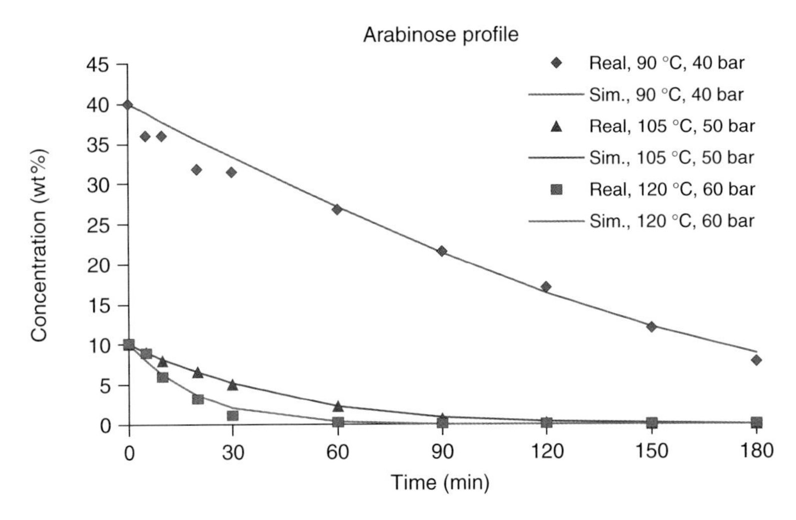

Figure 8.5 Hydrogenation of arabinose to arabinitol on Ru/C [26].

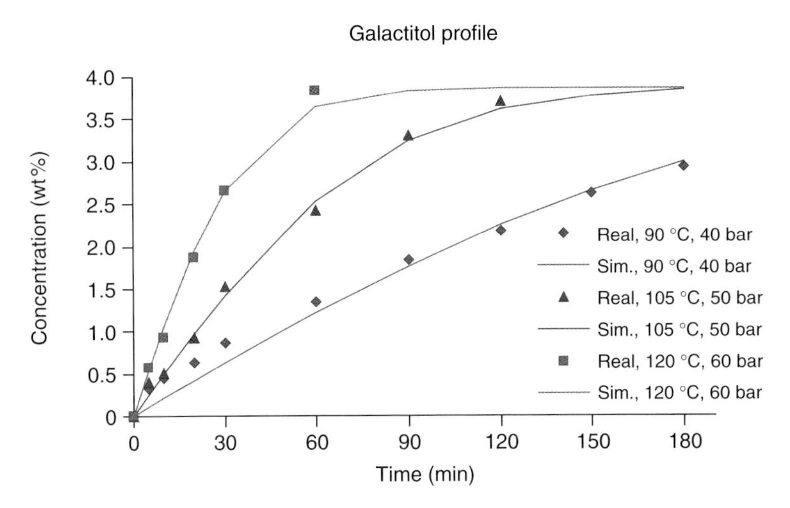

Figure 8.6 Hydrogenation of galactose on Ru/C – concentrations of the main product, galactitol [26].

Preliminary product distribution plots were prepared on the basis of experimental data representing intrinsic kinetics. The hydrogenation of arabinose and galactose over Ru/C was straightforward: arabinitol and galactitol were the absolutely dominating products with the selectivity close to 100%. The kinetics was described with a Langmuir–Hinshelwood model including rapid competitive reactant adsorption and rate control by the surface reaction between the adsorbed sugar molecule and hydrogen. The kinetic parameters were determined with regression analysis.

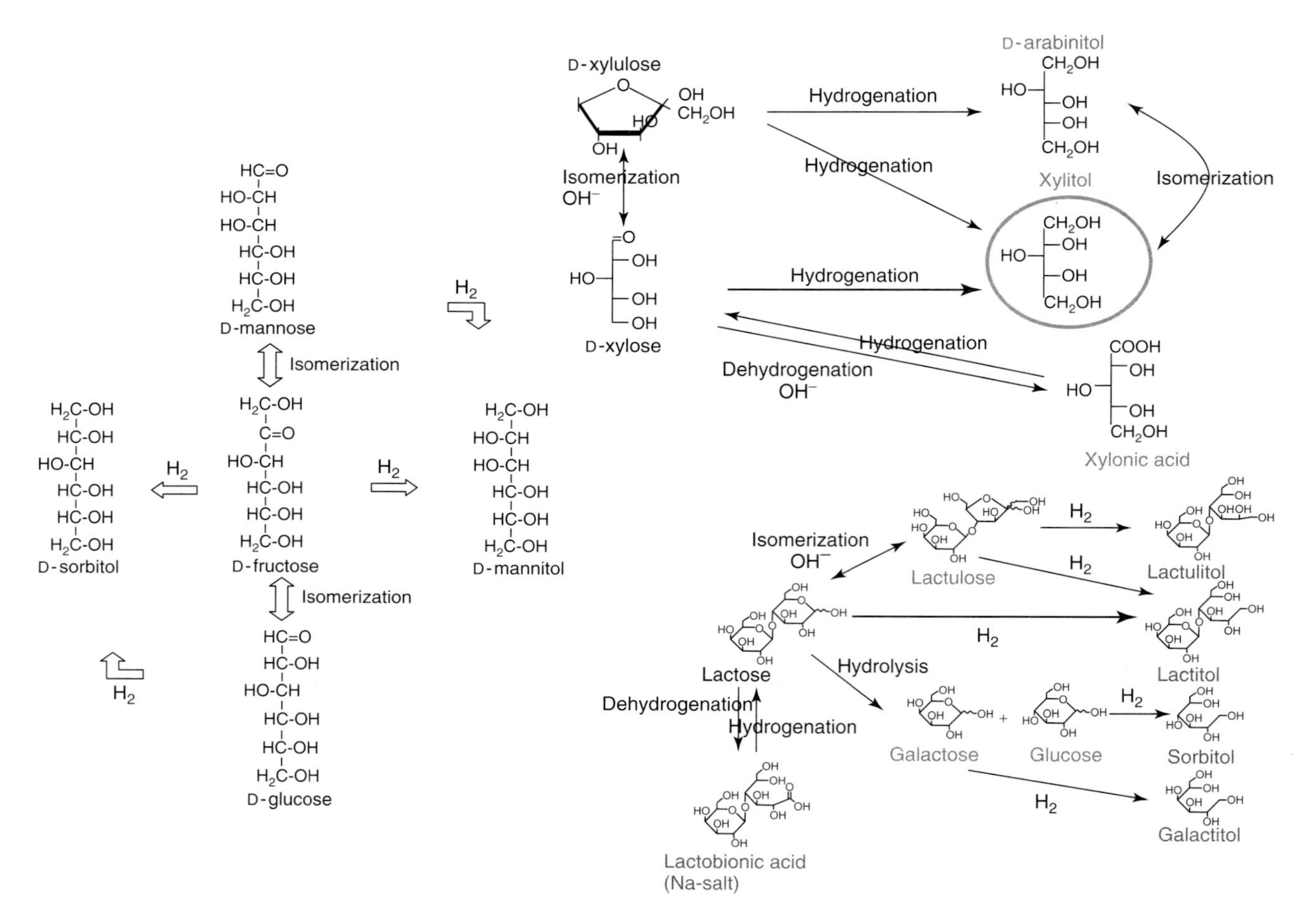

Figure 8.7 Reaction schemes for hydrogenation of D-fructose, D-xylose, and D-lactose.

Examples of the fit of the model are provided by Figures 8.5 and 8.6 [26]. The model completely describes the progress of the reaction.

The case of lactose hydrogenation is more complicated, since it is a disaccharide. The following facts were revealed: the dominant product on Ni and Ru catalysts is lactitol, while some lactulose and lactitulitol are formed through a consecutive route. In addition, formation of sorbitol and galactitol was noticed. They can principally originate either from a hydrolysis-hydrogenation route of lactose (lactose gives glucose and galactose which are hydrogenated to sorbitol and galactitol) or from hydrolysis of lactitol. After a very precise HPLC analysis, lactobionic acid was detected, but it was partially hydrogenated away. For xylose hydrogenation on Ni and Ru, the picture was simpler: the main product was xylitol, but because of the lack of hydrogen, considerable amounts of the isomerization product, xylulose, were formed, which was further hydrogenated to arabinitol and xylitol. For fructose hydrogenation on Cu-based catalysts, a parallel reaction scheme was evident: both sorbitol and mannitol were formed, the selectivity remaining constant during the reaction.

Preliminary kinetic analysis revealed that the reactions mentioned for various sugars were close to first order with respect to the organic reactant, while the reaction order with respect to hydrogen varied between 0.5 and 2.2, being 0.7 for hydrogenation of lactose on sponge nickel and about 2 for fructose hydrogenation on CuO/ZnO.

Rate expressions based on the principle of an ideal surface, rapid adsorption and desorption, but with rate-limiting hydrogenation steps were derived. The competitiveness of adsorption and the state of hydrogen on the catalyst surface has been the subject of intensive discussions in the literature (see e.g., [40]), but in this case we adopted the simplest possible model, noncompetitive adsorption of hydrogen and organics along with molecular adsorption of hydrogen. Previous comparisons [40] have shown that this approach gives an adequate description of the kinetics, with a minimum number of adjustable parameters. A thorough comparison of experimental data with modeling results indicated that the model for intrinsic kinetics is sufficient and can be used for simulation of the reaction–diffusion phenomena in larger catalyst particles.

Some simulation results for large catalyst particles by using the model for porous catalyst layer, Eqs. (1–6) are provided by Figure 8.8. As the figure reveals, the process becomes easily diffusion-limited, not only by hydrogen diffusion but also by that of the organic educts and products. For particles with diameters more than 0.03 mm, diffusion resistance is visible. This is valid not only for the main components, such as lactose and lactitol, but also for by-products, such as lactulose and lactulitol. The effect of deactivation is illustrated in the lower part of Figure 8.8, where the concentration front moves toward the center of the particle, since the outer layer of the particle deactivates. As the reaction progresses to high conversions, the role of diffusion resistance diminishes, because all of the reaction rates become low.

For "small" catalyst particles used in sugar hydrogenation (slurry reactors), one would intuitively conclude that the system is safely within the kinetic regime.

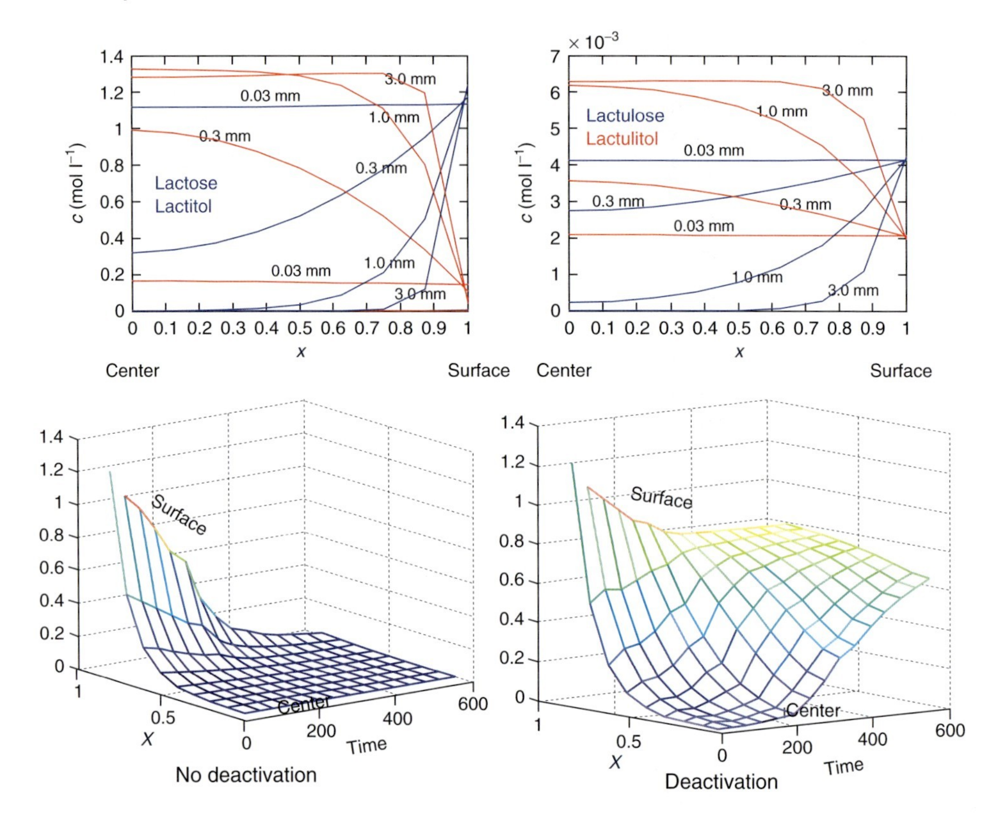

Figure 8.8 Internal mass transfer resistance and catalyst deactivation: concentration profiles inside a catalyst particle – lactose hydrogenation to lactitol and by-products (sponge Ni).

The simulation results obtained for sugar hydrogenation strongly suggest that this is not the case. The organic molecules are large, having molecular diffusion coefficients of the magnitude 4.0×10^{-10} m^2 s^{-1}. The porosity-to-tortuosity factor can be rather small ($\ll 0.5$); thus the effective diffusion coefficient becomes small, and internal diffusion resistance becomes an important factor even for thin catalyst layers.

In addition, external diffusion resistance is often a crucial factor in sugar hydrogenation: due to inappropriate stirrer design and high viscosity, the risk for lack of hydrogen in the liquid phase is high. A simulation of the hydrogen content in the liquid phase indicated that a complete saturation is achieved at high conversions, while the hydrogen concentration is below saturation at the beginning of the experiment. By increasing the stirring efficiency, the k_L-values can be improved and the role of external mass transfer resistance is suppressed. The hydrogen concentration in the liquid phase plays a crucially important role for the production: in the case of external mass transfer limitation of hydrogen,

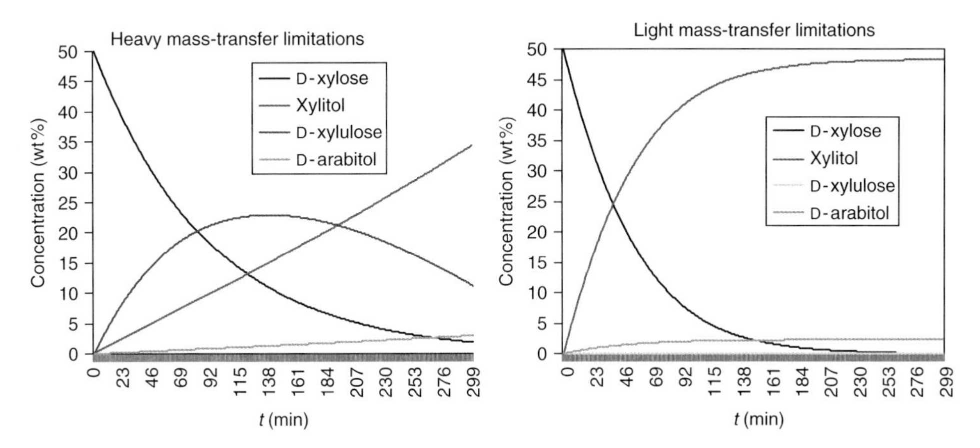

Figure 8.9 External mass transfer resistance – xylose hydrogenation and isomerization to xylitol and by-products on sponge Ni (based on the results of Mikkola *et al.* [22]).

the isomerization reaction yielding xylulose, which does not need any hydrogen, is favored (Figure 8.9).

8.3.7 Hydrogenation of Double Bond – from Sitosterol to Sitostanol

Raw materials originating from natural sources often contain impurities, which tend to adsorb irreversibly on the catalyst thus occupying active sites by poisoning. This is the case, for instance, in the preparation of the cholesterol-suppressing agent sitostanol through catalytic hydrogenation of sitosterol. Small amounts of campesterol are usually present as coreactants in the system. The double bond of the sterol molecule is hydrogenated to get the corresponding stanol. Some side reactions can also be present yielding sitostane and campestane and plant reactor data are influenced by external mass transfer resistance as discussed by Wärnå *et al.* [54]. A simplified reaction scheme is displayed below in Figure 8.10.

Palladium is known to be a metal that works catalytically in the system. Various supports can be used for Pd, such as active carbon, mesoporous materials, and polymers. All of them deactivate in the sitosterol hydrogenation, most probably because of sulfur and phosphorus impurities present in the raw material, which originates from the tall oil production, a side process of chemical pulping.

The sitosterol hydrogenation and deactivation kinetics was determined in a shaking constant-pressure batch reactor by using the new type of synthetic support material (mesoporous carbon Sibunit) for palladium (4 wt% Pd) [55].

Preliminary kinetic studies revealed that the catalyst deactivated; it was difficult to achieve a complete conversion of the substrate (sitosterol) in a batch experiment, but the conversion approached a limiting value, depending on the amount of catalyst added. By the addition of higher amounts of the catalyst, higher conversions were obtained; even so the standard plots of the substrate conversion versus the mass

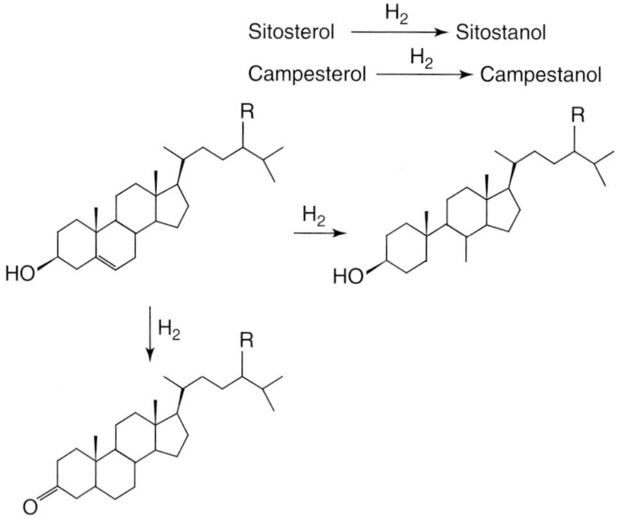

Sitosterol $\xrightarrow{H_2}$ Sitostanol

Campesterol $\xrightarrow{H_2}$ Campestanol

Figure 8.10 Reaction scheme in the hydrogenation of sitosterol to sitostanol.

of catalyst × time/liquid volume did not show the behavior expected for the mass transfer resistance, but an increasing efficiency was observed with an increasing amount of the catalyst. This was taken as a clear evidence for the fact that the increasing amount of the supported catalyst was able to adsorb the poison and still leave vacant sites for the hydrogenation reaction.

A mathematical model was developed, starting from mechanistic aspects, to describe the adsorption, desorption, and surface reaction steps on the catalyst. The empirical approach to the catalyst activity description was totally abandoned, and the deactivation was treated as an irreversible adsorption step on the surface. The rate model was coupled to a model for a well-mixed batch reactor. A full numerical solution of the model is in principle possible, but in order to construct a more robust algorithm for the parameter estimation, semianalytical solutions were developed for the surface species and they were incorporated in the mass balances of the bulk components (sitosterol, campesterol, and the hydrogenation products). Sitosterol was assumed to be the most abundant surface species among the organic molecules, and the adsorption of the organic molecules and hydrogen was assumed to be noncompetitive due to the large size difference of hydrogen and the organic molecules. The reaction mechanism presumed can be summarized as follows (A = sitosterol, R = product, P = poison):

$$
\begin{aligned}
&A + {}^{*} \leftrightarrow A^{*} \\
&A^{*} + H_2{}^{*\prime} \rightarrow R + {}^{*} + {}^{*\prime} \\
&P + {}^{*} \rightarrow P^{*}
\end{aligned}
\tag{8.8}
$$

The rates of the elementary steps can be formulated in a conventional manner, and the quasi-steady state hypothesis is applied to the adsorbed substrate (A^{*}). The mass balances of the surface intermediates are given by $dc^*{}_j/dt = r_j$ and, for the organic bulk-phase species in the batch reactor, $dc_i/dt = \rho_B r_i$ (kinetic regime).

A detailed mathematical treatment gave the following models for the main substrate (A) and product (R):

$$\begin{aligned} dc_A/dt &= -dc_R/dt \\ &= -[\rho_B\, k' c_H c_A/(1 + K_H c_H)/(1 + K_A c_A)] \\ &\quad \times \exp\left(-\int k_d c_{OP}(c_A/c_{0A})\delta t/(1 + K_A c_A)\right) \end{aligned} \tag{8.9}$$

The ODE-model was solved numerically during the estimation of the kinetic, adsorption, and deactivation parameters.

At a later stage, the basic model was extended to comprise several organic substrates. An example of the data fitting is provided by Figure 8.11, which shows a very good description of the data. The parameter estimation statistics (errors of the parameters and correlations of the parameters) were on an acceptable level. The model gave a logical description of all the experimentally recorded phenomena.

It should be emphasized that the approach proposed is by no means limited to the actual case, hydrogenation of sitosterol, but is a general one for the reaction scheme, which is very common in catalyst poisoning. The methodology with semianalytical solutions of the surface species turned out to be very robust in the

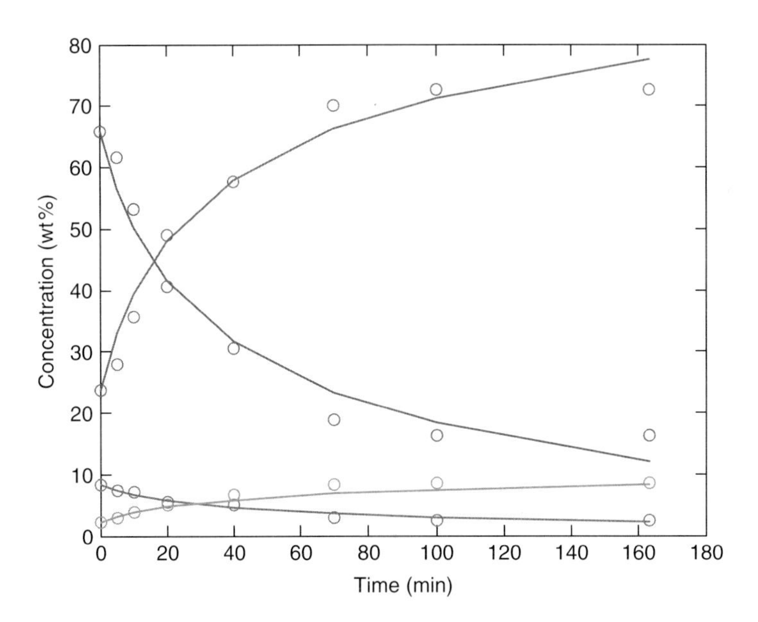

Figure 8.11 Example of data fitting results – concentrations of sitosterol, campesterol, and the products in the hydrogenation of sitosterol to sitostanol (increasing curve) on Pd/Sibunit. The lower curves represent campesterol and campestanol.

parameter estimation, which suggests that this approach should be used for plant data, too.

8.4 Conclusions and Future Perspectives

Catalytic treatment of chemicals from biomass provides many challenges and opportunities. Important platform molecules, such as various sugars and sugar alcohols, can be obtained from biomass, such as cellulose, hemicelluloses, and starch. These platform molecules can be hydrogenated, oxidized, and isomerized to valuable products, to fine and specialty chemicals, health-promoting components, and food additives. Sugar alcohols, the hydrogenation products of sugars, can be exposed to aqueous reforming to produce liquid fuel components. Sugars can be fermented to bioethanol, which has many potential applications, from fuels to the production of polymers (polyethene). Extractives from wood can be used as solvents, as fragrances, and as various specialty chemicals. Fatty and resin acids are – after some treatment – excellent components in biodiesel.

In all of the transformations indicated, catalysts, particularly heterogeneous catalysts, play the key role. It has turned out that few precious metals (Ru, Pd, Pt, and Au) can do most of the transformations needed, but the development of catalyst supports and structures is a challenge. In future, the discontinuous slurry reactor technology should be replaced by continuous reactor technology, either in the form of fixed beds with egg-shell catalysts or structured reactors, such as catalyst monoliths, foams, and fibers. These structures enable suppressing the internal diffusion resistance which is inevitable in the treatment of large organic molecules originating from biomass. In the design of future biorefineries, it is better to start from the biomass as such – from the macromolecules – and search for technologically and economically feasible solutions instead of trying desperately to recover components from existing chemical pulping processes.

Acknowledgment

This work is part of the activities at the Åbo Akademi Process Chemistry Centre within the Finnish Centre of Excellence Programmes (2000–2011) by the Academy of Finland. T. Salmi is grateful for the Academy Professor grant by the Academy of Finland. The authors are grateful to the current and former students, who have performed the experimental work: Andreas Bernas, Heidi Bernas, Manuel Flores Geant, Alexey Kirilin, Bright Kusema, Jyrki Kuusisto, Krista Laaksonen, Gerson Martin Rivero, Eloisa Mateos Sevilla, Elena Murzina, Oluwamuyiwa Olatunde Oladele, Pablo Perez Casal, Victor Sifontes Herrera, Bartosz Rosmyszlowicz, Irina Simakova, Olga Simakova, Mathias Snåre, Mona Sparv, Anton Tokarev, Blanka Toukoniitty, Esa Toukoniitty, and Hanna Vainio. The discussions with Professor Bjarne Holmbom have been a continuous source of information and inspiration for us.

List of Symbols

a catalyst activity factor
a_{P} catalyst particle mass transfer area to reactor volume relation
a_{GL} gas–liquid mass transfer area to reactor volume relation
c concentration
D diffusion coefficient
f metal distribution function in egg-shell catalysts
k rate constant
k' merged rate constant
K equilibrium constant
n reaction order
N flux (mol m^{-2} s)
r component reaction rate
r radius (m)
R reaction rate
s catalyst particle shape factor
t time (min)
δ parameter in catalyst deactivation model
ε_p porosity
ρ_{B} catalyst particle density
ρ_p catalyst particle density
ν stoichiometric coefficient, dimensionless

Subscripts and Superscripts

B bulk property
d deactivation
e effective property
G gas phase
i component index
j reaction index
L liquid phase
p catalyst particle property
* equilibrium conditions

References

1. Salmi, T., Mikkola, J.-P., and Wärnå, J. (2009) *Chemical Reaction Engineering and Reactor Technology*, Taylor & Francis, Boca Raton.
2. Cybulski, A. and Moulijn, J.A. (eds) (2006) *Structured Catalysts and Reactors*, Taylor & Francis, Boca Raton.
3. van Santen, R.A. (2007) Renewable catalytic technologies – a perspective, in *Catalysis for Renewables* (eds G. Centi and R.A. van Santen), Wiley-VCH Verlag GmbH, Weinheim.
4. Gallezot, P. (2007) Process options for the catalytic conversion of renewables

into bioproducts, in *Catalysis for Renewables* (eds G. Centi. and R.A. van Santen), Wiley-VCH Verlag GmbH, Weinheim.

5. Sjöström, E. (1989) *Puukemia – Teoreettiset Perusteet Ja Sovellutukset*, Otakustantamo, Espoo.
6. Allahverdiev, A.I., Gündüz, G., and Murzin, D.Yu. (1998) Kinetics of α-pinene isomerization. *Ind. Eng. Chem. Res.*, **37**, 2373–2377.
7. Vital, J., Ramos, A.M., Silva, I.F., and Castanheiro, J.E. (2001) The effect of α-terpienol On the hydration of α-pinene over zeolites dispersed in polymeric membranes. *Catal. Today*, **67**, 217–223.
8. Corma, C.A., Eduardo Domine, M., Susarte, R.M., and Rey, G.F. (2002) MCM-41 type microporous materials containing titanium and their utilization as catalysts in α- pinene oxidation, Patent WO0054880.
9. Michalson, E.T. and Devore, J.D. (2004) Process for catalytically hydrogenating Phytosterols. US 6673951.
10. Ekblom, J. (2005) Process for the preparation of stanol esters. US 6855837.
11. Frankel, E. (1970) Conversion of polyunsaturates in vegetable oils to cis Monounsaturates by homogeneous hydrogenation catalyzed with chromium carbonyls. *J. Am. Oil Chem. Soc.*, **47**, 11–14.
12. DeJarlais, W. and Gast, W.L. (1971) Conjugation of polyunsaturated fats: Methyl linoleate with tris(triphenylphosphine) chlororhodium. *J. Am. Oil Chem. Soc.*, **48**, 21–24.
13. Mukesh, D., Narasimham, S., Deshpande, V.M., and Ramnarayan, K. (1988) Isomerization of methyl linoleate on supported ruthenium-nickel catalyst. *Ind. Eng. Chem. Res.*, **27**, 409–414.
14. Bernas, A., Laukkanen, P., Kumar, N., Mäki-Arvela, P., Väyrynen, J., Laine, E., Holmbom, B., Salmi, T., and Murzin, D.Yu. (2002) A new heterogeneously catalytic pathway for isomerization of linoleic acid over Ru/C and Ni/H-MCM-41. *J. Catal.*, **210**, 354–366.
15. Snåre, M., Kubickova, I., Mäki-Arvela, P., Eränen, K., and Murzin, D.Yu. (2006) Heterogeneous catalytic deoxygenation of stearic acid for production of biodiesel. *Ind. Eng. Chem. Res.*, **45**, 5708–5715.
16. Abatzoglou, N. and Chornet, E. (1998) *Acid hydrolysis of hemicellulose and cellulose.* Theory and Applications, in Polysaccharides, Marcel Dekkes, New York, pp. 1007–1045.
17. Willför, S., Sjöholm, R., Laine, C., and Holmbom, B. (2002) Structural features of water-soluble arabinogalactans from Norway spruce and Scots pine heartwood. *J. Wood Sci. Technol.*, **36**, 101–110.
18. Willför, S. and Holmbom, B. (2004) Isolation and characterization of water-soluble polysaccharides from Norway spruce and Scots pine. *J. Wood Sci. Technol.*, **38**, 173–179.
19. Xu, C., Pranovich, A., Vähäsalo, L., Hemming, J., Holmbom, B., Schols, H.A., and Willför, S. (2008) Kinetics of acid hydrolysis of water-soluble spruce O-acetyl galactoglucomannans. *J. Agric. Food. Chem.*, **56**, 2429–2435.
20. Kusema, B.T., Mäki-Arvela, P., Xu, C., Willför, S., Holmbom, B., Salmi, T., and Murzin, D. (2010) Kinetics of acid hydrolysis of arabinogalactans. *Int. J. Chem. Reactor Eng.*, in press.
21. Maloney, M.T., Chapman, T.W., and Baker, A.J. (1986) An engineering analysis of the production of xylose by dilute acid hydrolysis of hardwood hemicellulose. *Biotechnol. Progr.*, **2**, 193.
22. Mikkola, J.P., Salmi, T., and Sjöholm, R. (1999) Modelling of kinetics and mass transfer in the hydrogenation of xylose over Raney nickel catalyst. *J. Chem. Technol. Biotechnol.*, **74**, 655–662.
23. Kuusisto, J., Mikkola, J.-P., Sparv, M., Heikkilä, H., Perälä, R., Väyrynen, J., and Salmi, T. (2006) Hydrogenation of lactose over sponge nickel catalysts – Kinetics and modeling. *Ind. Eng. Chem. Res.*, **45**, 5900–5910.
24. Gallezot, P., Nicolaus, N., Flèche, G., Fuertes, P., and Perrard, A. (1998) Glucose hydrogenation on ruthenium catalysts in a trickle-bed reactor. *J. Catal.*, **180**, 51–55.

25. Kuusisto, J., Tokarev, A.V., Murzina, E.V., Roslund, M.U., Mikkola, J.-P., Murzin, D., and Salmi, T. (2007) From renewable raw materials to high-value added fine chemicals – catalytic hydrogenation and oxidation of D-lactose. *Catal. Today*, **121**, 92–99.
26. Sifontes, V. (2008) Continuous reactor technology in the production of alternative sweetners, Annual Report, Graduate School in Chemical Engineering, Turku/Åbo, Finland, pp. 263–270.
27. Holmbom, B., Eckerman, C., Hemming, J., Reunanen, M., Sundberg, K., and Willför, S. (2002) A method for isolating phenolic substances or juvabiones from wood comprising knot-wood. WO 02/098830.
28. Holmbom, B., Eckerman, C., Eklund, P., Hemming, J., Nisula, L., Reunanen, M., Sjöholm, R., Sundberg, A., Sundberg, K., and Willför, S. (2004) Knots in trees – a new rich source of lignans. *Phytochem. Rev.*, **2**, 331–340.
29. Eklund, P., Lindholm, A., Mikkola, J.-P., Smeds, A., Lehtilä, R., and Sjöholm, R. (2003) Synthesis of (-)-matairesinol, (-)-enterolactone, and (-)-enterodiol from natural lignan hydroxy-matairesinol. *Org. Lett.*, **52**, 491–493.
30. Sjöholm, R., Eklund, P., Mikkola, J.-P., Lehtilä, R., Södervall, M., and Kalapudas, A. (2003) Preparation of matairesinol from hydroxymatairesinol, WO 03/057209.
31. Markus(Bernas), H., Mäki-Arvela, P., Kumar, N., Heikkilä, T., Lehto, V.P., Sjöholm, R., Holmbom, B., Salmi, T., and Murzin, D.Yu. (2006) Reactions of hydroxymatairesinol over supported palladium catalysts. *J. Catal.*, **238**, 301–308.
32. Blecker, C., Fougnies, C., van Herck, J.-C., Chevalier, J.-P., and Paquot, M. (2002) Kinetic study of the acid hydrolysis of various oligofructose samples. *J. Agric. Food. Chem.*, **50**, 1602–1607.
33. Mäki-Arvela, P., Holmbom, B., Salmi, T., and Murzin, D. (2007) Recent progress in synthesis of fine and specialty chemicals from wood and other biomass by heterogeneous catalytic processes. *Catal. Rev. Sci. Eng.*, **49**, 197–340.
34. Augustine, R.L. and Reardon, E.J. Jr. (1969) Palladium catalyzed hydrogenation of cholesterol. *Org. Prep. Proced.*, **1**, 107–109.
35. Wenkin, M., Touillaux, R., Ruiz, P., Delmon, B., and Devillers, M. (1996) Influence of metallic precursors on the properties of carbon-supported bismuth-promoted palladium catalysts for the selective oxidation of glucose to gluconic acid. *Appl. Catal., A*, **148**, 181–199.
36. Tokarev, A., Murzina, E.V., Mikkola, J.-P., Kuusisto, J., Kustov, L.M., and Murzin, D. (2007) Application of in-situ catalyst potential measurements for estimation of reaction performance: D-lactose oxidation over Au and Pd catalysts. *Chem. Eng. J.*, **134**, 153–161.
37. Haakana, T., Kolehmainen, E., Turunen, I., Mikkola, J.-P., and Salmi, T. (2004) The development of monolith reactors: general strategy with a case study. *Chem. Eng. Sci.*, **59**, 5629–5635.
38. Mikkola, J.-P. and Salmi, T. (1999) In-situ ultrasonic catalyst rejuvenation in three-phase hydrogenation of xylose. *Chem. Eng. Sci.*, **54**, 1583–1588.
39. Toukoniitty, B., Toukoniitty, E., Kuusisto, J., Mikkola, J.-P., Salmi, T., and Murzin, D. (2006) Suppression of catalyst deactivation by means of acoustic irradiation – application on fine and specialty chemicals. *Chem. Eng. J.*, **120**, 91–98.
40. Salmi, T., Murzin, D., Mikkola, J.-P., Wärnå, J., Mäki-Arvela, P., Toukoniitty, E., and Toppinen, S. (2004) Advanced kinetic concepts and experimental methods for catalytic three-phase processes. *Ind. Eng. Chem. Res.*, **43**, 4540–4550.
41. Cubylski, A., Sharma, M.M., Moulijn, J.A., and Sheldon, R.A. (2001) *Fine Chemicals Manufacture, Technology and Engineering*, Elsevier, Amsterdam.
42. Murzin, D. and Salmi, T. (2005) *Catalytic Kinetics*, Elsevier, Amsterdam.
43. Thiele, E.W. (1939) *Ind. Eng. Chem.*, **31**, 916.
44. Frank-Kamenetskii, D.A. (1955) *Diffusion and Heat-Transfer in Chemical Kinetics*, Princeton University Press.

45. Sandelin, F., Salmi, T., and Murzin, D. (2006) Dynamic modelling of catalyst deactivation in fixed bed reactors: skeletal isomerization of 1-pentene on ferrierite. *Ind. Eng. Chem. Res.*, **45**, 558–566.
46. Sandelin, F., Oinas, P., Salmi, T., Paloniemi, J., and Haario, H. (2006) Dynamic modelling of catalytic liquid-phase reactions in fixed beds – kinetics and catalyst deactivation. *Chem. Eng. Sci*, **61**, 4528–4539.
47. Wilke, C.R. and Chang, P. (1955) Correlation of diffusion coefficients in dilute solutions. *AIChE J.*, **1**, 264.
48. Hayduk, W. and Minhas, B.S. (1982) Correlations for prediction of molecular diffusivities in liquids. *Can. J. Chem. Eng.*, **60**, 295.
49. Wild, G. and Charpentier, J.-C. (1987) Diffusivité des gaz dans les liquids, Techniques de l'ingenieur, P615, 1-13, Paris.
50. Schiesser, W.E. (1991) *The Numerical Method of Lines: Integration of Partial Differential Equations*, Academic Press, San Diego.
51. Hindmarsh, A.C. (1983) ODEPACK–A systematized collection of ODE solvers, in *Scientific Computing* (eds R. Stepleman *et al.*), IMACS/North Holland Publishing Company, pp. 55–64.
52. Haario, H. (2007) MODEST User's Guide 6.0, Profmath Oy, Helsinki.
53. Marquardt, D.W. (1963) An algorithm for least squares estimation of nonlinear parameters. *J. Soc. Ind. Appl. Maths*, **11**, 431–441.
54. Wärnå, J., Flores Geant, M., Salmi, T., Hamunen, A., Orte, J., Hartonen, R., and Murzin, D. (2006) Modelling and scale-up of sitosterol hydrogenation process: from laboratory slurry reactor to plant scale. *Ind. Eng. Chem. Res.*, **45**, 7067–7076.
55. Martin Curvelo, G. (2008) Hydrogenation of sitosterol to sitostanol on Pd catalysts. M.Sc. Thesis, Åbo Akademi, Turku/Åbo Finland.

9
Structured Reactors, a Wealth of Opportunities

Andrzej Cybulski† and Jacob A. Moulijn

9.1
Introduction

Catalytic reactors can roughly be classified as random and structured reactors. In random reactors, catalyst particles are located in a chaotic way in the reaction zone, no matter how carefully they are packed. It is not surprising that this results in a nonuniform flow over the cross-section of the reaction zone, leading to a nonuniform access of reactants to the outer catalyst surface and, as a consequence, undesired concentration and temperature profiles. Not surprisingly, this leads, in general, to lower yield and selectivity. In structured reactors, the catalyst is of a well-defined spatial structure, which can be designed in more detail. The hydrodynamics can be simplified to essentially laminar, well-behaved uniform flow, enabling full access of reactants to the catalytic surface at a low pressure drop.

A wealth of structures exists and can be found in the literature [1–3]. Figure 9.1 shows examples of monoliths and arrayed catalysts. Monoliths (Figure 9.1a) consist of parallel channels, whereas arrayed catalysts are built from structural elements that are similar to monolithic structures but containing twisted (zig-zag or skewed) passages and/or interconnected passages (Figure 9.1b,c) or arrays of packets of conventional catalyst particles located in the reaction zone in a structured way, whereby the position of particles inside the packets is random (Figure 9.1d). The latter are mainly used for catalytic distillation and are not discussed further in this chapter.

Recently, many papers have been published on fiber catalysts and foam structures (Figure 9.2). Although, strictly speaking, fibers and foams might not be considered as structured systems, beds of such catalysts exhibit typical features of structured catalysts, namely, low pressure drop, uniform flow, a good and uniform access to the catalytic surface, and they are definitely nonrandom. Therefore, we have included them in this chapter.

Autocatalysts, based on monoliths, are probably the most extensively used catalytic reactors: around a hundred million have been installed and are performing well in car exhaust systems [10–12]. Reduction of volatile organic carbon (VOC) emissions [13] and removal of NO_x from stationary sources [14, 15] are also

Novel Concepts in Catalysis and Chemical Reactors: Improving the Efficiency for the Future.
Edited by Andrzej Cybulski, Jacob A. Moulijn, and Andrzej Stankiewicz

ISBN: 978-3-527-32469-9

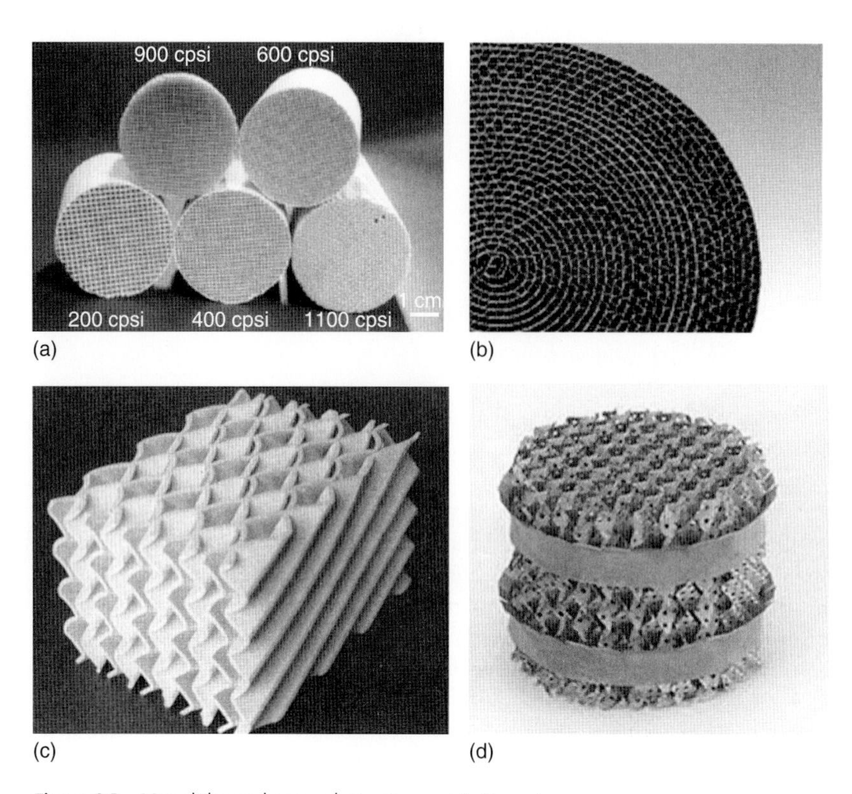

Figure 9.1 Monoliths and arrayed structures. (a) Ceramic monoliths with parallel channels. (Reprinted from [4].) (b) Metallic structure. (Reprinted from [5].) (c) Ceramic Sulzer structure, Katapak®. (Reprinted from [6].) (d) Mellapak Sulzer structure.

Figure 9.2 Fiber and foam structures. (a) Knitted silica fibers catalyst. (Reprinted from [7].) (b) Woven active carbon fiber catalyst. (Reprinted from [8].) (c) Aluminum foam. (Reprinted from [9].)

commonly based on structured catalysts. Applications of structured catalysts in other areas are scarce, although the topic has drawn a lot of attention in academia and industrial R&D. Attractive novel structures are presented for gas-phase reactions in [16, 17] and for three-phase reactions in [9, 18]. Only a selection of papers is discussed here. A complete literature survey was not our intention.

9.2 Monoliths

Monolithic catalysts can be produced by extrusion of a paste containing support material (often cordierite, but popular catalyst carrier materials such as alumina and titania are also used), and possibly catalyst particles (e.g., zeolites, V-based catalysts), allowing high catalyst loading (incorporated type). Alternatively, catalyst supports and active materials can be put as a thin layer on a monolithic support structure ("washcoating"). An interesting example is the carbon monolith support. It can be produced by extrusion with a precursor for the final product (e.g., polymers) but also via coating procedures, for example, silica coating followed by impregnation of Ni salt, oxidation, reduction, and growth of carbon fibers [19]. Ceramic materials are most popular, but metallic materials are also widely used, in particular for automotive applications. The choice for a certain catalyst type will strongly depend on the balance between maximizing the catalyst inventory and catalyst effectiveness. For slow reactions a high catalyst loading is desired and the incorporated-type monolith is a good choice, while for fast reactions with relatively slow internal and external diffusion rates a thin washcoated structure with a maximum geometric area is preferred.

9.2.1 Single-Phase Applications

Because of the huge car market, monolithic catalysts are produced in very large numbers. This leads to the attractive situation that, although they are sophisticated structures, the standard systems are of high quality and commercially available at reasonable costs. It should be noted that when a specific type is desired, an extensive development program might be unavoidable. Monolithic reactors for cleaning of exhaust gases are state of the art, while for most other uses, commercial applications are scarce and a large R&D program might be required.

9.2.1.1 Environmental Applications: Search for the Holy Grail for Diesel Exhaust Cleaning

Control of emissions of CO, VOC, and NO_x is high on the agenda. Heterogeneous catalysis plays a key role and in most cases structured reactors, in particular monoliths, outperform packed beds because of (i) low pressure drop, (ii) flexibility in design for fast reactions, that is, thin catalytic layers with large geometric surface area are optimal, and (iii) attrition resistance [17]. For power plants the large flow rates and the high reaction rate that have to be accommodated make monoliths a logical choice. Furthermore, for VOC, a monolith suggests itself; however, dependent on the flow rates and reaction rates, packed beds might also do. The potential areas of applications are countless; they include surface coating, printing and converting, plants for manufacturing of solvents, cooking (deodorization), greenhouses, soil remediation, and hazardous waste sites.

In automotive exhaust gas purification so-called three-way catalysts based on monoliths are the state-of-the-art technology. The term *three-way catalysts* refers to the three reactions catalyzed, that is, oxidation of CO, oxidation of hydrocarbons, and reduction of NO_x. The three-way catalyst for gasoline engines is a big success. What is the situation for diesel exhaust gas? In the past, it was thought that diesel exhaust gases were clean, because their CO and NO_x emissions were modest compared with gasoline exhaust gases (before the introduction of the successful three-way catalyst). Nowadays, however, the NO_x, CO, and fine particle levels in diesel exhaust are unacceptable, although, in the past, they were an order of magnitude higher! Several innovative ideas have emerged [20–24]. The challenge is to simultaneously reduce NO_x, and to oxidize CO, hydrocarbons, and soot particles. The problem is how to find the Holy Grail, catalyzing the soot/O_2 reaction with a solid, while reducing NO_x.

The reduction of NO_x cannot be performed by three-way catalysis because O_2 strongly inhibits the catalytic activity, but selective catalytic reduction (SCR with ammonia or urea) technology can be used. The obvious disadvantage of SCR is the need to add ammonia or urea. An alternative is a system consisting of a NO_x trapping and a regeneration step. The chemistry is based on the fact that some nitrate salts, for instance, $BaNO_3$, are stable under oxidative conditions, whereas under reducing conditions they decompose into N_2, among others. During normal operation, the conditions are oxidic, allowing the formation of $BaNO_3$:

$$BaO + NO_x(+O_2) \rightarrow BaNO_3 \tag{9.1}$$

Reduction into N_2 is provoked by (short periods of) rich conditions:

$$BaNO_3(+CO) \rightarrow BaO + \frac{1}{2}N_2(+CO_2) \tag{9.2}$$

The oxidation of soot is a slow process, owing to its refractory character. Therefore, the soot particles are trapped in a filter, thus increasing the reaction time. However, catalysis is still required. The design of a catalytic filter is a challenge mainly because solid/solid contact is too poor for efficient catalysis [23]. Several ideas have been put forward.

The first is to add to the diesel fuel a catalytic additive, which accumulates in the soot particles formed. The catalyst is thus brought to the right place and there is contact.

The second is a neat idea coming from Johnson Ma tthey. They invented the so-called continuously regenerating trap (CRT) consisting of a monolithic preoxidizer and a particulate trap, see Figure 9.3 [24]. The first monolith (containing Pt) oxidizes hydrocarbons and CO to CO_2 and NO into NO_2, which is very reactive

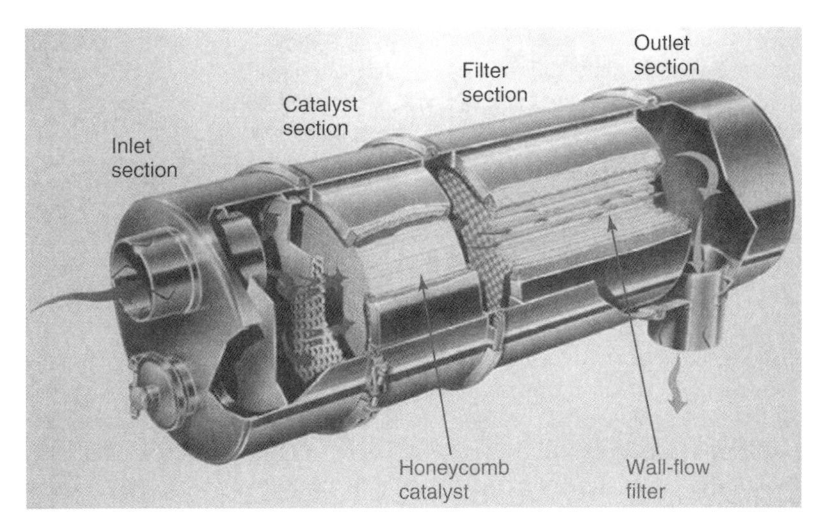

Figure 9.3 Continuously regenerating trap. (Reprinted from [22].)

and gasifies the soot deposited on the trap. Because soot is continuously removed, temperature runaways within the trap are at least partly avoided, which prolongs the life expectation of the trap. The CRT is an elegant technology but the search for the Holy Grail for diesel exhaust gas cleaning is still going on.

In all these innovative processes, monoliths and monolith-based filters are used. As temperature runaways can occur special materials are used. SiC appears to be the best choice because of its excellent heat conductivity.

9.2.1.2 **Combustion of Fuels**

Industrial and municipal installations for combustion of fuels all generate emissions. Natural gas predominates as the fuel for power stations, although coal and (fractions of) crude oil are also widely used. Emission arises from incomplete combustion (CO, UHC, soot) and reactions inherent to combustion (NO_x, CO_2). The dilemma is as follows: the higher the temperature, the lower the emission of unburnt hydrocarbons but the higher the NO_x emission, owing to the high activation energy of the oxidation reaction of N_2. Heterogeneous catalysis has potential in both aspects: complete combustion can be realized at relatively low temperature. On top of this, catalysis enables a stable combustion at lean conditions, again having the benefit of a low NO_x production.

In the engineering of catalytic combustion, the stability of the material is of prime concern. Taking a gas turbine as an example, three different regimes can be distinguished: an ignition zone at low temperature (350–550 °C), an intermediate zone at mid-temperature (600–1000 °C), and a high-temperature zone (1000–1400 °C). In particular, material stability is a point of concern in the last zone. In hybrid systems, part of the fuel reacts over the catalyst and the rest downstream in a homogeneous combustion zone, reducing the temperature of the catalyst. Ceramic or metal monoliths can be used. The thermal expansion coefficient of metals is very high but damage of the catalyst can be prevented by an appropriate configuration of the substrate. Washcoating the substrate with alumina is standard but γ-alumina is unstable above 1000 °C. Zirconia is a better choice for high-temperature applications. Active catalytic species are usually selected from noble metals or transition metal oxides. For instance, Pd oxide was applied in the first commercial catalytic combustion system [27]. A search for better catalysts is still going on.

As of now, only one catalytic combustion system has been implemented on a full scale: the XONON™ Cool Combustion technology, developed by Catalytica Energy Systems [25, 26]. The system is operated as follows: Fuel from a lean mix preburner and main fuel together with compressed air is passed through the catalyst module (palladium oxide catalyst deposited on corrugated metal foil) in which the gas reaches a temperature $\leq$1350 °C. The UHC and CO are combusted to essentially full conversion downstream of the catalyst in the homogenous combustion zone. The guaranteed emission levels are <3 ppm of NO_x, <6 ppm of CO, and <6 ppm of UHC. The technology was tested successfully for 8000 hours, using a natural gas-fueled 1.5-MW M1A-13X turbine of Kawasaki Heavy Industries Ltd. The first commercial XONON-equipped M1A-13X gas turbine entered operation at a 120-building campus in 2002. This installation has consistently performed well for over 13 000 hours.

9.2.1.3 **Industrial Processes with a High Thermal Effect**

The potential of monoliths in industry is generally recognized and many reactions are studied, for instance, short contact time processes (oxidative dehydrogenation, partial oxidation to synthesis gas, etc.), dehydrogenation processes, methanation, methanol-to-gasoline process, Fischer–Tropsch Synthesis, steam reforming, steam cracking, and methanol-to-formaldehyde oxidation. Structured catalysts that can be used in multitubular reactor configurations, be it before, in, or after the main reaction zone seem to be the closest to full-scale implementation.

Metal monoliths show good thermal characteristics. A typical support with herringbone channels made from Fecralloy® performed satisfactory in automotive applications [27]. Modeling showed that overall heat transfer was about 2 times higher than for conventional pellets [28, 29]. Hence, there is potential for structured catalysts for gas-phase catalytic processes in multitubular reactors.

Scientists from Politecnico di Milano and Ineos Vinyls UK developed a tubular fixed-bed reactor comprising a metallic monolith [30]. The walls were coated with catalytically active material and the monolith pieces were loaded lengthwise. Corning, the world leader in ceramic structured supports, developed metallic supports with straight channels, zig-zag channels, and wall-flow channels. They were produced by extrusion of metal powders, for example, copper, tin, zinc, aluminum, iron, silver, nickel, and mixtures and alloys [31]. An alternative method is extrusion of softened bulk metal feed, for example, aluminum, copper, and their alloys. The metal surface can be covered with carbon, carbides, and alumina, using a CVD technique [32]. For metal monoliths, it is to be expected that the main resistance lies at the interface between reactor wall and monolith. Corning patented a metallic support whereby the gap distance was minimized (<250 μm) [33]. Metal monoliths of the Corning type were studied for the partial oxidation of *o*-xylene to phthalic anhydride (PA) [34]. Modeling [35] predicted lower hot spots for monolithic catalyst than conventional packed beds. Investigations were performed using a pilot reactor consisting of a jacketed tube with 16 Al monoliths washcoated with a catalyst load of 62 kg m^{-3}. Loading and unloading of the monoliths appeared to be easy in spite of the small gap (0.1 mm) between the monolith and the tube wall. The reactor was operated continuously for over 1600 hours. The improved thermal characteristics of the monolithic catalyst allow for process intensification by increasing the *o*-xylene concentration or by using tubes of greater diameter.

A general frustration in the field of catalysis is the occurrence of deactivation. A consortium of Elektrochemische Industrie GmbH, Wacker Chemie GmbH, and Lurgi Öl Gas Chemie GmbH developed a postreactor technology based on an adiabatically operated monolith for PA plants that allows for longer operations between subsequent catalyst replacements [36]. The monolithic postreactor was installed in India with a plant of capacity 60 000 tons of PA per annum and was run successfully. New PA plants of Wacker/Lurgi are equipped with this postreactor system, which is considered as standard. The postreactor is also suited for retrofitting.

9.2.2
Multiphase Reactions

In the design of optimal catalytic gas–liquid reactors, hydrodynamics deserves special attention. Different flow regimes have been observed in co- and countercurrent operation. Segmented flow (often referred to as *Taylor flow*) with the gas bubbles having a diameter close to the tube diameter appeared to be the most advantageous as far as mass transfer and residence time distribution (RTD) is concerned. Many reviews on three-phase monolithic processes have been published [37–40].

Because of significant advantages of monoliths in three-phase processes, several studies have been carried out; catalytic hydrogenation in bulk and fine chemistry processes, hydrodesulfurization, hydrodenitrification, oxidation (also for environmental processes), acylation, esterification, and bioprocesses with immobilized enzymes. Column reactors with reactant downflow were used in these studies. Up to now, only one monolithic process was implemented on an industrial scale (200 ktons/year): hydrogenation of alkylanthraquinones in H_2O_2 manufacture [41–44]. The gas–liquid mixture flows downward in the Taylor regime. The reactor was proven to work successfully for a prolonged period of time. In process development studies the monolithic reactor was compared with conventional slurry and packed-bed reactors. Productivity of the monolith reactor was higher by more than an order of magnitude compared to the slurry reactor and almost two times greater than that for packed-bed technology. Other reactor designs for three-phase processes are discussed in the following sections.

9.2.2.1 Countercurrent Reactors

Countercurrent flow has advantages in product and thermodynamically limited reactions. Catalytic packings (see Figure 9.1d) are commonly used in that mode of operation in catalytic distillation. Esterification (methyl acetate, ethyl acetate, and butyl acetate), acetalization, etherification (MTBE), and ester hydrolysis (methyl acetate) were implemented on an industrial scale.

Typical monoliths with parallel channels, possibly internally finned, were shown to be advantageous. Film flow is possible both for cocurrent and countercurrent flow because of the low interaction between both phases in this regime [45]. The feasibility of film flow monolithic reactors has been demonstrated for an esterification reaction. At high conversion levels, countercurrent operation outperforms cocurrrent operation, while for low conversion, the difference is small. An optimal design might be to use a countercurrent monolith reactor as the finishing reactor. When comparing film flow monolithic reactors with conventional catalytic packings, it can be concluded that the critical hydrodynamic characteristics (hydraulic capacity, pressure drop, volumetric mass transfer rates) are similar but monoliths have distinct advantages: higher flexibility, easier scale-up, the surface is better suited for coating procedures, and advances in flooding control enable the use of very small channels, allowing efficient catalyst utilization.

9.2.2.2 Monolithic Loop Reactor (MLR)

A novel monolithic loop reactor (MLR) has been developed at Air Products and Chemicals, see Figure 9.4 [46]. The reactor contains a monolithic catalyst operating under cocurrent downflow conditions. As the residence time in the monolith is short and the heat of reaction has to be removed, the liquid is continually circulated via an external heat exchanger until the desired conversion is reached. The above concept was patented for the hydrogenation of dinitrotoluene to toluenediamine [47]. The reactor performance was compared with traditional slurry operation for several catalytic hydrogenation processes [48]. It was found that the monolith productivity was 2–3 orders of magnitude higher than for the slurry catalyst. Running costs can be lower for the MLR since the high costs of processing of the slurry catalyst are avoided. Savings on the replacement of a Pt slurry catalyst with a Pt monolith were evaluated for a typical hydrogenation process for fine chemicals and pharmaceuticals that was carried out in a 60-l tank. Net savings were $600–800 per batch. For the hydrogenation of glucose to sorbitol (29 000 tons/year) the MLR resulted in annual savings per reactor ranging from $70 000 to 370 000, depending on the monolithic catalyst applied. High catalyst stability was a prerequisite.

The MLR has become a subject of marketing agreement between Air Products and Chemicals Inc. and Johnson Matthey [49].

Boger *et al.* [50] analyzed the performance of MLRs with internal density-driven circulation (IMLR). They found the gas–liquid mass transfer superior and the overall mass transfer performance comparable with those for slurry reactors.

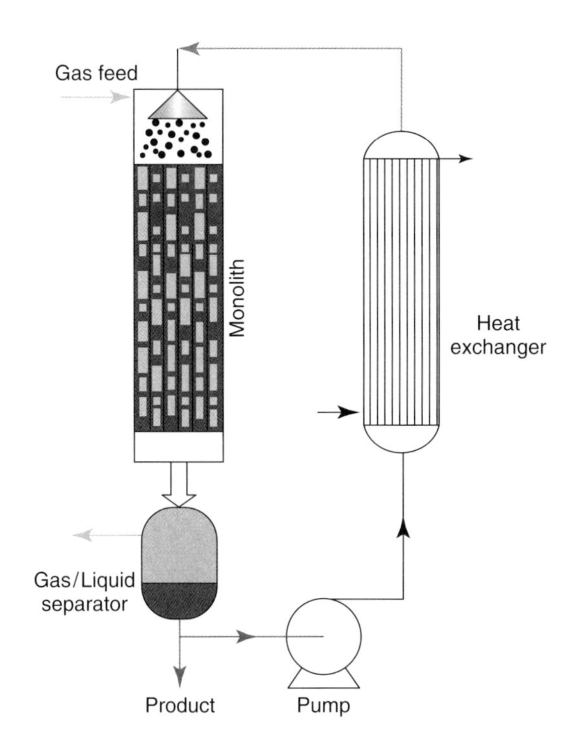

Figure 9.4 Monolithic loop reactor (MLR). (Reprinted from [4].)

9.2.2.3 **Stirred Tank Reactor with a Monolithic Agitator**

A monolith reactor that might be particularly useful in fine chemicals manufacture and biotechnology was developed in The Netherlands [51, 52]. Monolithic structures in this reactor are mounted on the stirrer shaft replacing conventional impeller blades (see Figure 9.5). The monolithic stirrers can be mounted on a vertical shaft or on a horizontal one and more than one set of stirrers can be placed on the shaft. The so-called ROTACAT reactor was tested in the hydrogenation of 3-methyl-1-pentyn-3-ol [53]. The performance was comparable to that of a conventional slurry reactor. Another process tested is the hydroformylation of 1-octene [54]. The reactor performed very well, showing an overall selectivity of 90% for the linear aldehyde. The ROTACAT reactor was also used in bioprocesses with enzymes immobilization [55]. The concept was proven with acylation of butanol with vinylacetate in the presence of immobilized lipase. Compared to conventional stirrers the monolith impellers have a much larger geometric catalytic surface area and generate a much larger mixing zone of low intensity.

9.3 Other Structured Catalysts

A wealth of other regular, mainly metallic structures has been reported. Many of them are derived from static mixers. A general comment has to be made that the relatively low catalyst loading often makes them impractical for processes occurring

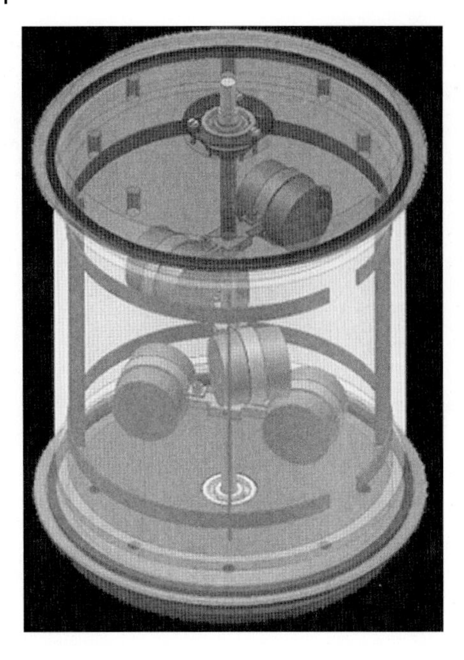

Figure 9.5 Monolithic stirrer reactor (MSR).

in the kinetic regime. To solve this, a structured system is occasionally used to encapsulate the catalyst present in the form of particles.

9.3.1 Arrays of Structural Elements

While conventional monoliths contain parallel channels, in practice, systems are often made from alternate layers that allow lighter structures with better mass transfer characteristics in gas-phase applications, see Figure 9.6 showing interconnected flow paths. They are usually made from metal, mostly Fecralloy®, Kanthal®, or stainless steel, and widely used in autocatalysts and in environmental

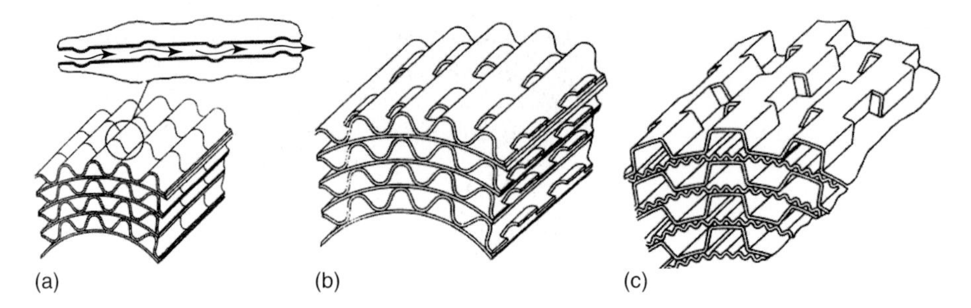

Figure 9.6 Structural metal monoliths: (a) transversal structure, (b) SM design, and (c) LS design of EMITEC. (Reprinted from [12].)

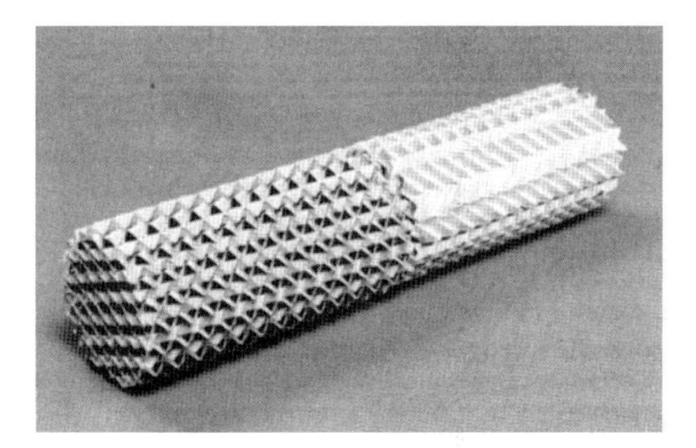

Figure 9.7 The Sulzer OCFS metal structure with ceramic washcoat. (Reprinted from [6].)

processes (deNOxification, SCR, incineration of VOC). The potential for multiphase application has been recently reviewed [9].

Sulzer Chemtech [56] developed a successful structure, often referred to as *open crossflow structure* (OCFS), see Figure 9.7. It contains superimposed individual corrugated sheets with the corrugations in opposed orientation. The main stream of fluid(s) is divided into a number of substreams that are recombined, intensifying the turbulence of the mass and heat transfer within the structure and (to some extent) between the fluid and the reactor wall. In radial direction, the effective thermal conductivity can be up to 2 times greater than that for conventional packed beds. The pressure drop is 10–100 times lower than in conventional packed beds. The OCFSs might have potential for highly exothermic reactions [9]. The geometric surface area ranges from 300 to 1800 $m^2\ m^{-3}$, void fraction is about 90%. They are made from metal or ceramics (Figure 9.1c), the latter being relatively easy to use as catalyst support. At present, no industrial applications of OCFS as catalyst support have been reported.

A promising novel structure for highly exothermic reactions is shown in Figure 9.8. The void fraction of this structure exceeds 90% and the geometric

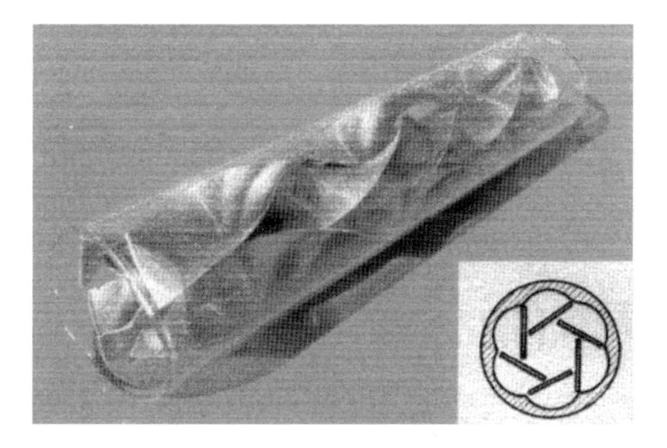

Figure 9.8 A structured support consisting of twisted ribbons.

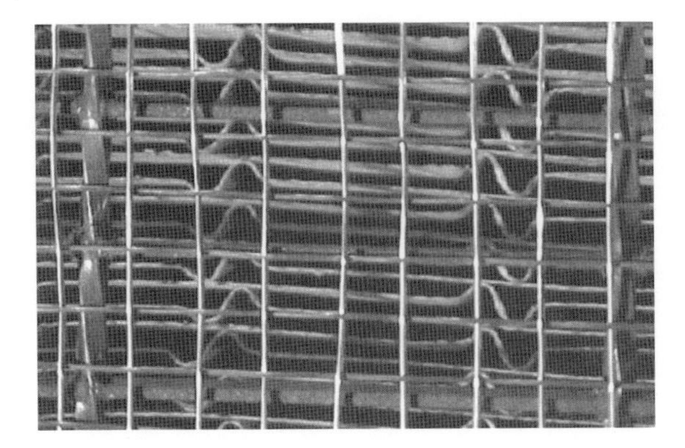

Figure 9.9 Plate-type catalyst (PTC) of Babcock-Hitachi. (Reprinted from [59].)

surface area of the elements is about 300–350 $m^2\ m^{-3}$. The radial mixing is very intensive and the heat transfer with the tube wall is much better than that for fixed beds of conventional catalysts [57]. In catalyst loading, it is helpful to stiffen the structure by rings. Combining many structures in a bundle by means of a grid would make loading and discharging easier.

Plate-type catalysts (PTCs) consist of metal sheets, metal net, or perforated metal plates with the catalytic species deposited onto, assembled in modules that are inserted into the reactor in layers [58, 59]. Several innovative structures have been reported. Figure 9.9 gives an example of a structure permitting some vibration of the individual plates, reducing possible blockage; the plates are very thin, reducing the pressure drop.

Babcock-Hitachi developed a plate-like structure covered with fabric of inorganic fibers [60]. Kawasaki Heavy Industries, Ltd. [61] developed an integrated unit comprising a plurality of foamed metallic plates covered with titania-supported catalysts. Allied Signal [62] developed a catalytic converter comprising a large number of fins arranged in an axial succession of offset fin rows. Nippon Steel [63] developed a support consisting of plain and corrugated plates, both made of nonwoven fabric of metal fiber, alternately superposed over each other and rolled or laminated. Recently, a microstructured plate reactor and microstructured reactor/heat exchanger assembled from aluminum plates appropriately grooved was described [64]. It showed nearly isothermal operation for ammonia oxidation.

A vast majority of structured catalysts and reactors are used in flue gas cleaning. However, due to proprietary reasons, not much information is provided. Babcock Power Environmental reported [65] that in the United States they have over 30 000 MWs of SCRs commissioned, in design or evaluation, and under construction. BP's licensor has over 26 000 MWs of SCRs operating in Europe. The German company EON operates 44 SCR systems, totaling 12 300 MWs. It should be noted that not all SCRs plants are based upon structured catalysts.

9.3.2 Fibrous Structured Catalysts

Fibrous structured catalysts consist of threads that are knitted or woven into fabrics, felts, cloths, and so on or are formed by chemical or mechanical processing of ceramic or metal sheets. Figure 9.2a,b shows typical examples. A review can be found in [66].

Voidage of fibrous beds ranges from 0.7 to 0.9, resulting in low pressure drop. Owing to the small size of building units diffusion limitations are minimized, making them attractive catalyst supports. They are effective as filters that retain even fine particles. The downside is that they are more prone to clogging, more so when a catalytic coating is applied. Moreover, satisfactory washcoating will be relatively difficult to achieve. In order to reduce channeling, uniform packing is crucial.

Commercial glass fiber structured catalysts consist of threads (about 0.3–1 mm in diameter) made from a bundle of elementary filaments of size of 2–100 μm. The surface area of the pristine material is low, $1{-}2\,m^2\,g^{-1}$. They are usually leached with HCl solution to remove nonsilica components increasing the surface area up to $75{-}400\,m^2\,g^{-1}$. Glass-fiber- based systems have been modified with TiO_2, ZrO_2, and Al_2O_3 to increase thermostability and to impart suitable catalytic properties.

Sintered metal fibers with filaments of uniform size (2–40 μm), made of SS, Inconel, or Fecralloy®, are fabricated in the form of panels. Gauzes based on thicker wires (100–250 μm) are made from SS, nickel, or copper. They have a low surface area of about $10^{-3}\ m^2\ g^{-1}$. Several procedures are used to increase the surface area, for example, leaching procedures, analogous to the production of Ra-Nickel, and electrophoretic deposition of particles or colloid suspensions. The porosity of structures formed from metal fibers range from 70 to 90%. The heat transfer coefficients are high, up to 2 times larger than for random packed beds [67].

The surface area of activated carbon filaments (ACFs) can be very large, up to $3000\,m^2\ g^{-1}$. They are available commercially as woven or knitted cloths or nonwoven felts and papers. Fabrics are made typically from threads about 0.5 mm in diameter, consisting of bundles of elementary filaments 3–5 μm in diameter. Although ACFs are commercially available, many are prepared in the laboratory using cellulose, viscose rayon, polyacrylic resins, and so on. A complex protocol is needed, consisting of precarbonization (pyrolysis at about 400 °C), carbonization (about 1200 °C), activation (e.g., by steam or CO_2 treatment at elevated temperature), and if desired, further oxidation in order to generate oxygen-containing functional groups at the surface.

Noble metal catalysts such as Pd, Pt, and Ru and transition metals are deposited on or incorporated in the fibers by methods similar to those used in catalyst preparation. Zeolitic (e.g., TS-M, Na-X, ZSM-type) crystals have been grown on the surfaces.

Investigations of gas-phase processes in the presence of fiber catalysts focused on oxidation, SCR, and synthesis of chemicals. Oxidation studies of CO [68–71], methane [72, 73, 77], propane [74–76], toluene, terpenes, naphthalene, lower alkanols, and flue gases showed their potential. Fibrous SCR catalysts, mostly activated carbon fiber (ACF)-based, performed satisfactorily [78–82]. Conversion of NO to N_2 using CO was carried out in the presence of Rh/Cu–TS-1/mullite fibers with excellent activity and low light-off temperature [83]. Excellent conversions, up to 99% to cis- and trans-isomers of 1,2-dimethylcyclohexane were reached in hydrogenation of *o*-xylene in the presence of Pt/silica fiber [84]. Rh on sintered metal fibers showed a very high activity and selectivity in the hydrogenation of 1,3-cyclohexadiene [85].

Hydrogenation reactions, particularly for the manufacture of fine chemicals, prevail in the research of three-phase processes. Examples are hydrogenation of citral (selectivity $> 80\%$ [86–88]) and 2-butyne-1,4-diol (conversion $> 80\%$ and selectivity $> 97\%$ [89]). For Pt/ACF the yield to D-sorbitol in hydrogenation of D-glucose exceeded 99.5% [90]. Water denitrification via hydrogenation of nitrites and nitrates was extensively studied using fiber-based catalysts [91–95]. An attempt to use fiber-structured catalysts for wet air oxidation of organics (4-nitrophenol as a model compound) in water was successful. TOC removal up to 90% was achieved [96].

The fibrous catalysts were investigated mostly as slurry catalysts in autoclaves. Scientists from Swiss Federal Institute of Technology in Lausanne studied scale-up by applying a bubble column (ID 24–140 cm), containing a series of fibrous catalytic panels (woven fabrics) with a certain distance between them. The gas–liquid mixture flows upward and is separated after leaving the reactor; the liquid is recirculated via an external loop. Different flow regimes were observed [97, 98]. Hydrodynamic properties depend mainly on the distance between layers and the distance between threads inside the layer. Pressure drop was found to be dependent mainly on the gas velocity. Residence time distribution is narrow; no backmixing occurs between interlayer spaces and perfect mixing takes place in those spaces.

9.4 Foams

Foam-type structured catalysts are 3D cellular materials with interconnected pores resembling the inverse of a packed bed, see Figure 9.1d. They combine a high porosity (up to 97%) with a large surface area. They exhibit very good heat transfer characteristics and modest pressure drop. Foams are produced in a large variety of materials (carbon, metals, ceramics, etc.). They have not drawn much interest in the industrial R&D, but the major engineering data needed for design are available. *A priori*, it is expected that it is not easy to deposit a catalyst, but washcoating of Fecralloy with Pd/γ-alumina has been reported. Methods of foam catalysts preparation, hydrodynamics, and transport properties were mainly studied using CO oxidation as a test reaction [99–103].

Recently, hydrodynamic aspects of multiphase applications have been studied in detail for the cocurrent and countercurrent flow regimes. Useful correlations were determined and it was found that foams combine high rates and low pressure drop, proving their high potential in multiphase applications [9, 107, 108].

9.5
Why are Industrial Applications of Structured Reactors so Scarce?

At present, all autocatalysts are structured catalysts and this will not change. The other environmental applications of structured catalysts and reactors will expand further with emerging new (and cheaper) technology and stricter environmental standards. Investments will continue in combustion processes for power stations and local heating stations where fuels of lower grade are used. Monoliths will prevent emissions of harmful components in flue gases and enable advanced combustion technology such as lean burning. There is a bright future for structured catalysts, in particular, monoliths, in these areas. However, this does not seem to apply to the chemical industry. What is the reason for this?

Firstly, there are technical reasons concerning catalyst and reactor requirements. In the chemical industry, catalyst performance is critical. Compared to conventional catalysts, they are relatively expensive and catalyst production and standardization lag behind. In practice, a robust, proven catalyst is needed. For a specific application, an extended catalyst and washcoat development program is unavoidable, and in particular, for the fine chemistry in-house development is a burden. For coated systems, catalyst loading is low, making them unsuited for reactions occurring in the kinetic regime, which is particularly important for bulk chemistry and refineries. In that case, incorporated monolithic catalysts are the logical choice. Catalyst stability is crucial. It determines the amount of catalyst required: for a batch process, the number of times the catalyst can be reused, and for a continuous process, the run time.

When considering structured reactors, novel engineering solutions are required in several respects. Gas-phase catalytic reactions are conventionally carried out in adiabatic column reactors or multitubular reactors. It is essential to provide uniform flow through the reactor. Structured catalysts with their low pressure drop are prone to nonuniformity of flow, in particular at the inlet and the outlet of the reactor. Nonideal distribution could result in backflow within the structured catalyst. Chemical reactions proceed usually with evolution or consumption of large amounts of heat, leading to undesired temperature profiles in ceramic structured catalysts. This can lead to an undesired course of the catalytic process, particularly for gas-phase reactions. Success has been noted for oxidation of *o*-xylene to PA where an aluminum monolith of good thermal characteristics was used. Although this is promising, confirmation is required in a demonstration plant with a multitubular reactor. Between the monolithic catalyst and the tube, the gap should be minimal, requiring tubes of a very small tolerance. A satisfactory catalyst-loading protocol into the large number of tubes is needed. It is worthwhile to test the twisted-ribbon structure for gas-phase reactions. In general, efficient deflectors at the inlet and collectors at the outlet of the reactor are needed. A monolithic postreactor seems to be mature technology that requires only a dedicated catalyst. Such a relatively cheap reactor can be installed when retrofitting the plant or designed for new plants.

Sufficient engineering data for designing reactors for three-phase processes are available. A column reactor with gravitational liquid downflow was industrially proven. An MLR with forced liquid downflow with ejector was also well studied. Dedicated catalysts for particular processes must be, however, worked out.

Secondly, investments during the last decade were rather limited because of slow developments in (petro)chemical industries and because of overcapacity in the fine chemicals sector. For new plants, capital required for structured reactors and slurry reactors is similar in magnitude. When retrofitting a plant, new investment is required: external devices (piping, pump, heat exchanger) and internal arrangements (ejector, housing for the catalyst). In such a situation, management prefers processes that have been well proven on a full scale. This is not the case for structured reactors: no detailed reports have been published on demonstration plants with structured catalysts that have run for a prolonged period of time. If any industrial company were to decide to use a structured catalyst for a highly exothermic process, this would constitute a breakthrough and further applications would become open.

Extensive research programs have been carried out to meet these challenges. The industrial use of structured catalysts in the field of three-phase catalytic processes is still limited to the hydrogenation step in the alkylanthraquinone method for production of hydrogen peroxide. Results of physical and process studies of the MLR are very promising for applications in fine chemicals manufacture. One of the main obstacles is that commercial monolithic supports with a washcoat of activated carbon, a very popular support in fine chemicals manufacture, are not commercially available. The cooperation of Air Products and Chemicals with Johnson Matthey might lead to commercially available washcoated monoliths. Methods that are effective in washcoating were recently reviewed [104, 105]. We think that the MLR is a potential candidate to replace slurry reactors in the manufacture of fine chemicals. We expect that more processes based upon monolithic reactors will be implemented in the industry if AkzoNobel publishes a detailed technical and economical evaluation of their process. An interesting niche in the field of three-phase processes is the catalytic wet air oxidation [106]. Nippon Shokubai successfully tested monolithic catalysts for such oxidation.

Results of R&D are promising for industrial applications of fibrous structured catalysts. There are, however, constraints in applications. On the one hand, filtering ability of fiber structures causes clogging and, consequently, maldistribution and increasing pressure drop. On the other hand, separation of particles from flue gases can be an advantage if cleaning of fibrous filters became easy. Fibrous structured supports are not easy in manufacturing a well-behaved coated catalyst system.

9.6 Concluding Remarks

Structured catalysts and reactors are in common use for environmental applications: autocatalysts, VOCs incinerators, denoxification plants, and the like. Chemical industries prefer technologies that have been proven at a full scale. The only monolithic process that was implemented in an industrial plant is the hydrogenation step of the alkylanthraquinone method for the manufacture of hydrogen peroxide (AkzoNobel). Two promising monolithic technologies are close to implementation: (i) MLR for hydrogenation in the fine chemicals manufacture (Air Products and Chemicals in cooperation with Johnson Matthey) and (ii) monolithic catalysts for carrying fast and highly exothermic gas-phase reactions (Corning, Polynt and Milan Polytechnic). The use of structured reactors seems to have potential for several processes, including catalytic wet air oxidation.

References

1. (a) Cybulski, A. and Moulijn, J.A. (eds) (1998) *Structured Catalysts and Reactors*, 1st edn, Marcel Dekker, New York; (b) Cybulski, A. and Moulijn, J.A. (eds) (2006) *Structured Catalysts and Reactors*, 2nd edn, CRC Taylor & Francis, Boca Raton.
2. Irandoust, S. and Andersson, B. (1988) *Catal. Rev. Sci. Eng.*, **30** (3), 341.
3. Cybulski, A. and Moulijn, J.A. (1994) *Catal. Rev. Sci. Eng.*, **36** (2), 179.
4. DeDeugd, R.M., Kapteijn, F., and Moulijn, J.A. (2003) *Top. Catal.*, **26** (1-4), 29–39.
5. Emerachem Leaflet *http://www.emerachem.com/*.
6. Stringaro, J.-P., Collins, P., and Bailer, O. (1998) Open cross-flow catalysts and supports, in *Structured Catalysts and Reactors*, 1st edn, Chapter 14 (eds A. Cybulski and J.A. Moulijn), Marcel Dekker, New York, p. 393.
7. Toukoniitty, E., Mäki-Arvela, P., Kalantar Neyestanaki, A., Salmi, T., and Murzin, D.Yu. (2002) *Appl. Catal. A: Gen.*, **235** (1-2), 125–138, Fig 9.1.
8. Aumo, J., Oksanen, S., Mikkola, J.-P., Salmi, T., and Murzin, D.Yu. (2005) *Catal. Today*, **102-103**, 128–132.
9. Pangarkar, K., Schildhauer, T.J., van Ommen, J.R., Nijenhuis, J., Kapteijn, F., and Moulijn, J.A. (2008) *Ind. Eng. Chem. Res.*, **47** (10), 3720–3751.
10. Heck, R.M., Farrauto, R.J., and Gulati, S.T. (eds) (2002) *Catalytic Air Pollution Control: Commercial Technology*, 2nd edn, Wiley/Interscience, New York.
11. (a) Gulati, S.T. (1998) Ceramic catalyst supports for gasoline fuel, in *Structured Catalysts and Reactors*, 1st edn, Chapter 2 (eds A. Cybulski and J.A. Moulijn), Marcel Dekker, New York, p. 15; (b) Gulati, S.T. (2006) Ceramic catalyst supports for gasoline fuel, in *Structured Catalysts and Reactors*, 2nd edn, Chapter 2 (eds A. Cybulski and J.A. Moulijn), CRC Taylor & Francis, Boca Raton, p. 21.
12. (a) Twigg, M.V. and Wilkins, A.J.J. (1998) Autocatalysts – past, present, and future, in *Structured Catalysts and Reactors*, 1st edn, Chapter 4 (eds A. Cybulski and J.A. Moulijn), Marcel Dekker, New York, p. 91; (b) Twigg, M.V. and Wilkins, A.J.J. (2006) Autocatalysts – past, present, and future, in *Structured Catalysts and Reactors*, 2nd edn, Chapter 4 (eds A. Cybulski and J.A. Moulijn), CRC Taylor & Francis, Boca Raton, p. 109.
13. Kolaczkowski, S. (2006) Treatment of volatile organic carbon

(VOC) emissions from stationary sources, in *Structured Catalysts and Reactors*, 2nd edn, Chapter 5 (eds A. Cybulski and J. A. Moulijn), CRC Taylor & Francis, Boca Raton, p. 147.

14. (a) Beretta, A., Tronconi, E., Groppi, G., and Forzatti, P. (1998) Monolithic catalysts for selective reduction of NOx with NH_3 from stationary sources, in *Structured Catalysts and Reactors*, 1st edn, Chapter 5 (eds A. Cybulski and J.A. Moulijn), Marcel Dekker, New York, p. 121; (b) Nova, I., Beretta, A., Groppi, G., Lietti, L., Tronconi, E., and Forzatti, P. (2006) Monolithic catalysts for NOx removal from stationary sources, in *Structured Catalysts and Reactors*, 2nd edn, Chapter 6 (eds A. Cybulski and J.A. Moulijn), CRC Taylor & Francis, Boca Raton, p. 171.
15. Tomašić, V. (2007) *Catal. Today*, **119** (1-4), 106–113.
16. Voecks, G.E. (1998) Unconventional utilization of monolithic catalysts for gas-phase reactions, in *Structured Catalysts and Reactors*, 1st edn, Chapter 7 (eds A. Cybulski and J.A. Moulijn), Marcel Dekker, New York, p. 179; (b) Groppi, G., Beretta, A., and Tronconi, E. (2006) Structured catalysts for gas-phase syntheses of chemicals, in *Structured Catalysts and Reactors*, 2nd edn, Chapter 8 (eds A. Cybulski and J.A., Moulijn), CRC Taylor & Francis, Boca Raton, p. 243.
17. Farrauto, R.J., Liu, Y., Ruettinger, W., Ilinich, O., Shore, L., and Giroux, T. (2007) *Catal. Rev.*, **49**, 141–196.
18. Irandoust, S., Cybulski, A., and Moulijn, J.A. (1998) The use of monolithic reactors for three-phase reactions, in *Structured Catalysts and Reactors*, 1st edn, Chapter 9 (eds A. Cybulski and J.A. Moulijn), Marcel Dekker, New York, p. 239; (b) Cybulski, A., Edvinsson-Albers, R., and Moulijn, J.A. (2006) Monolithic catalysts for three-phase processes, in *Structured Catalysts and Reactors*, 2nd edn, Chapter 10 (eds A. Cybulski and J.A. Moulijn), CRC Taylor & Francis, Boca Raton, p. 355.
19. de Lathouder, K.M., Lozano-Castelló, D., Linares-Solano, A., Wallin, S.A., Kapteijn, F., and Moulijn, J.A. (2007) *Microporous Mesoporous Mater.*, **99** (1-2), 216–223.
20. van Setten, B.A.A.L., Makkee, M., and Moulijn, J.A. (2001) *Catal. Rev. Eng.*, **43** (4), 489–564.
21. Gulati, S.T., Makkee, M., and Setiabudi, A. (2006) Ceramic catalysts, supports and filters for diesel exhaust aftertreatment, in *Structured Catalysts and Reactors*, 2nd edn, Chapter 19 (eds A. Cybulski and J.A. Moulijn), CRC Taylor & Francis, Boca Raton, p. 663.
22. Fino, D. (2007) *Sci. Technol. Adv. Mater.*, **8**, 93–100.
23. Neeft, J.P.A., Makkee, M., and Moulijn, J.A. (1996) *Fuel Process. Technol.*, **47**, 1–69.
24. Twigg, M.V. and Wilkins, A.J.J. (2006) Autocatalysts: past, present, and future, in *Structured Catalysts and Reactors*, 2nd edn, Chapter 4 (eds A. Cybulski and J.A. Moulijn), CRC Taylor & Francis, Boca Raton, p. 109.
25. McNamara, W. (2005) *Power Eng.*, **109** (8), 18.
26. Caron, T., Lemons, S., Nickolas, S., Tuet, P., and Gilbert, G.W.; Catalytica Energy Systems, Inc. (2005) Xonon ultra low-nox combustion applied to multi-combustor gas turbines. Report for PIER, California Energy Commission, November 2005.
27. Cybulski, A. and Moulijn, J.A. (1994) *Chem. Eng. Sci.*, **49** (1), 19–27.
28. (a) Groppi, G. and Tronconi, E. (2005) *Catal. Today*, **105**, 297; (b) Groppi, G. and Tronconi, E. (2001) *Catal. Today*, **69**, 307.
29. Boger, T. and Heibel, A.K. (2005) *Chem. Eng. Sci.*, **60** (7), 1823–1835.
30. Carmello, D., Marsella, A., Forzatti, P., Tronconi, E., and Groppi, G. (2001) EP1110605; US2002038062 (2002).
31. Cutler, W.A., Lin, He, Olszewski, A., and Sorensen, C.M. (2003) US2003100448; US6881703 (2005).
32. Abbott, J.H. and Boger, T. (2006) CN1832820; WO2005011889 (2005).
33. Amsden, J.M., Boulc, H.G., Heibel, A., and Partridge, N.E., (2005)

US2005142049; WO2005065187 (2005); EP1699552 (2006).

34. (a) Groppi, G., Tronconi, E., Cruzolin, F., Cortelli, C., Leanza, R., and Marsaud, S. (2008) paper presented at Eurokin 10th Anniversary Meeting, May 20, 2008, IFP Lyon.; (b) Groppi, G., Tronconi, E., Cruzolin, F., Cortelli, C., Leanza, R., and Marsaud, S. (2008) Paper No. 300496-1, presented at ISCRE 20, September 7-10, 2008, Kyono.
35. Boger, T. and Menegola, M. (2005) *Ind. Eng. Chem. Res.*, **44**, 30.
36. (a) Eberle, H.-J., Breimair, J., Domes, H., and Gutermuth, T. (2000) *Petrochemicals Gas Process.*, **Q3**, 129–133; (b) Eberle H.-J., Helmer, O., Stocksiefen, K.H., Trinkhaus, S., Wecker, U., and Zeitler, N. (2001) WO0103832; EP1181097 (A1); US6730631 (B1); KR20020027472 (A); EP1181097 (A0); DE19931902 (A1).
37. Kapteijn, F., Heiszwolf, J.J., Nijhuis, T.A., and Moulijn, J.A. (1999) *CAT-TECH*, **3**, 24–41.
38. Kreutzer, M.T., Kapteijn, F., and Moulijn, J.A. (2006) *Catal. Today*, **111**, 111–118.
39. Roy, S., Bauer, T., Al-Dahhan, M., Lehner, P., and Turek, T. (2004) *AIChE J.*, **50** (11), 2918–2938.
40. Kapteijn, F., Nijhuis, T.A., Heiszwolf, J.J., and Moulijn, J.A. (2001) *Catal. Today*, **66** (2-4), 133–144.
41. Berglin, T. and Herrmann, W. (1984) Swedish Patent No. 431,532; Eur. Patent No. EP 102,934 (1986).
42. Berglin, T. and Schöön, N.-H. (1983) *Ind. Eng. Chem. Process Des. Develop.*, **22**, 150.
43. Irandoust, S., Andersson, B., Bengtsson, E., and Siverström, M. (1989) *Ind. Eng. Chem. Res.*, **28**, 1489–1493.
44. Edvinsson Albers, R., Nyström, M., Siverström, M., Sellin, A., Dellve, A.-C., Andersson, U., Hermann, W., and Berglin, Th. (2001) *Catal. Today*, **69**, 247–252.
45. Heibel, A.K. and Lebens, P.J.M. (2006) Film flow monolith reactors, in *Structured Catalysts and Reactors*, 2nd edn, Chapter 13 (eds A. Cybulski and J.A. Moulijn), CRC Taylor & Francis, Boca Raton, p. 479.
46. Broeckhuis, R.R., Machado, R.M., and Nordquist, A.F. (2001) *Catal. Today*, **69**, 87–93.
47. Machado, R.M., Parrillo, D.J., Boehme, R.P., and Broekhuis, R.R. (1999) US6005143.
48. Machado, R.M., Broekhuis, R.R., Nordquist, A.F., Roy, B.P., and Carney, S.R. (2005) *Catal. Today*, **105** (3-4), 305–317.
49. Johnson, M. (2003) *Focus Catal.*, (January issue), 2; Johnson, M. (2003) *Focus Catal.*, (March issue), 7; (2003) *Chem. Eng. (New York)*, **110** (1), 17.
50. Boger, T., Roy, S., Heibel, A.K., and Borchers, O. (2003) *Catal. Today*, **79-80**, 441–451.
51. Edvinsson, R.K. and Moulijn, J.A. (1998) WO9830323.
52. Edvinsson Albers, R.K., Houterman, M.J.J., Vergunst, Th., Grolman, E., and Moulijn, J.A. (1998) *AIChE J.*, **44**, 2459–2464.
53. Hoek, I., Nijhuis, T.A., Stankiewicz, A.I., and Moulijn, J.A. (2004) *Chem. Eng. Sci.*, **59** (22-23), 4975–4981.
54. Sandee, A.J., Ubale, R.S., Makkee, M., Reek, J.N.H., Kamer, P.C.J., Moulijn, J.A., and van Leeuwen, W.N.M. (2001) *Adv. Synth. Catal.*, **341** (2), 201–206.
55. de Lathouder, K.M., Marques Flo, T., Kapteijn, F., and Moulijn, J.A. (2005) *Catal. Today*, **105**, 443–447.
56. Sulzer Chemtech (2006) *Mixing and Reaction Technology*, Sulzer Chemtech, Winterthur.
57. Werner, K. and Cybulski, A. (1987) *Chem. Eng. Apparatus (Inżynieria i Aparatura Chemiczna, Polish)*, **26** (1), 11.
58. Yokoyama, K., Yoshida, N., and Kato, Y. (1998) JP10028871.
59. Morita, I., Ogasahara, T., and Franklin, H.N. (2002) Recent Experience with Hitachi Plate Type SCR Catalyst, The Institute of Clean Air Companies Forum '02, February 12-13.
60. Yokoyama, K., Kato, Y., and Miyamoto E. (1999) JP11319583.
61. Niwa, S., Sasaki, T., Shinto, M., and Imamura, N. (1984) US4446250.

62. Lester, G.R. and Homeyer, S.T. (2001) US6203771.
63. Okazaki, Y. and Shinoda, T. (2002) JP2002113798.
64. Rebrov, E.V., de Croon, M.H.J.M., and Schauten, J.C. (2001) *Catal. Today*, **69** (1-4), 183–192.
65. *http//WWW.bbabcockpower/index.php?option=products&task=viewproduct&coid=17&proid=45* (last accessed May 2010).
66. Matatov-Meytal, Yu. and Sheintuch, M. (2002) *Appl. Cat. A: Gen.*, **231**, 1–16.
67. Cahela, D.R. and Tatarchuk, B.J. (2001) *Catal. Today*, **69**, 33–39.
68. Kiwi-Minsker, L., Yuranov, I., Siebenhaar, B., and Renken, A. (1999) *Catal. Today*, **54**, 39–46.
69. Bulushev, D.A., Kiwi-Minsker, L., Yuranov, I., Suvorova, E.I., Buffat, P.A., and Renken, A. (2002) *J. Catal.*, **210** (1), 149–159.
70. Ahlström-Silversand, A.F. and Odenbrand, C.U.I. (1999) *Chem. Eng. J.*, **73** (3), 205–216.
71. Kalantar Neyestanaki, A. and Lindfors, L.-E. (1998) *Fuel*, **77** (15), 1727–1734.
72. Maione, A., André, F., and Ruiz, P. (2007) *Appl. Catal. B: Environ.*, **75**, 59–70.
73. Klvana, D., Kirchnerová, J., Chaouki, J., Delval, J., and Yaïci, W. (1999) *Catal. Today*, **47** (1-4), 115–121.
74. Kiwi-Minsker, L., Yuranov, I., Slavinskaia, E., Zaikovskii, V., and Renken, A. (2000) *Catal. Today*, **59**, 61–68.
75. Yuranov, I., Renken, A., and Kiwi-Minsker, L. (2003) *Appl. Catal. B: Environ.*, **43**, 217–227.
76. Yuranov, I., Dunand, N., Renken, A., and Kiwi-Minsker, L. (2002) *Appl. Catal. B: Environ.*, **36**, 183–191.
77. Klvana, D., Kirchnerová, J., Gautheer, P., Delval, J., and Chaouki, J. (1997) *Can. J. Chem. Eng.*, **75**, 509–519.
78. Lu, Y., Fu, R., Chen, Y., and Zeng, H. (1994) *Mater. Res. Sci. Soc., Symp. Proc.*, **344**, 95.
79. Yoshikawa, M., Yasutake, A., and Mochida, I. (1998) *Appl. Catal. A: Gen.*, **173**, 239–245.
80. Marbán, G. and Fuertes, A.B. (2001) *Appl. Catal. B: Environ.*, **34** (1), 43–53.
81. Marbán, G. and Fuertes, A.B. (2001) *Appl. Catal. B: Environ.*, **34** (1), 55–71.
82. Marbán, G., Antuña, R., and Fuertes, A.B. (2003) *Appl. Catal. B: Environ.*, **41** (3), 323–338.
83. Petrov, L., Soria, J., Dimitrov, L., Cataluna, R., Spasov, L., and Dimitrov, P. (1996) *Appl. Catal. B: Environ.*, **8**, 9–31.
84. KalantarNeyestanaki, A. Maki-Arvela, P. Backman, H. Karhu, H. Salmi, T. Vayrynen, J., and Murzin, D.Yu. (2003) *Ind. Eng. Chem. Res.*, **42** (14), 3230–3236.
85. Ruta, M., Yuranov, I., Dyson, P.J., Laurenczy, G., and Kiwi-Minsker, L. (2007) *J. Catal.*, **247** (2), 269–276.
86. Salmi, T., Mäki-Arvela, P., Toukoniitty, E., Kalantar Neyestanaki, A., Tiainen, L.-P., Lindfors, L.-E., Sjöholm, R., and Laine, E. (2000) *Appl. Catal. A: Gen.*, **196** (1), 93–102.
87. Auomo, J., Oksanen, S., Mikkola, J.-P., Salmi, T., and Murzin, D.Yu. (2005) *Catal. Today*, **102-103**, 128–132.
88. Mikkola, J.-P., Auomo, J., Murzin, D.Yu., and Salmi, T. (2005) *Catal. Today*, **105**, 325–330.
89. Joannet, E., Horny, C., Kiwi-Minsker, L., and Renken, A. (2002) *Chem. Eng. Sci.*, **57** (16), 3453–3340.
90. Perrard, A., Gallezot, P., Joly, J.-P., Durand, R., Baljou, C., Coq, B., and Trens, P. (2007) *Appl. Catal. A: Gen.*, **331**, 100–104.
91. Höller, V., Yuranov, I., Kiwi-Minsker, L., and Renken, A. (2001) *Catal. Today*, **69**, 175–181.
92. Matatov-Meytal, Yu., Barelko, V., Yuranov, I., and Sheintuch, M. (2000) *Appl. Catal. B: Environ.*, **27** (2), 127–135.
93. Matatov-Meytal, Yu., Barelko, V., Yuranov, I., Kiwi-Minsker, L., Renken, A., and Sheintuch, M. (2001) *Appl. Catal. B: Environ.*, **31** (4), 233–240.
94. Matatov-Meytal, Yu., Shindler, Yu., and Sheintuch, M. (2003) *Appl. Catal. B: Environ.*, **45** (2), 127–134.
95. Matatov-Meytal, Yu. and Sheintuch, M. (2005) *Catal. Today*, **102-103**, 121–127.
96. Wang, J., Zhu, W., and Yang, S. (2008) *Appl. Catal. B: Environ.*, **78** (1/2), 30–37.

97. Höller, V., Wegricht, D., Kiwi-Minsker, L., and Renken, A. (2000) *Catal. Today*, **60**, 51–56.
98. Grasemann, M., Semagina, N., Renken, A., and Kiwi-Minsker, L. (2007) *Ind. Eng. Chem. Res.*, **46** (25), 8602–8606.
99. Richardson, J.T., Peng, Y., and Remue, D. (2000) *Appl. Catal. A: Gen.*, **204**, 19–32.
100. Richardson, J.T., Remue, D., and Hungl, J.-K. (2003) *Appl. Catal. A: Gen.*, **250**, 319–329.
101. Sirijaruphan, A., Goodwin, J.G., Rice, R.W., Wei, D., Butcher, K.R., Roberts, G.W., and Spiveye, J.J. Jr (2005) *Appl. Catal. A: Gen.*, **281**, 1–9.
102. Groppi, G., Giani, L., and Tronconi, E. (2005) *Ind. Eng. Chem. Res.*, **44**, 4993–5002.
103. Groppi, G., Giani, L., and Tronconi, E. (2007) *Ind. Eng. Chem. Res.*, **46**, 3955–3958.
104. Xu, X. and Moulijn, J.A. (2006) Transformation of a structured carrier into a structured catalyst, in *Structured Catalysts and Reactors*, 2nd edn, Chapter 21 (eds A. Cybulski and J.A. Moulijn), CRC Taylor & Francis, Boca Raton, p. 751.
105. Vergunst, T., Linders, M.J.G., Kapteijn, F., and Moulijn, J.A. (2001) *Catal. Rev. Sci. Eng.*, **43** (3), 291–314.
106. Cybulski, A. (2007) *Ind. Eng. Chem. Res.*, **46**, 4007–4033.
107. Stemmet, C.P., Jongmans, J.N., van der Schaaf, J., Kuster, B.F.M., and Schouten, J.C. (2005) *Chem. Eng. Sci.*, **60**, 6422–6429.
108. Stemmet, C.P., Meeuwse, M., van der Schaaf, J., Kuster, B.F.M., and Schouten, J.C. (2007) *Chem. Eng. Sci.*, **62**, 5444–5450.

10
Zeolite Membranes in Catalysis: What Is New and How Bright Is the Future?

Johan van den Bergh, Norikazu Nishiyama, and Freek Kapteijn

10.1
Introduction

Membranes in catalysis can be used to improve selectivity and conversion of a chemical reaction, improve stability and lifetime of the catalyst, and improve the safety of operation. The most well-known example is *in situ* removal of products of an equilibrium-limited reaction. However, many more ways of application of a membrane can be thought of [1–3], such as using the membrane as a reactant distributor to control the reactant concentration levels in the reactor, or performing catalysis inside the membrane and having control over reactant feed and product removal.

In this chapter, we limit ourselves to the topic of zeolite membranes in catalysis. Many types of membranes exist and each membrane has its specific field where it can be applied best. Comparing polymeric and inorganic membranes reveals that for harsher conditions and high-temperature applications, inorganic membranes outperform polymeric membranes. In the field of heterogeneous catalYsis, elevated temperatures are quite common and therefore this is a field in which inorganic membranes could find excellent applications.

The main advantage of zeolite membranes compared to other porous membranes is their uniform pore size due to the crystalline nature of the material. This may lead to molecular sieving effects. Currently known zeolites have pore sizes ranging from approximately 0.26 (6-ring window) to 0.74 nm (14-ring window), yielding an extensive library of materials to match a desired application. Dense membranes, such as metal- or perovskite-based ones, are often restricted to transport of only one component (e.g., H_2 or O_2). This is a limitation on the one hand, but implies an absolute separation for a limited number of very relevant (reactive) separations on the other. More pros and cons of inorganic membranes will be discussed in detail in Section 10.5.

In this chapter, the focus is on how zeolite membranes can be applied in the field of catalysis and to what extent this is successful. The latter is illustrated by reviewing some commonly studied zeolite membrane applications. Finally, the current hurdles that impede industrial application are discussed and some remarks

Novel Concepts in Catalysis and Chemical Reactors: Improving the Efficiency for the Future.
Edited by Andrzej Cybulski, Jacob A. Moulijn, and Andrzej Stankiewicz

ISBN: 978-3-527-32469-9

on the status of zeolite membranes in catalysis are made. But first, the properties of zeolite (membranes) are explored in detail to acquaint the reader with this interesting class of materials.

10.2
Zeolites: a Versatile, Well-Defined Class of Materials

Zeolite catalysts and adsorbents are widely accepted in industry. Commercial adsorbents based on synthetic aluminosilicates zeolite A and X became available in 1948 [4]. Zeolite Y as FCC catalyst became commercially available in 1964 [5].

Zeolites are crystalline aluminosilicates with well-defined pores. These pores are of molecular dimensions and can be classified as small (8-ring), medium (10-ring), and large pore (12-ring) zeolites. Each zeolite has its specific pore connectivity forming a one-, two-, or three-dimensional pore network. This interior network is well defined and can consist of combinations of channels, cages, intersections, or side pockets each of which has its influence on the adsorptive, catalytic, diffusive, and molecular sieving properties of the zeolite. The aforementioned properties can be varied by selecting the type of zeolite, the chemical composition of the framework (e.g., Si–Al ratio), or the counterion present in the structure to balance the charge of the framework aluminum atoms. The framework types of some well-known zeolites are presented in Figure 10.1. Zeolite structures can have anisotropic pore networks. A well-known example of this is the MFI pore network that consists of straight and sinusoidal channels with slightly different pore diameters of ~0.51–0.54 and 0.54 nm respectively. Owing to this anisotropy molecules enter a MFI crystal from either type of channels depending on the crystal facet.

Other materials, closely related to zeolites, with zeo-type structures are silico-aluminophosphates (SAPOs) and aluminophosphates (ALPOs) [6].

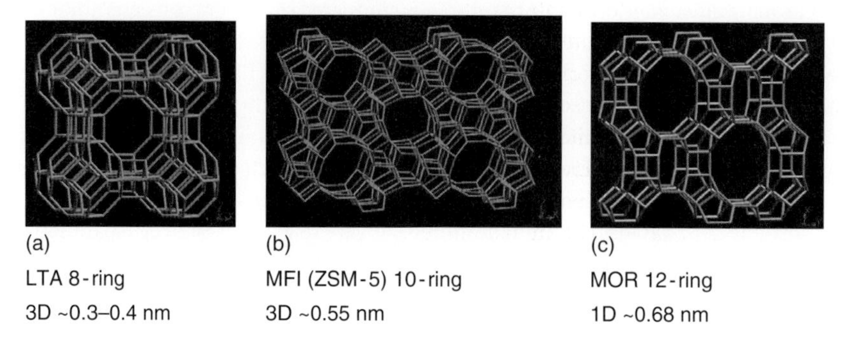

Figure 10.1 Framework types of some well-known zeolites with their specific pore size and pore network dimensions [6].

10.2.1
Zeolite Catalysis

Aluminum-containing zeolites are inherently catalytically active in several ways. The isomorphic substituted aluminum atom within the zeolite framework has a negative charge that is compensated by a counterion. When the counterion is a proton, a Brønsted acid site is created. Moreover, framework oxygen atoms can give rise to weak Lewis base activity. Noble metal ions can be introduced by ion exchanging the cations after synthesis. Incorporation of metals like Ti, V, Fe, and Cr in the framework can provide the zeolite with activity for redox reactions. A well-known example of the latter type is titanium silicalite-1 (TS-1), a redox molecular sieve catalyst [7].

It is not the catalytic activity itself that make zeolites particularly interesting, but the location of the active site within the well-defined geometry of a zeolite. Owing to the geometrical constraints of the zeolite, the selectivity of a chemical reaction can be increased by three mechanisms: reactant selectivity, product selectivity, and transition state selectivity. In the case of reactant selectivity, bulky components in the feed do not enter the zeolite and will have no chance to react. When several products are formed within the zeolite, and only some are able to leave the zeolite, or some leave the zeolite more rapidly, we speak about product selectivity. When the geometrical constraints of the active site within the zeolite prohibit the formation of products or transition states leading to certain products, transition state selectivity applies.

10.2.2
Zeolite Membranes

Because of their remarkable molecular sieving properties, it is not surprising that many attempts have been made to make zeolite-based membranes or films. The first zeolite membranes were reported about 20 years ago. The first systems were mixed matrix systems [8] and, in 1987, the concept that is now most commonly applied to make zeolite membranes was defined: an intergrown layer of zeolite crystals [9]. Such a zeolite membrane consists of a macroporous support layer with a thin zeolite layer in the order of a micrometer on top. The support layer gives the membrane the required mechanical strength. Figure 10.2a shows a typical cross-section of a thin zeolite film on top of a support layer and Figure 10.2b shows the top view of the same membrane. The quality of zeolite membranes improved over time and, the first industrial-scale application of a zeolite membrane was accomplished by Mitsui [10] in 2001. They developed a large-scale pervaporation plant based on NaA zeolite membranes to dewater alcohols producing $530\,l\,h^{-1}$ of solvents.

Examples of zeolites that have been prepared as membrane are: MFI (ZSM-5/Silicalite-1) [12], LTA (zeolite A) [13], FAU (Faujasite) [14], MOR (Mordenite) [15], DDR (DD3R) [16, 17], CHA (Chabasite, SAPO-34) [18], BEA (Beta) [19], FER (Ferrierite) [20], and SOD (Sodalite) [21].

Figure 10.2 NaA zeolite membrane (a) cross-section and (b) top view [11]. (Reprinted with permission of Elsevier.)

The composition of the framework (Si–Al ratio) is a very important factor regarding the application of a zeolite membrane and influences the membrane properties in various ways. Zeolites with low Si–Al ratios are very hydrophilic and, therefore, often used for water separation. An advantage of these types of membranes in such applications is that possible membrane defects are sealed off by strongly adsorbing water. But, it is found that membranes synthesized with higher Si–Al ratios have fewer defects and are of higher quality [22]. Therefore, for gas separation applications, high-silica zeolite membranes are the preferred choice. As a final remark, the chemical and hydrothermal stability of a zeolite is high in general, but zeolites with low Si–Al ratios are the least stable. An example of this is the low-silica zeolite, a membrane that is very suitable for water removal from alcohols, but is unstable in the presence of acids [23]. All silica zeolites are very stable and, therefore, much effort goes into synthesizing zeolite membranes with this composition.

10.3 Application Options

Membranes can be applied to catalysis in different ways. In most of the literature reports, the membrane is used on the reactor level (centimeter to meter scale) enclosing the reaction mixture (Figure 10.3). In most cases, the membrane is used as an inert permselective barrier in an equilibrium-limited reaction where at least one of the desired products is removed *in situ* to shift the extent of the reaction past the thermodynamic equilibrium.

A second option is to apply the membrane on the particle level (millimeter scale) by coating catalyst particles with a selective layer. As a third option, application at the microlevel (submicrometer scale) is distinguished. This option encompasses, for example, zeolite-coated crystals or active clusters (e.g., metal nanoparticles). Advantages of the latter two ways of application are that there are no sealing issues, it is easy to scale-up, the membrane area is large per unit volume, and, if there is a defect in the membrane, this will have a very limited effect on the overall reactor performance. Because of these advantages, it is believed that using a zeolite

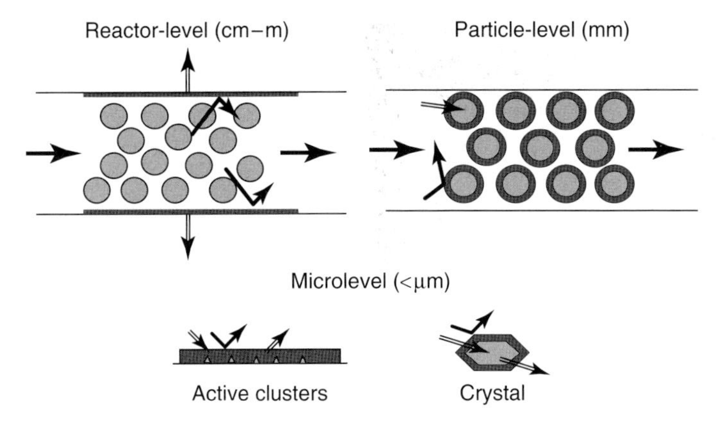

Figure 10.3 Membranes can be applied at the reactor-, particle-, and microlevels.

membrane on the particle or microlevel will be much easier than application at the reactor level.

10.3.1 Reactor Level

10.3.1.1 Membrane Reactors: Nomenclature

Membrane reactors are defined here based on their membrane function and catalytic activity in a structured way, predominantly following Sanchez and Tsotsis [2]. The acronym used to define the type of membrane reactor applied at the reactor level can be set up as shown in Figure 10.4. The membrane reactor is abbreviated as MR and is placed at the end of the acronym. Because the word "membrane" suggests that it is permselective, an "N" is included in the acronym in case it is nonpermselective. When the membrane is inherently catalytically active, or a thin catalytic film is deposited on top of the membrane, a "C" (catalytic) is included. When catalytic activity is present besides the membrane, additional letters can be included to indicate the appearance of the catalyst, for example, packed bed (PB) or fluidized bed (FB). In the case of an inert and nonpermselective

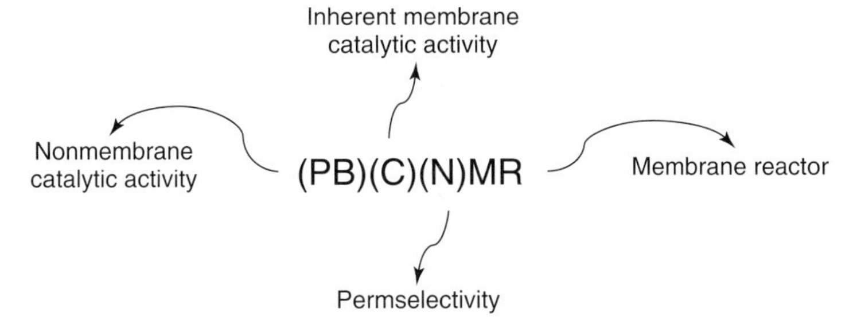

Figure 10.4 Meaning of acronyms used to define types of membrane reactors at the reactor level. (After [2]).

membrane, the nonmembrane catalytic activity is often not mentioned in the acronym.

Before going into detail on the different MR configurations, it should be realized that the main function of the membrane can be different for each configuration:

- **Extractor**: The membrane is used to selectively remove components from a reaction mixture.
- **Distributor**: The membrane is used to selectively feed components to a reaction mixture.
- **Contactor**: The Membrane is used to create a reaction front. When all reactants are fed from one side of the membrane, it is called a *flow-through contactor* (or reactor), when reactants are fed from opposite sides of the membrane it is called an *interfacial contactor*.

10.3.1.2 **Packed-Bed Membrane Reactor**

Most research reports involve an inert, selective membrane that encloses a PB of catalyst particles, a packed-bed membrane reactor (PBMR). It must be noted that the catalyst bed can also be fluidized or fixed, but types other than PBs are rarely found in literature. The following are the advantages of this type of reactor:

1) increase of single-pass conversion due to selective removal of products in an equilibrium-limited reaction;
2) increase of selectivity by selective removal of desired intermediate products or by selectively distributing one of the reactants and thereby suppressing undesired side reactions; and
3) improving safety by selective feeding of reactant and thereby preventing formation of, for example, explosive mixtures.

Typical examples of PBMRs where the membrane is used as extractor to shift the equilibrium conversion are dehydrogenation of alkanes [24], esterification, and etherification [23]. An interesting point is that, for example, in dehydrogenation, a low operating pressure is selected to obtain a higher conversion. In an MR, however, with increasing operating pressure both the reaction rate as well as membrane permeation rate can be increased without sacrificing a high conversion level, since this is ultimately not restricted to the equilibrium conversion of the feed composition. In this way, a MR can lead to process intensification [25], but integration of the reaction and separation steps leads to a decrease in degrees of freedom to optimize the overall process performance.

Zeolite membranes can also be envisaged as distributors. Gora *et al.* [26] used a silicalite-1 tubular membrane to selectively feed linear C6 alkanes from a mixed isomers feed to a platinum-containing chlorinated alumina fixed-bed catalyst. By combining the separation and reaction step into one unit the ongoing isomerization reaction of linear C6 alkanes provides a driving force for the separation of linear and branched isomers. Furthermore, combining these process steps might lead to higher energy efficiency, better process control, and lower energy consumption. An MFI membrane was used by Cruz-Lopez *et al.* [27] in the selective oxidation of butane to maleic anhydride, allowing the use of much higher butane feed concentrations while straying out of the flammability region.

10.3.1.3 Catalytic Membrane Reactor

In the case of a catalytic membrane reactor (CMR), the membrane is (made) intrinsically catalytically active. This can be done by using the intrinsic catalytic properties of the zeolite or by making the membrane catalytically active. When an active phase is deposited on top of a membrane layer, this is also called a *CMR* because this becomes part of the composite membrane. In addition to the catalytic activity of the membrane, a catalyst bed can be present (PBCMR). The advantages of a CMR are as follows:

1) improved selectivity due to good control of contact time and thereby limiting the occurrence of unwanted consecutive reactions;
2) reduction of mass transfer limitations;
3) shift in chemical equilibrium conversion due to a reduced concentration level of faster diffusing components in the membrane; and
4) safer operation when the membrane is used as an interfacial contactor keeping certain reactants segregated and preventing the formation of flammable or explosive mixtures.

Hasegawa *et al.* [28] selectively oxidized CO in a hydrogen-rich mixture using a Pt-loaded Y-type zeolite membrane. The envisaged application is to protect fuel cell electrodes from CO poisoning by zeolite films. In the methanol-to-olefins (MTO) process, methanol is converted by consecutive reactions into olefins and it can further react to form paraffins. Masuda *et al.* [29] applied a H–ZSM5 membrane in this reaction. The pressure drop over the membrane was used to control the contact time and a high selectivity (80–90%) at high conversions (60–98%) was obtained.

10.3.1.4 Nonselective Membrane Reactors

The last two configurations that involve a nonselective membrane are, to the best of our knowledge, not encountered in zeolite MRs. Zeolite membranes are almost always selective and suffer from the trade-off that high selectivities are typically combined with low fluxes. Therefore, when a nonselective membrane process is envisaged, a meso- or macroporous membrane would be the preferred choice. In the case of a catalytic nonpermselective membrane reactor (CNMR) the stoichiometric feeding of reactants can be controlled. A well-known example of such a membrane is presented by Sloot *et al.* [30] for the oxidation of H_2S to elemental sulfur and water. While the reaction is fast compared to mass transport in the membrane, a thin reaction zone is found in the membrane. The location of the reaction front is determined by the fluxes of the components to the reaction front. In this way, the system self-regulates stoichiometric feeding of reactants.

A packed-bed nonpermselective membrane reactor (PBNMR) is presented by Diakov *et al.* [31], who increased the operational stability in the partial oxidation of methanol by feeding oxygen directly and methanol through a macroporous stainless steel membrane to the PB. Al-Juaied *et al.* [32] used an inert membrane to distribute either oxygen or ethylene in the selective ethylene oxidation. By accounting for the proper kinetics of the reaction, the selectivity and yield of ethylene oxide could be enhanced over the fixed-bed reactor operation.

10.3.2 Particle Level

10.3.2.1 Basic Concepts

Figure 10.5 shows the basic concept of the particle-level MR that gives (i) selective addition of reactants to the reaction zone and (ii) selective removal of products from the reaction zone. In the first case, if the diffusivity of one reactant (A) is much higher than that of the other components (B), the reactant (A) selectively diffuses into a catalyst particle through a membrane. Undesired reactions or the adsorption of poisons on the catalysts can be prevented. In the second case, the reaction has a limited yield or is selectivity controlled by thermodynamics. The selective removal of the desired product from the catalyst particle gives enhancement of selectivity when the diffusivity of one product (R) is much greater than that of the other products (S).

10.3.2.2 Reactant-Selective Reactions

Nishiyama *et al.* [33, 34] have developed a method of coating silicalite-1 on spherical Pt/TiO_2 particles (silicalite/Pt/TiO_2). The silicalite/Pt/TiO_2 particles were used for hydrogenation of linear and branched alkenes. The composite silicalite-1/Pt/TiO_2 catalyst showed 1-hexene (1-Hex)/3,3-dimethylbut-1-ene (3,3-DMB) hydrogenation selectivities of 12–20 at 50 °C and 18–30 at 100 °C owing to the selective permeation of the reactant 1-Hex into Pt/TiO_2 particles through

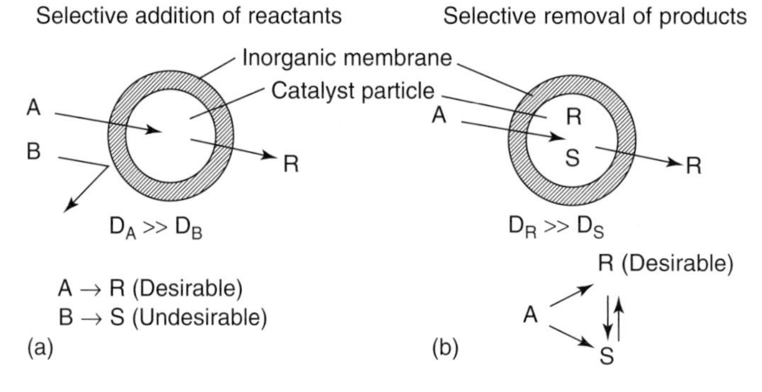

Figure 10.5 Principle of operation of a catalyst particle coated with a permselective membrane; (a) selective addition of reactants, (b) selective removal of products.

the silicalite-1 layer. Deactivation of the catalyst was also reduced, probably by protection against poisoning impurities in the feed. Zhong *et al.* [35] reported defect-free zeolite-4A membranes coated on Pt/γ-Al_2O_3 particles. Oxidation of a mixture of CO and *n*-butane over this composite catalyst has demonstrated the concept of reactant selectivity. The conversion of CO was over 90%, while *n*-butane hardly reacted. Since a trace amount of CO often coexists in hydrocarbon feed streams and CO can poison the catalysts of desired reactions, the zeolite-coated catalyst is attractive for the removal of CO from hydrocarbon streams.

10.3.2.3 **Product-Selective Reactions**

As an example of the selective removal of products, Foley *et al.* [36] anticipated a selective formation of dimethylamine over a catalyst coated with a carbon molecular sieve layer. Nishiyama *et al.* [37] demonstrated the concept of the selective removal of products. A silica–alumina catalyst coated with a silicalite membrane was used for disproportionation and alkylation of toluene to produce *p*-xylene. The product fraction of *p*-xylene in xylene isomers (para-selectivity) for the silicalite-coated catalyst largely exceeded the equilibrium value of about 22%. As illustrated in Figure 10.6, the high para-selectivity in the toluene disproportionation is caused by the selective removal of *p*-xylene from the silica–alumina particles, which leads to an apparent equilibrium shift between the xylene isomers.

The zeolite membrane has also been used as a catalytic membrane. Tsubaki *et al.* [38–40] reported a "capsule" catalyst for isoparaffin synthesis based on the FT reaction (Figure 10.6). H–ZSM-5 membrane was coated onto the surface of a preshaped Co/SiO_2 pellet. Syngas passed through the zeolite membrane to reach the Co/SiO_2 catalyst to be converted, and all straight-chain hydrocarbons formed left the particle through the zeolite channels undergoing hydrocracking as well as isomerization. A narrow, non-Anderson–Schultz–Flory product distribution was obtained. Contrary to a physical mixture of H–ZSM-5 and Co/SiO_2, C_{10+} hydrocarbons were suppressed completely on this catalyst, and the selectivity to middle isoparaffins was considerably improved.

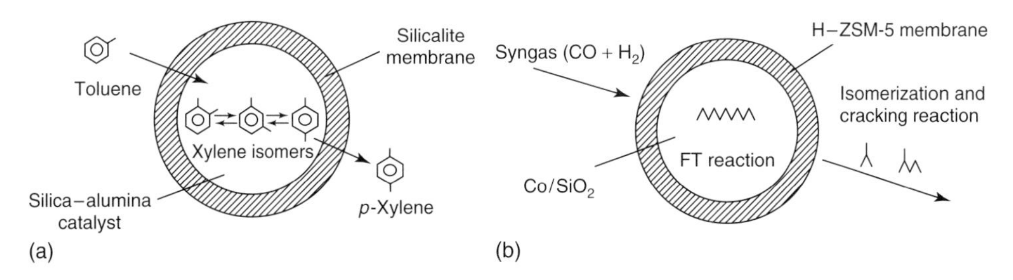

Figure 10.6 (a) Selective permeation of *p*-xylene though the silicalite-1 membrane coated on silica–alumina catalyst particles (b) Fischer–Tropsch reaction and hydrocracking over H–ZSM-5 membrane/Co/SiO_2 catalyst.

10.3.3
Micro Level

The principles of application of zeolite membranes at the microlevel can be very similar to those on the particle level, but now at the crystal (micrometer) scale, enclosing the active catalytic material.

As described in the previous section, the silica–alumina catalyst covered with the silicalite membrane showed excellent *p*-xylene selectivity in disproportionation of toluene [37] at the expense of activity, because the thickness of the silicalite-1 membrane was large (40 μm), limiting the diffusion of the products. In addition, the catalytic activity of silica–alumina was not so high. To solve these problems, Miyamoto *et al.* [41–43] have developed a novel composite zeolite catalyst consisting of a zeolite crystal with an inactive thin layer. In Miyamoto's study [41], a silicalite-1 layer was grown on proton-exchanged ZSM-5 crystals (silicalite/H–ZSM-5) [42]. The silicalite/H–ZSM-5 catalysts showed excellent para-selectivity of >99.9%, compared to the 63.1% for the uncoated sample, and independent of the toluene conversion.

The excellent high para-selectivity can be explained by the selective escape of *p*-xylene from the H-ZSM-5 catalyst and inhibition of isomerization on the external surface of catalysts by silicalite-1 coating. In addition to the high para-selectivity, toluene conversion was still high even after the silicalite-1 coating because the silicalite-1 layers on H–ZSM-5 crystals were very thin.

High catalytic activity and selectivity of silicalite-1/H–ZSM-5 composites must be caused by the direct pore-to-pore connection between H–ZSM-5 and silicalite-1 as revealed by Fe-SEM and TEM [43]. The silicalite-1 crystals were epitaxially grown on the surface of the H–ZSM-5 crystals.

The zeolite overgrowth has been reported for FAU on EMT zeolite [44] and MCM-41 on FAU zeolite [45]. On the other hand, in this study, zeolite layers were grown on the zeolite with the same framework structure, resulting in high coverage of ZSM-5 crystals with silicalite layers and high para-selectivity. The zeolite crystals with oriented thin layer on their external surface are expected to form a new class of shape-selective catalysts.

Apart from the above described core-shell catalysts, it is also possible to coat active phases other than zeolite crystals, like metal nanoparticles, as demonstrated by van der Puil *et al.* [46]. More examples of applications on the micro level are given in Section 10.5, where microreactors and sensor applications are discussed.

10.4
Potential Applications

In this section, an attempt is made to sketch the current status of zeolite MRs with respect to specific applications. The application of zeolite MRs is strongly related to the development status of zeolite membranes. Topics that are discussed are the most often studied reactions for zeolite membrane applications: dewatering, (de)hydrogenation, and isomerization (e.g., xylenes), and some special applications such as zeolite membranes in microreactors and sensor applications.

10.4.1 Dehydration

In situ removal of water of can be beneficial by shifting the equilibrium conversion of, for example, esterification, etherification, or condensation polymerization reactions. The state of development of zeolite membranes for this type of separation processes is high: the first commercial application of a zeolite membrane process has been accomplished by Mitsui [10]. Water-selective zeolite MRs are not commercially available, but several research groups have explored this type of application. *In situ* dehydration of esterification reactions have been recently investigated with Zeolite A [23, 47], T-type [48], MOR [23], and H–ZSM5 [49] membranes. The H–ZSM5 was particularly interesting because there was no catalytic activity in the membrane. In all cases, a proof of principle has been demonstrated that the MR yields a higher conversion than the PB reactor without a membrane. Owing to the very high membrane selectivity toward water, high conversions of 80–95% are found.

Zeolite A is a very successful membrane for separation of water from alcohols, but it suffers from stability issues under acid conditions [23]. Usually, a liquid phase should be avoided and, for this reason, vapor permeation is preferred. Recent developments show that the hydrophilic MOR [23] and PHI [50] membranes are more stable under acidic conditions in combination with a good membrane performance.

The preferred choice of a water-selective membrane up to now has been hydrophilic membranes because of their high water affinity. However, recently Kuhn *et al.* reported an all-silica DDR membrane for dehydration of ethanol and methanol with high fluxes (up to 20 kg m^{-2} h^{-1}) and high selectivities (H_2O/ethanol ~1500 and H_2O/methanol ~70 at 373 K) in pervaporation operation. The separation is based on molecular sieving with water fluxes comparable to well-performing hydrophilic membranes [51].

Recently, high-quality SOD membranes for water separation have been developed by Khajavi *et al.* [21, 52]. These zeolite membranes should allow an absolute separation of water from almost any mixture since only very small molecules such as water, hydrogen, helium, and ammonia can theoretically enter through the six-membered window apertures. Water/alcohol separation factors $\gg$10 000 have been reported with reasonable water fluxes up to 2.25 kg m^{-2} h^{-1} at 473 K in pervaporation experiments.

Furthermore, the application of the SOD membrane in a FT reaction has been investigated. The advantages of water removal in a FT reaction are threefold: (i) reduction of H_2O-promoted catalyst deactivation, (ii) increased reactor productivity, and (iii) displaced water gas shift (WGS) equilibrium to enhance the conversion of CO_2 to hydrocarbons [53]. Khajavi *et al.* report a mixture of H_2O/H_2 separation factors $\gg$10 000 and water fluxes of 2.3 kg m^{-2} h^{-1} under

FT conditions for their HSOD membrane. A conceptual process design indicated that application of SOD membranes in the FT reaction could be economically viable [54].

10.4.2 Dehydrogenation

The proof of principle that hydrogen-selective zeolite membranes can increase the conversion compared to a classical PB reactor in a dehydrogenation reaction has been demonstrated by various groups. Examples are the dehydrogenation of isobutane in combination with a ZSM5 membrane [24, 55], dehydrogenation of cyclohexane with an FAU membrane [56], and dehydrogenation of ethylbenzene to styrene with an Fe–ZSM5 [57], Al–ZSM5 [57], and a silicalite membrane [58]. Note that in the cyclohexane dehydrogenation, the FAU membrane removed both hydrogen and benzene selectively from the reaction mixture. The conversion increase in that case was attributed mostly to the removal of benzene, which had the highest permeance across the membrane. Table 10.1 summarizes the results of the zeolite MRs in terms of conversion increase, hydrogen permeance, and hydrogen separation performance.

The separation factors are relatively low and consequently the MR is not able to approach full conversion. With a molecular sieve silica (MSS) or a supported palladium film membrane, an (almost) absolute separation can be obtained (Table 10.1). The MSS membranes however, suffer from a flux/selectivity trade-off meaning that a high separation factor is combined with a relative low flux. Pd membranes do not suffer from this trade-off and can combine an absolute separation factor with very high fluxes. A favorable aspect for zeolite membranes is their thermal and chemical stability. Pd membranes can become unstable due to impurities like CO, H_2S, and carbonaceous deposits, and for the MSS membrane, hydrothermal stability is a major concern [62]. But the performance of the currently used zeolite membranes is insufficient to compete with other inorganic membranes, as was also concluded by Caro *et al.* [63] for the use of zeolite membranes for hydrogen purification.

Interestingly, the currently used zeolite membranes are all 10-membered (or larger) ring zeolites. It can be expected that smaller pore zeolites, like six- or eight-membered ring zeolites, would yield significantly higher separation factors or even absolute separation. Examples of such zeolites are: (i) the small-pore zeolite DDR that yields ideal separation factors of H_2 with respect to CH_4 and n-C_4 of >100 and >1000, respectively at 298 K [16] and (ii) the nanoblock zeolite membrane of Nishiyama *et al.*, which, they claimed, had a pore size smaller than the eight-membered ring zeolites and could separate H_2 from CH_4 with a selectivity of more than 1200 [61]. Although the fluxes for these membranes are still relatively low, they appear to be promising candidates with a significantly better performance than the currently used zeolite membranes in hydrogen-selective MRs.

Table 10.1 A comparison of recent examples of zeolite membranes in dehydrogenation reactions.

Membrane type	T (K)	Dehydrogenated HC	H_2 permeance (mol $m^{-2} s^{-1} Pa^{-1}$)	~H_2/HC selectivity	PBC[a] (%)	MRC[b] (%)	Reference
ZSM5 (MFI)	783	*i*-Butane	1.1×10^{-7}	70 – H_2/*i*-butane	29.1	48.6	[24]
ZSM5 (MFI)	730	*i*-Butane	–	9 – H_2/*i*-butane	~10	~55	[55]
FAU	473	Benzene	~8×10^{-7}	10 benzene/cyclohexene 4 H_2/cyclohexane	32.2	72.1	[56]
Fe–ZSM5 (MFI)	873	Ethylbenzene	20×10^{-7}	25.8 H_2/propane	45.1	60.1	[57]
Silicalite (MFI)	883	Ethylbenzene	6×10^{-7}(573 K)	15.4 H_2/*n*-butane (573 K)	67.5	74.8	[58]
MSS[c]	623	–	9×10^{-7}	13 (H_2/CO_2)	–	–	[59]
MSS	623	–	0.5×10^{-7}	>1000 (H_2/CO_2)	–	–	[59]
Pd film	773	–	40.9×10^{-7}	Infinite	–	–	[60]
DDR	298	–	0.2×10^{-7}	~100 (H_2/CH_4[d])	–	–	[16]
DDR	298	–	0.2×10^{-7}	>1000 (H_2/*n*-C_4[d])	–	–	[16]
Nanoblock	298	–	0.69×10^{-7}	>1200 (H_2/CH_4)	–	–	[61]

[a] Packed-bed conversion.
[b] Membrane reactor conversion.
[c] Molecular Sieve Silica.
[d] Permselectivity.

10.4.3
Isomer Separation

Separation of isomers is an application where zeolite membranes could be specifically interesting because of their well-defined pores that lead to molecular sieving effects. An application that is often considered is the xylene isomerization and related reactions.

Several research groups have attempted to improve the conversion and selectivity of the xylene isomerization (Table 10.2). Van Dyk *et al.* [64] carried out an isomerization of *m*-xylene in a silicalite-1 PBMR and compared it with the performance of a classical PB reactor. The membrane had a very high *para*-to-*ortho* selectivity that led to 100% selectivity in the permeate stream of the MR. The overall *para*-selectivity (retentate + permeate) was enhanced from 58 to 65% and the yield was increased from 21 to 23%. Haag *et al.* [65] showed, however, how difficult it is to compare data in an experimental study. A comparison was made between a conventional PB reactor and a CMR (H–ZSM5). In the CMR, the only catalytic activity was introduced by the intrinsic activity of the membrane. It was found by a kinetic study that the activity of the conventional H–ZSM5 catalyst was not comparable to the activity of the membrane. Furthermore, conversions well below the calculated equilibrium conversions were obtained, making it difficult to draw conclusions based on the obtained data. No convincing improvements were achieved mainly because of the poor membrane performance. Tarditty *et al.* [66] used a CMR (Pt–ZSM5) and a PBMR (Ba–ZSM5) to perform an *m*-xylene isomerization reaction. In both cases, the MRs showed a reasonable improvement of the selectivity and yield as compared to a conventional PB reactor.

These three studies show that MFI membranes can enhance yield and selectivity in a xylene isomerization process, but the improvements obtained are not yet convincing regarding commercial application. An approach that seems more promising is an MR applied at the particle level. Van Vu *et al.* [42] demonstrated that *para*-xylene can be synthesized from toluene and methanol with very high selectivities (>99.9%), which sustained even at higher conversions (Figure 10.7). In their studies H–ZSM5 catalyst particles were coated with an inert silicalite layer. The increase in selectivity may be attributed to the selective mass transport resistance introduced by the silicalite layer. Additionally, the coated catalyst particles deactivated more slowly compared to the uncoated particles, probably because of a reduction in coke formation due to a reduction of surface acid groups by the coating.

10.4.4
Microreactors

The following are some of the reasons that microreactors can be be used: (i) reduced mass and heat transfer limitations, (ii) high area to volume ratio, (iii) safer operation, and (iv) ease of scaling up by numbering out. The advantages of scaling down zeolite membranes are that it could be easier to create defect-free membranes and

Table 10.2 Performance of several zeolite membrane reactors in the xylene isomerization reaction.

Membrane type	*T* (K)	*p*-Xylene permeance (mol m^{-2} s^{-1} Pa^{-1})	~*p*/*o* Membrane selectivity	PBS^a (%) (*para*)	MRS^b (%)	PBY^c (%)	MRY^d (%)	Reference
Silicalite (MFI)	577	6×10^{-9}	Infinite	58	65	21	23	[64]
H–ZSM5 (MFI)	573	7.4×10^{-7}	2.2	52.3	63.6	3.1	4.4	[65]
Ba–ZSM-5 (MFI)	643	6.5×10^{-8}	19.0	52	69	21	25	[66]
Pt–ZSM5 (MFI)	643	7.2×10^{-8}	4.9	n.a.	72	18	22	[66]

n.a., not applicable,
[a] Packed-bed selectivity.
[b] Membrane reactor selectivity.
[c] Packed bed yield.
[d] Membrane reactor yield.

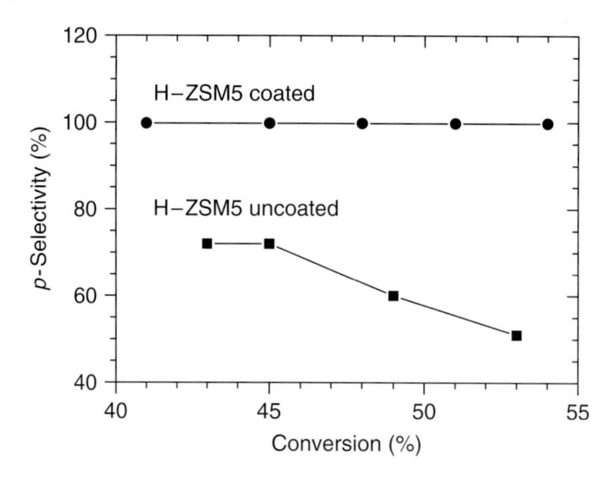

Figure 10.7 *Para*-selectivity as a function in the conversion of the alkylation of methanol and toluene to xylene by bare and silicalite-coated H–ZSM5 catalyst particles.

that by using single-crystal membranes, the advantages of oriented crystals may be more efficiently exploited. Studies on zeolite membranes applied in microreactors are scarce and most of them focus on demonstrating that zeolite films can very well be synthesized on the microreactor scale.

Yeung *et al.* have shown that as in the macroscale, the yield and selectivity of equilibrium-limited reactions in a microzeolite MR can be similarly improved. For the Knoevenagel condensation, *in situ* water removal in a microreactor using a NaA [67] and a ZSM5 [68] zeolite membrane increased the yield as compared to a conventional microreactor. A typical yield increase from ~60 (equilibrium) to ~85% was demonstrated [68]. A slight yield increase and a reduction in catalyst deactivation could be achieved by application of a micro-MR in the oxidation of aniline. These advantages are attributed to the *in situ* removal of water by the ZSM5 zeolite membrane [69].

The application of zeolite membranes in microreactors is still in an early stage of development, and suffers sometimes from unexpected problems arising from template removal [70]. However, several application examples of zeolite membranes in microstructured devices have been demonstrated yielding similar advantages as were to be expected from experiences on the macroscale. Because of the high surface to volume ratio of microreactors, the application of zeolite membranes in these systems has great potential.

10.4.5
Chemical Sensors

Adequate techniques to accurately and selectively detect concentration levels are available, but are mostly expensive, slow, and complex. Cheaper sensors are available, but the working principle is usually not very selective. A wealth of different sensing techniques exists where typical sensor principles rely on changes in mass (microbalance), electrical properties (e.g., capacitance), or optical properties (e.g., IR). A good sensor has a fast response, high sensitivity, and high selectivity. Additionally, for several applications, sensors need to withstand harsh conditions (e.g., exhaust gases) for long periods. Zeolites are potential candidates to improve the sensitivity and selectivity by exploiting their molecular sieving, selective adsorption, and catalytic properties.

Three different ways in which a zeolite membrane can contribute to a better sensor performance can be distinguished: (i) the add-on selective adsorption or molecular sieving layer to the sensor improves selectivity and sensitivity, (ii) the zeolite layer acts as active sensing material and adds the selective adsorption and molecular sieving properties to this, and (iii) the zeolite membrane adds a catalytically active layer to the sensor, improving the selectivity by specific reactions.

A very recent example of the first case is presented by Vilaseca *et al.* [71] where an LTA coating on a micromachined sensor made the sensor much more selective to ethanol than methane. Moos *et al.* [72, 73] report H–ZSM5 NH_3 sensor based on impedance spectroscopy using the zeolite as active sensing material. At elevated temperatures (>673 K) NH_3 still adsorbs significantly in contrast to CO_2, NO, O_2, and hydrocarbons, leading to a NH_3 selective sensor. Arguments in favor of using a zeolite-based sensor in NH_3 detection for automotive applications are low cost, high temperature stability, and suitability for use in thick-film technology, of common use in the automotive industry [72]. The sensors were tested on an engine test bank and the authors claim that "the sensor itself meets all the technological and economical demands of the automotive industry" [73].

When NO_x levels are measured electrochemically, NO and NO_2 can lead to opposing signals because NO is oxidized and NO_2 tends to be reduced. Moreover, it is preferred to obtain a "total" NO_x measurement instead of only one of the constituents. The latter can be achieved by catalytically equilibrating the feed with oxygen before contact with the sensor by coating an active zeolite layer on top or placing a active catalyst bed in front of the sensor. Both approaches have been demonstrated successfully with a Pt–Y zeolite as active catalyst [74, 75]. The additional advantage of the filter bed is a reduction in the cross-sensitivity with CO due to CO oxidation above 673 K.

In a similar manner, Sahner *et al.* [76, 77] utilized a Pt–ZSM-5 layer to reduce the cross-sensitivity of a hydrocarbon (propane) sensor toward CO, propene, H_2, and NO at 673 K. The zeolite layer was put on the sensor as a paste. The improved cross-sensitivity is attributed to selective oxidation of all considered components except propane. Trimboli *et al.* [78] demonstrated the same concept by using a Pt–Y zeolite for the CO oxidation, maintaining the sensitivity for propane.

The added value, variety of use, and methods to apply zeolite coatings or films in sensor applications has been convincingly demonstrated. Although current trends focus on miniaturization of sensors and creating smaller zeolite crystals and thinner films, to decrease the response time of the sensor [79], often thick-film technology is sufficient to apply zeolite films for this type of application. Some sensor materials cannot withstand the high temperatures necessary for template removal by air calcination. Recent work demonstrated that ozonication yields sufficient removal at lower temperatures (~470 K), extending the application of templated zeolites for thermosensitive sensors materials [80–82]. The long-term stability has not been demonstrated but, considering the stability of zeolites in general, this is not expected to be a limiting factor. A bright future is anticipated for the application of zeolite films in sensor applications.

10.5 Current Hurdles

It is evident that zeolite membranes can have added value in catalysis, but industrial applications are still lacking. Thus, the question remains what the major factors are that stand in the way of application. The following factors are considered troublesome and are discussed in more detail:

1) Cost considerations
2) Synthesis of thin (<1 μm) defect-free and stable membranes
3) Scale-up and reproducibility
4) Cheap high-temperature sealing
5) Understanding of transport phenomena
6) Catalyst development
7) Reaction and membrane integration.

10.5.1 Cost Considerations

The estimated cost of a zeolite membrane is about €1000–5000 m^{-2} [83–86] (I. Voigt, personal communication) at this moment, including a full module design. A large part of the costs are related to the module and support, and only 10–20% [83] to 50% (I. Voigt, personal communication) to the membrane itself.

Although cost is the predominant factor regarding application, feasibility studies for zeolite membrane-based processes are scarce. The cost that would make a zeolite membrane process profitable is strongly related to the targeted process. Meindersma and de Haan [84] estimated that for an industrial-scale separation process of aromatic hydrocarbons, a cost of €200 m^{-2} would be profitable; in addition, a minimum selectivity of 40 and an aromatic hydrocarbon flux 25 times higher than reported are additional requirements. According to Tennison [85], detailed flow-sheeting studies pointed out that very few processes could tolerate installed membrane costs of more than €1000 m^{-2}. As an example, the dehydrogenation process of butane is mentioned where, with the fluxes reported in the late 1990s and with a hydrogen separation factor of 40, costs should be below €1000 m^{-2} to make the process profitable. Although, a recent report by Khajavi *et al.* [54] indicates that using SOD membranes for water removal in an FT process, a viable zeolite-membrane reactor-based process could be within reach, the overall picture is that the current price level of zeolite membranes is too high for profitable application. On the other hand, the energy savings and/or CO_2 emission reductions that might be achieved is an aspect that gains increasing importance and may shift the break even price for membrane investment. Considering that the major part of the costs of supported membranes are related to the support leads to the conclusion that this is the part where significant cost reduction is required to permit widespread industrial application of zeolite membranes.

The considerations above apply to zeolite membranes as applied on the macrolevel (e.g., PBMR). Zeolite membranes applied on the particle level or smaller might lead to a more optimistic outlook since this type of application neither involves expensive modules and supports nor expensive sealing material.

10.5.2 Synthesis of Thin Defect-Free Membranes

Tremendous progress has been made in the last few decades in zeolite membrane synthesis, and, as shown in Section 10.3, high-quality thin membranes of a growing number of different zeolites can be produced. But, considering that zeolite membranes have a relative low flux, "ultra" thin ($<$1-μm) membranes are desired for a more widespread application. Hedlund *et al.* have been very active in this field and have been able to make high-flux MFI membranes with a thickness of 0.5 μm in a very reproducible way [87]; however, such a membrane thickness remains at a lower limit which is not common practice. Thin MFI membranes on a titania support, to avoid incorporation of any destabilizing Al from a support, were synthesized for butene isomer separation. The goal of high fluxes was achieved, but selectivity relevant for practice was moderate and decreased at higher pressures [88].

10.5.3 Scale-up and Reproducibility

For installation of large membrane areas, module designs with high surface to volume ratios are desired. Geometries that are under consideration are multitubular, monolithic (or multichannel), capillaries, and hollow fibers. Scaling up by a multitubular geometry is applied by Mitsubishi Chemicals (formerly by BNRI) in their medium-scale isopropanol and large-scale ethanol dehydration plants ([10]) and has been demonstrated to be a viable technology. Inocermic also is involved in a new ethanol dehydration plant [22]. For high temperatures, sealing is an important issue and an interesting novel concept is the all-ceramic multitubular support. Although not demonstrated for zeolite membranes yet, successful H_2/N_2 separation experiments with a silica membrane on this type of support at 773 K and a pressure drop of 9.5 bar have been demonstrated for 1000 h of operation [89].

Although hollow fibers are thought to be an excellent candidate to be used as support – they are cheap and have a very high surface area to volume ($>$1000 $m^2\ m^{-3}$) – very few reports on hollow-fiber-supported zeolite membranes exist in the open literature. For zeolite membranes, ceramic hollow fibers are preferred because of their mechanical and thermal stability. Recently, Alshebani

Figure 10.8 (a) Ceramic hollow-fiber module of Hyflux [91]; (b) ceramic multichannel supports of Inopor GmbH [93]; and (c) the multichannel membrane design of NGK insulators.

et al. [90] obtained a high-quality MFI membrane supported by an alumina hollow fiber via a pore-plugging synthesis. The fluxes and binary separation factors of a H_2/n-butane mixture were comparable to high-quality MFI membranes prepared on tubular supports. Zeolite membranes based on ceramic hollow fibers appear to be a promising option, moreover, because ceramic hollow-fiber modules are commercially available [91] (Figure 10.8). Richter *et al.* [92] report the synthesis of MFI membranes supported by small tubes and capillaries. Reasonable selectivities, better than Knudsen selectivity, are demonstrated.

Bowen *et al.* [94] made a B–MFI membrane on a monolithic support. The pervaporation fluxes and selectivities of several alcohol/water mixtures were comparable to similar tubular-based B–MFI membranes, demonstrating the scale-up, although, for pervaporation, the quality requirements are much more forgiving. Kuhn *et al.* tested a multichannel high-silica MFI membrane for ethanol/water separation. The membrane was supplied by NGK Insulators and, also, in this case, the multichannel membrane measures up to its tubular counterparts [95] (Figure 10.8).

An important driver for zeolite membrane applications has been the commercialization of the NaA membranes for dehydration. However, for these membranes, the quality required is not as high as compared to gas-phase molecular sieving applications, since in the case of dehydration, defects or imperfections are blocked by water adsorption, resulting in high selectivities, even if gas-phase separations yield only Knudsen selectivity values. Natural gas purification by zeolite membranes (e.g., DD3R) seems to be a viable application [16, 17], and this application could strongly catalyze the development of large-surface-area gas permeation zeolite membranes.

The reproducibility of zeolite membranes is sometimes questioned. But, taking into account the large surface areas that are produced for the mentioned application examples, for well-studied zeolite membrane syntheses this does not appear to be a limiting factor.

10.5.4 High-Temperature Sealing

Sealing can be expensive when zeolite membranes are used at high temperatures. Cheap polymer sealings can be used up to around 400–500 K. Higher temperatures require much more expensive polymers or graphite. Brazed or glazed tube ends can withstand high temperatures, but suffer from mismatches in thermal expansion coefficients with the support and module, which can lead to thermally induced stress. Innovative methods to simplify sealing are, for example, the synthesis of the membrane on a steel support [96] that allows welding of the support onto a module. An innovative and promising approach is the all-ceramic multitubular support [89, 97] in which several tubes are connected to a dense alumina endplate by a glass-based sealant prior to the synthesis of the membrane. No additional sealing is required.

10.5.5 Transport Phenomena

In order to design a zeolite membrane-based process a good model description of the multicomponent mass transport properties is required. Moreover, this will reduce the amount of practical work required in the development of zeolite membranes and MRs. Concerning intracrystalline mass transport, a decent continuum approach is available within a Maxwell–Stefan framework for mass transport [98–100]. The well-defined geometry of zeolites, however, gives rise to microscopic effects, like specific adsorption sites and nonisotropic diffusion, which become manifested at the macroscale. It remains challenging to incorporate these microscopic effects into a generalized model and to obtain an accurate multicomponent prediction of a "real" membrane.

In the case of supported membranes also, the support can play an important role in the separation performance of the membrane in the gas as well as in the liquid phase [101–103]. Transport in these support pores can be accurately described by the "Dusty Gas Model" [100, 104] although it is put forward by Kerkhof and Geboers that their "Binary Friction Model" is physically more correct [105].

Modeling on the reactor level, which is needed in designing the zeolite membrane reactor, could receive some more attention, since the number of studies in this particular field are few [56, 106].

10.5.6 Catalyst Development

An important aspect of a MR is the catalyst [107]. Replacing a conventional reactor by a MR can change the operating conditions considerably (e.g., operating temperature, concentration levels), which might be rather different from the catalyst it was originally designed for. Furthermore, for a MR system with a high-performance membrane, the catalyst can become the factor limiting the reactor's effectiveness [108].

When considering dehydrogenation reactions, an extractor-type PBMR can lead to higher conversion by hydrogen removal, than a conventional fixed-bed reactor. Further, the operation temperature can be reduced without sacrificing product yield. But, such a temperature reduction leads to a reduced catalytic activity and the extraction of hydrogen can lead to increased coke formation. Thus, for this application, more active catalysts with reduced coke formation would be required to ultimately arrive at an optimal reactor system. The main message is that, with increasing membrane performance, it should not be overlooked that eventually another part of the system will limit the reactor performance and this can very well be the catalyst developed for normal steady-state operation.

10.5.7 Reaction and Membrane Integration

A trivial point, which arises from the previous subsection and is, without saying, assumed throughout this chapter, is that the operational regime of the membrane and of the catalyst should overlap to be able to combine catalysis and membrane permeation. Furthermore, the performance of the catalyst and the membrane, in terms of productivity and permeability, respectively, should match and be integrated in an appropriate way [109]. Otherwise, either the catalyst or the membrane will be poorly utilized. In practice, the reactor space time yield (STY, mol m^{-3} s^{-1}) is matched with the membrane areal time yield (ATY, mol m^{-2} s^{-2}) by application of a support geometry with the required membrane surface area to volume ratio (A/V, m^2 m^{-3}). As pointed out by van de Graaf *et al.*, A/V values of porous ceramic membranes range from 20 to 5000 m^{-1}, which appears sufficient to match typical STY values of 1–10 mol m^{-3} s^{-1} making application of zeolite membranes in catalysis feasible from this perspective [110].

Integration of the separation and reaction step has several advantages, but an inherent downside of such a process intensification step is the loss of degrees of freedom for process design and process control (Figure 10.9).

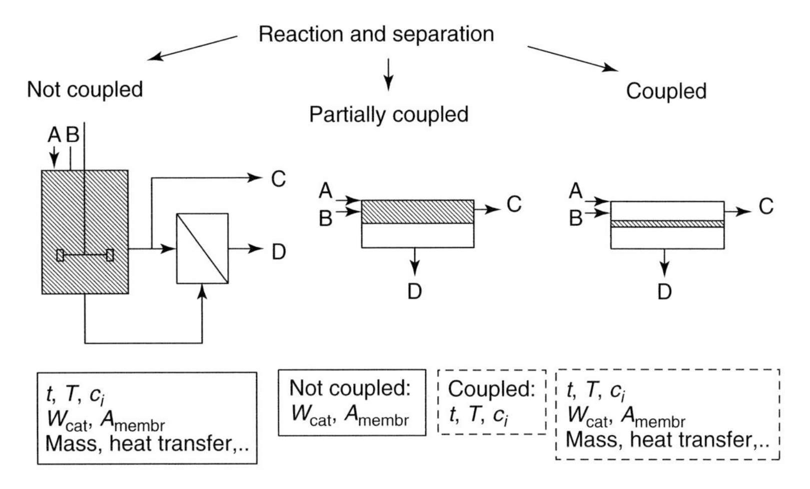

Figure 10.9 Combining reactions and separation steps leads to a loss in process design degrees of freedom.

10.5.8
Concluding Remarks

The consideration that many zeolite types exist, each with many tunable properties (e.g., pore size and alumina content), leads not only to a wealth of options but also to a high level of complexity. Owing to this complexity and limited understanding of zeolite formation and permeation behavior, a lot of experimental effort is required in this field, slowing down developments toward successful application.

A general remark regarding industrial implementation of any new technology is that the process industry is conservative and new technologies need to be well demonstrated before they are accepted, particularly when high investment costs are involved. Moreover, application of a MR in an existing plant will not be straightforward since this MR will presumably be designed for a different conversion level than the original reactor, which will affect downstream processing. Therefore, introduction of a MR is not likely to be a simple change of reactor, but requires a considerable plant redesign. A completely new process layout has a better outlook but requires considerable investments. Rather, membrane add-on units are expected to have a larger chance of introduction in existing plants to improve performance, debottlenecking, or increasing capacity.

As is obvious, many potential hurdles discussed in the previous sections do not apply to application of zeolite membranes at the micro- and particle levels. Issues like scale-up and high-temperature sealing do not play a role here. Additionally, coated catalyst particles do not require a change of reactor, but only replacement of the catalyst. Application of zeolite membranes at these levels is therefore considered to be easier and their implementation will probably occur earlier.

10.6 Concluding Remarks and Future Outlook

In the early 1990s, when the first zeolite membranes were developed, the expected time to successful application in separation and catalytic processes was underestimated. The main reasons for this were the great complexity – for instance, the challenging synthesis of thin defect-free membranes and a good understanding of multicomponent transport in zeolite membranes – and the high installation costs. But, reasons to use zeolite membranes – their thermal and chemical stability, catalytic and separation properties – are without doubt well-founded and industrial application appears to be technically feasible. However, the currently estimated installed membrane cost (€1000–5000 m^{-2}) appears to be too high and can be considered as the predominant hurdle with respect to successful widespread industrial application. For current supported zeolite membrane technology, a dominant cost factor is the support price, which should be reduced.

Zeolite membranes are developing into a more mature technology, which is emphasized by the industrial application of zeolite membranes in alcohol dehydration. Yet it remains a technology with high potential, since further applications are still lacking. However, taking into account the solid progress made in the last decade and the demonstrated technological benefits, application of zeolite membranes in the field of catalysis has a serious prospect.

References

1. Falconer, J.L., Noble, R.D., and Sperry, D.P. (1995) *Catalytic Membrane Reactors, in Membrane Separations Technology; Principles and Applications*, Chapter 14 (eds R.D.Noble and S.A. Stern), Elsevier, Amsterdam, pp. 669–712.
2. Sanchez, J. and Tsotsis, T.T. (2002) *Catalytic Membranes and Membrane Reactors*, Wiley-VCH Verlag GmbH, Weinheim.
3. Hsieh, H.P. (1996) *Inorganic Membranes for Separation and Reaction*, Elsevier, Amsterdam.
4. Corma, A. (2003) *J. Catal.*, **216**, 298–312.
5. Plank, C.J., Hawthorne, W.P., and Rosinski, E.J. (1964) *Ind. Eng. Chem. Prod. Res. Dev.*, **3**, 165–169.
6. Baerlocher, C. and McCusker, L.B. (2008) Database of zeolite structures, *www.iza-structure.org/databases/* (visited November 2008).
7. Taramasso, M., Perego, G., and Notari, B. (1983) Preparation of porous crystalline synthetic material comprised of silicon and titanium oxides. US Patent 4,410,501.
8. Kulprathipanja, S. and Charoenphol, J. (1986) Mixed matrix membrane for separation of gases. US Patent 6,726,744.
9. Suzuki, H. (1987) Composite membrane having a surface layer of an ultrathin film of cage-shaped zeolite and processes for production thereof. US Patent 4,699,892.
10. Morigami, Y., Kondo, M., Abe, J., Kita, H., and Okamoto, K. (2001) *Sep. Purif. Technol.*, **25**, 251–260.
11. Aguado, S., Gascon, J., Jansen, J.C., and Kapteijn, F. (2009) *Microporous Mesoporous Mater.*, **120**, 170–176.
12. Bakker, W.J.W., Kapteijn, F., Poppe, J., and Moulijn, J.A. (1996) *J. Membr. Sci.*, **117**, 57–78.

13. Okamoto, K., Kita, H., Horii, K., Tanaka, K., and Kondo, M. (2001) *Ind. Eng. Chem. Res.*, **40**, 163–175.
14. Lassinantti, M., Hedlund, J., and Sterte, J. (2000) *Microporous Mesoporous Mater.*, **38**, 25–34.
15. Matsukata, M., Sawamura, K.I., Shirai, T., Takada, M., Sekine, Y., and Kikuchi, E. (2008) *J. Membr. Sci.*, **316**, 18–27.
16. Tomita, T., Nakayama, K., and Sakai, H. (2004) *Microporous Mesoporous Mater.*, **68**, 71–75.
17. van den Bergh, J., Zhu, W., Gascon, J., Moulijn, J.A., and Kapteijn, F. (2008) *J. Membr. Sci.*, **316**, 35–45.
18. Li, S.G., Falconer, J.L., and Noble, R.D. (2004) *J. Membr. Sci.*, **241**, 121–135.
19. Maloncy, M.L., van den Berg, A.W.C., Gora, L., and Jansen, J.C. (2005) *Microporous Mesoporous Mater.*, **85**, 96–103.
20. Matsufuji, T., Nakagawa, S., Nishiyama, N., Matsukata, M., and Ueyama, K. (2000) *Microporous Mesoporous Mater.*, **38**, 43–50.
21. Khajavi, S., Kapteijn, F., and Jansen, J.C. (2007) *J. Membr. Sci.*, **299**, 63–72.
22. Caro, J., Noack, M., and Kölsch, P. (2005) *Adsorpt. J. Int. Adsorpt. Soc.*, **11**, 215–227.
23. de la Iglesia, O., Mallada, R., Menéndez, M., and Coronas, J. (2007) *Chem. Eng. J.*, **131**, 35–39.
24. Illgen, U., Schäfer, R., Noack, M., Kölsch, P., Kühnle, A., and Caro, J. (2001) *Catal. Commun.*, **2**, 339–345.
25. Rezai, S.A.S. and Traa, Y. (2008) *J. Membr. Sci.*, **319**, 279–285.
26. Gora, L. and Jansen, J.C. (2005) *J. Catal.*, **230**, 269–281.
27. Cruz-López, A., Guilhaume, N., Miachon, S., and Dalmon, J.A. (2005) *Catal. Today*, **107-108**, 949–956.
28. Hasegawa, Y., Kusakabe, K., and Morooka, S. (2001) *J. Membr. Sci.*, **190**, 1–8.
29. Masuda, T., Asanuma, T., Shouji, M., Mukai, S.R., Kawase, M., and Hashimoto, K. (2003) *Chem. Eng. Sci.*, **58**, 649–656.
30. Sloot, H.J., Versteeg, G.F., and Van Swaaij, W.P.M. (1990) *Chem. Eng. Sci.*, **45**, 2415–2421.
31. Diakov, V. and Varma, A. (2002) *Chem. Eng. Sci.*, **57**, 1099–1105.
32. Al-Juaied, M.A., Lafarga, D., and Varma, A. (2001) *Chem. Eng. Sci.*, **56**, 395–402.
33. Nishiyama, N., Ichioka, K., Park, D.H., Egashira, Y., Ueyama, K., Gora, L., Zhu, W.D., Kapteijn, F., and Moulijn, J.A. (2004) *Ind. Eng. Chem. Res.*, **43**, 1211–1215.
34. Nishiyama, N., Ichioka, K., Miyamoto, M., Egashira, Y., Ueyama, K., Gora, L., Zhu, W.D., Kapteijn, F., and Moulijn, J.A. (2005) *Microporous Mesoporous Mater.*, **83**, 244–250.
35. Zhong, Y.J., Chen, L., Luo, M.F., Xie, Y.L., and Zhu, W.D. (2006) *Chem. Commun.*, 2911–2912.
36. Foley, H.C., Lafyatis, D.S., Mariwala, R.K., Sonnichsen, G.D., and Brake, L.D. (1994) *Chem. Eng. Sci.*, **49**, 4771–4786.
37. Nishiyama, N., Miyamoto, M., Egashira, Y., and Ueyama, K. (2001) *Chem. Commun.*, 1746–1747.
38. He, J.J., Xu, B.L., Yoneyama, Y., Nishiyama, N., and Tsubaki, N. (2005) *Chem. Lett.*, **34**, 148–149.
39. He, J.J., Yoneyama, Y., Xu, B.L., Nishiyama, N., and Tsubaki, N. (2005) *Langmuir*, **21**, 1699–1702.
40. He, J.J., Liu, Z.L., Yoneyama, Y., Nishiyama, N., and Tsubaki, N. (2006) *Chem. Eur. J.*, **12**, 8296–8304.
41. Miyamoto, M., Kamei, T., Nishiyama, N., Egashira, Y., and Ueyama, K. (2005) *Adv. Mater.*, **17**, 1985–1988.
42. van Vu, D., Miyamoto, M., Nishiyama, N., Egashira, Y., and Ueyama, K. (2006) *J. Catal.*, **243**, 389–394.
43. van Vu, D., Miyamoto, M., Nishiyama, N., Ichikawa, S., Egashira, Y., and Ueyama, K. (2008) *Microporous Mesoporous Mater.*, **115**, 106–112.
44. Yonkeu, A.L., Buschmann, V., Miehe, G., Fuess, H., Goossens, A.M., and Martens, J.A. (2001) *Cryst. Eng.*, **4**, 253–267.
45. Kloetstra, K.R., Zandbergen, H.W., Jansen, J.C., and van Bekkum, H. (1996) *Microporous Mater.*, **6**, 287–293.
46. van der Puil, N., Creyghton, E.J., Rodenburg, E.C., Sie, S.T., van Bekkum, H., and Jansen, J.C. (1996) *J. Chem. Soc., Faraday Trans.*, **92**, 4609–4609.

47. Jafar, J.J., Budd, P.M., and Hughes, R. (2002) *J. Membr. Sci.*, **199**, 117–123.
48. Tanaka, K., Yoshikawa, R., Ying, C., Kita, H., and Okamoto, Ki. (2001) *Catal. Today*, **67**, 121–125.
49. Bernal, M.P., Coronas, J., Menéndez, M., and Santamaría, J. (2002) *Chem. Eng. Sci.*, **57**, 1557–1562.
50. Kiyozumi, Y., Nemoto, Y., Nishide, T., Nagase, T., Hasegawa, Y., and Mizukami, F. (2008) *Microporous Mesoporous Mater.*, **116**, 485–490.
51. Kuhn, J., Yajima, K., Tomita, T., Gross, J., and Kapteijn, F. (2008) *J. Membr. Sci.*, **321**, 344–349.
52. Khajavi, S., Jansen, J.C., and Kapteijn, F. (2009) *J. Membr. Sci.*, **326**, 153–160.
53. Rohde, M.P., Schaub, G., Khajavi, S., Jansen, J.C., and Kapteijn, F. (2008) *Microporous Mesoporous Mater.*, **115**, 123–136.
54. Khajavi, S., Wong, V., Dokhale, N., Patil, D., Jansen, J.C., and Kapteijn, F. (2008) Proceedings ICIM10, Tokyo, Fischer Tropsch Synthesis with In-situ Water Removal: Industrial Implementation of Membranes and their Economic Feasibility.
55. Ciavarella, P., Casanave, D., Moueddeb, H., Miachon, S., Fiaty, K., and Dalmon, J.-A. (2001) *Catal. Today*, **67**, 177–184.
56. Jeong, B.H., Sotowa, K.I., and Kusakabe, K. (2003) *J. Membr. Sci.*, **224**, 151–158.
57. Xiongfu, Z., Yongsheng, L., Jinqu, W., Huairong, T., and Changhou, L. (2001) *Sep. Purif. Technol.*, **25**, 269–274.
58. Kong, C., Lu, J., Yang, J., and Wang, J. (2007) *J. Membr. Sci.*, **306**, 29–35.
59. Verweij, H., Lin, Y.S., and Dong, J.H. (2006) *MRS Bull.*, **31**, 756–764.
60. Tong, J.H., Kashima, Y., Shirai, R., Suda, H., and Matsumura, Y. (2005) *Ind. Eng. Chem. Res.*, **44**, 8025–8032.
61. Nishiyama, N., Yamaguchi, M., Katayama, T., Hirota, Y., Miyamoto, M., Egashira, Y., Ueyama, K., Nakanishi, K., Ohta, T., Mizusawa, A., and Satoh, T. (2007) *J. Membr. Sci.*, **306**, 349–354.
62. Lu, G.Q., Diniz da Costa, J.C., Duke, M., Giessler, S., Socolow, R., Williams, R.H., and Kreutz, T. (2007) *J. Colloid Interface Sci.*, **314**, 589–603.
63. Caro, J. and Noack, M. (2008) *Microporous Mesoporous Mater.*, **115**, 215–233.
64. van Dyk, L., Lorenzen, L., Miachon, S., and Dalmon, J.A. (2005) *Catal. Today*, **104**, 274–280.
65. Haag, S., Hanebuth, M., Mabande, G.T.P., Avhale, A., Schwieger, W., and Dittmeyer, R. (2006) *Microporous Mesoporous Mater.*, **96**, 168–176.
66. Tarditi, A.M., Horowitz, G.I., and Lombardo, E.A. (2008) *Catal. Lett.*, **123**, 7–15.
67. Zhang, X.F., Lai, E.S.M., Martin-Aranda, R., and Yeung, K.L. (2004) *Appl. Catal., A*, **261**, 109–118.
68. Lai, S.M., Ng, C.P., Martin-Aranda, R., and Yeung, K.L. (2003) *Microporous Mesoporous Mater.*, **66**, 239–252.
69. Wan, Y.S.S., Yeung, K.L., and Gavriilidis, A. (2005) *Appl. Catal., A*, **281**, 285–293.
70. den Exter, M.J., van Bekkum, H., Rijn, C.J.M., Kapteijn, F., Moulijn, J.A., Schellevis, H., and Beenakker, C.I.N. (1997) *Zeolites*, **19**, 13–20.
71. Vilaseca, M., Coronas, J., Cirera, A., Cornet, A., Morante, J.R., and Santamaría, J. (2008) *Sens. Actuators, B*, **133**, 435–441.
72. Moos, R., Müller, R., Plog, C., Knezevic, A., Leye, H., Irion, E., Braun, T., Marquardt, K.J., and Binder, K. (2002) *Sens. Actuators B*, **83**, 181–189.
73. Franke, M.E., Simon, U., Moos, R., Knezevic, A., Müller, R., and Plog, C. (2003) *Phys. Chem. Chem. Phys.*, **5**, 5195–5198.
74. Yang, J.C. and Dutta, P.K. (2007) *Sens. Actuators, B*, **123**, 929–936.
75. Szabo, N.F. and Dutta, P.K. (2003) *Sens. Actuators, B*, **88**, 168–177.
76. Sahner, K., Schönauer, D., Kuchinke, P., and Moos, R. (2008) *Sens. Actuators, B*, **133**, 502–508.
77. Sahner, K., Moos, R., Matam, M., Tunney, J.J., and Post, M. (2005) *Sens. Actuators, B*, **108**, 102–112.
78. Trimboli, J. and Dutta, P.K. (2004) *Sens. Actuators, B*, **102**, 132–141.
79. Biemmi, E. and Bein, T. (2008) *Langmuir*, **24**, 11196–11202.

80. Kuhn, J., Gascon, J., Gross, J., and Kapteijn, F. (2008) *Microporous Mesoporous Mater.*, **120**, 12–18.
81. Gora, L., Kuhn, J., Baimpos, T., Nikolakis, V., Kapteijn, F., and Serwicka, E.M. (2009) *Analyst*, **134**, 2118–2122.
82. Kuhn, J., Motegh, M., Gross, J., and Kapteijn, F. (2008) *Microporous Mesoporous Mater.*, **120**, 35–38.
83. Caro, J., Noack, M., Kolsch, P., and Schafer, R. (2000) *Microporous Mesoporous Mater.*, **38**, 3–24.
84. Meindersma, G.W. and de Haan, A.B. (2002) *Desalination*, **149**, 29–34.
85. Tennison, S. (2000) *Membr. Technol.*, **2000**, 4–9.
86. Caro, J. (2006) Workshop on Zeolite Membranes, ICIM9, Lillehammer.
87. Hedlund, J., Sterte, J., Anthonis, M., Bons, A.J., Carstensen, B., Corcoran, N., Cox, D., Deckman, H., De Gijnst, W., de Moor, P.P., Lai, F., McHenry, J., Mortier, W., Reinoso, J., and Peters, J. (2002) *Microporous Mesoporous Mater.*, **52**, 179–189.
88. Voss, H., Diefenbacher, A., Schuch, G., Richter, H., Voigt, I., Noack, M., and Caro, J. (2009) *J. Membr. Sci.*, **329**, 11–17.
89. Yoshino, Y., Suzuki, T., Taguchi, H., Nomura, M., Nakao, Si., and Itoh, N. (2008) *Sep. Sci. Technol.*, **43**, 3432–3447.
90. Alshebani, A., Pera-Titus, M., Landrivon, E., Schiestel, T., Miachon, S., and Dalmon, J.A. (2008) *Microporous Mesoporous Mater.*, **115**, 197–205.
91. Hyflux Ltd. (2008) *www.Hyflux.com* (visited November 2008).
92. Richter, H., Voigt, I., Fischer, G., and Puhlfürß, P. (2003) *Sep. Purif. Technol.*, **32**, 133–138.
93. Inopor GmbH (2008) *http://www.inopor.com/en/ membranes_e.html* (visited November 2008).
94. Bowen, T.C., Kalipcilar, H., Falconer, J.L., and Noble, R.D. (2003) *J. Membr. Sci.*, **215**, 235–247.
95. Kuhn, J., Gross, J., and Kapteijn, F. (2008) Proceedings ICIM10, 18-22 August, 2008, Tokyo, pp. 316–316.
96. van de Graaf, J.M., Kapteijn, F., and Moulijn, J.A. (1999) *Chem. Eng. Sci.*, **54**, 1081–1092.
97. Voigt, I., Fischer, G., Puhlfürß, P., Stahn, M., and Tusel, G.F. (2000) Sixth International Conference on Inorganic Membranes, Montpellier, Book of abstracts, pp. 43–43.
98. Kapteijn, F., Moulijn, J.A., and Krishna, R. (2000) *Chem. Eng. Sci.*, **55**, 2923–2930.
99. Krishna, R. and Baur, R. (2003) *Sep. Purif. Technol.*, **33**, 213–254.
100. Kapteijn, F., Zhu, W., Moulijn, J.A., and Gardner, T.Q. (2005) Zeolite Membranes: Modeling and Application, in *Structured Catalysts and Reactors*, Chapter 20 (eds A. Cybulski and J.A. Moulijn), Taylor & Francis Group, Boca Raton, pp. 701–747.
101. van de Graaf, J.M., van der Bijl, E., Stol, A., Kapteijn, F., and Moulijn, J.A. (1998) *Ind. Eng. Chem. Res.*, **37**, 4071–4083.
102. de Bruijn, F., Gross, J., Olujic, Ž., Jansens, P., and Kapteijn, F. (2007) *Ind. Eng. Chem. Res.*, **46**, 4091–4099.
103. de Bruijn, F.T., Sun, L., Olujic, Z., Jansens, P.J., and Kapteijn, F. (2003) *J. Membr. Sci.*, **223**, 141–156.
104. Coppens, M.O., Keil, F.J., and Krishna, R. (2000) *Rev. Chem. Eng.*, **16**, 71–197.
105. Kerkhof, P.J.A.M. and Geboers, M.A.M. (2005) *AIChE J.*, **51**, 79–121.
106. Casanave, D., Ciavarella, P., Fiaty, K., and Dalmon, J.-A. (1999) *Chem. Eng. Sci.*, **54**, 2807–2815.
107. Miachon, S. and Dalmon, J.A. (2004) *Top. Catal.*, **29**, 59–65.
108. van Dyk, L., Miachon, S., Lorenzen, L., Torres, M., Fiaty, K., and Dalmon, J.-A. (2003) *Catal. Today*, **82**, 167–177.
109. Dittmeyer, R., Svajda, K., and Reif, M. (2004) *Top. Catal.*, **29**, 3–27.
110. van de Graaf, J.M., Zwiep, M., Kapteijn, F., and Moulijn, J.A. (1999) *Appl. Catal. A*, **178**, 225–241.

11
Microstructures on Macroscale: Microchannel Reactors for Medium- and Large-Size Processes

Anna Lee Y. Tonkovich and Jan J. Lerou

11.1
Introduction

Microchannel reactors for medium- and large-scale processes offer great promise for the chemical and energy industry, where economic advantage is made by scaling to high throughput. Annualized production volumes measured in thousands of metric tons or greater represent the ultimate challenge for a technology based on harvesting value through small dimensions. The microchannel technology area offers a paradigm shift to reduce usage of costly raw materials, decrease energy consumption, and lower capital investment costs for plants whose investment capital can reach hundreds of millions of dollars. Realization of the potential economic benefits of microchannel reactors is built upon the practical reality of implementing the often but cavalierly cited numbering-up philosophy. In theory, a single channel will perform in the same manner as 10 000 equally sized channels – in practice, the challenges of manufacturing realities and the physics of fluid flow under extreme conditions at the microscale have made scale-up elusive.

While there are many technical and economic challenges associated with numbering up tens of thousands of microchannels to act in concert as if one, the solutions are not infeasible. Innovation for tough technical problems in the nineteenth and twentieth centuries was built upon an Edisonian approach – test, test, and test in the laboratory. Today, we have the luxury to test first on computers in a fraction of the time and for a fraction of the cost. Complex and interrelated physics make essential the need to use mathematical models to quantify behavior throughout the microchannel. Concurrent and integrated simulation of heat transfer, pressure drop, chemical reactions, mass transfer, mechanical design, and manufacturing variables is carefully choreographed to scale microchannel technology to medium- and large-scale processes.

11.2
Background on Medium- to Large-Scale Processes in Microchannels

Most microchannel work to date has focused on low-volume production of specialty and fine chemicals. However, a few pioneering firms and research

Novel Concepts in Catalysis and Chemical Reactors: Improving the Efficiency for the Future.
Edited by Andrzej Cybulski, Jacob A. Moulijn, and Andrzej Stankiewicz

ISBN: 978-3-527-32469-9

institutes, including Velocys, Battelle, IMM, Karlsruhe FZK, Heatric, Corning, Degussa/Evonik, Alfa Laval, Chart, and Compact GTL, have made forays into medium- and large-scale processes. The applications have varied from liquid-phase reactions, gas–liquid reactions, and hydrogen generation to the production of large gas-to-liquids (GTLs) and biofuels production facilities.

Scale-up of microchannel reactors is based on using the optimal channel dimensions rather than seeking the smallest or the largest microchannel. In some cases, the channels may range from 100 μm in hydraulic diameter to a few millimeters. The classification of a rigorous size range to designate a reactor as microchannel is not necessary.

Technology developers for medium- to large-scale microchannel technology are listed in Table 11.1.

11.3 Fundamental Challenges of Microchannel Scale-up

Microreactor scale-up is built upon the premise of numbering up channels, Figure 11.1. A single channel is demonstrated with the same geometry and fluid hydrodynamics as a full-scale reactor. Numbering up relies on creating a massively

Table 11.1 Microchannel technology developers for medium- to large-scale applications.

Organization (country)	Focus area
Heatric (UK)	Compact heat exchangers for offshore facilities, hydrogen production 4 tons per year [1]
Chart Industries (US)	Integrated heat exchangers and reactors for numerous applications
IMM (Germany)	Nitroglycerin production at 80 tons per year with Xi'an Chemical Company [2]; microfluidic solutions for bioanalytic and industrial analysis
Degussa/Evonik (Germany)	Falling film microreactors for ozonolysis and other chemical applications with IMM reactors [3] and DEMIS project collaboration [4, 5]
Alfa Laval (Sweden)	Plate reactors for producing pharmaceuticals and fine chemicals
Compact GTL (UK)	Offshore gas-to-liquids facilities. Developing steam reforming and Fischer–Tropsch reactor technologies [6]
Corning	Hazardous nitration reaction in collaboration with DSM at 800-tons-per-year capacity [7]
Karlsruhe FZK	Fine chemicals, collaboration with DSM for a 15 000-tons-per-year maximum feed [8, 9]
Battelle – Oxford Catalysts Group – Velocys	Offshore gas-to-liquids, biofuels, 50–50 000 tons per year [10, 11]

parallel fluid network with hundreds, thousands, or tens of thousands of channels operating identically. The simplicity of the numbering-up concept has proved elusive in practice, based on the challenges of manifold design.

The promise of equal performance at all scales of operation by numbering up channels is based upon the assumption that flow will be uniformly distributed from hundreds to tens of thousands of channels. The impact of flow maldistribution depends upon the chemistry or process design. For heat-transfer-only processes, flow maldistribution of 10 or 20% or more results in larger volumes, more metal, and in turn worse process economics – but the process will still run. Reactor processes by their very nature produce nonlinear responses to process variables, and, therefore, are more sensitive to flow maldistribution. Effects may range from a change in channel residence time and corresponding selectivity to a thermal imbalance across the reactor which can affect selectivity, mechanical integrity, and catalyst lifetime. A drop in performance as a process goes from a single to multiple channels is not desired and challenges commercial economic targets.

There are two primary approaches to numbering up microchannels – the first and more straightforward approach is a cross-flow device with a large open macromanifold. The second, and more challenging is the micromanifold approach, needed for counter- or coflow processes. For processes that can tolerate a cross-flow operation, maldistribution can be minimized by designing the reactor such that the majority of the system pressure drop occurs within the microchannels. Processes that are less temperature sensitive may utilize a cross-flow design. As the physical scale of the cross-flow reactor increases, the thermal-control challenge increases – for example, removing heat with a convective heat-transfer medium in a 0.01-m-length flow channel is fairly straightforward, but maintaining the same degree of heat removal and thermal control in a 0.1-m or even 1-m flow length is more challenging.

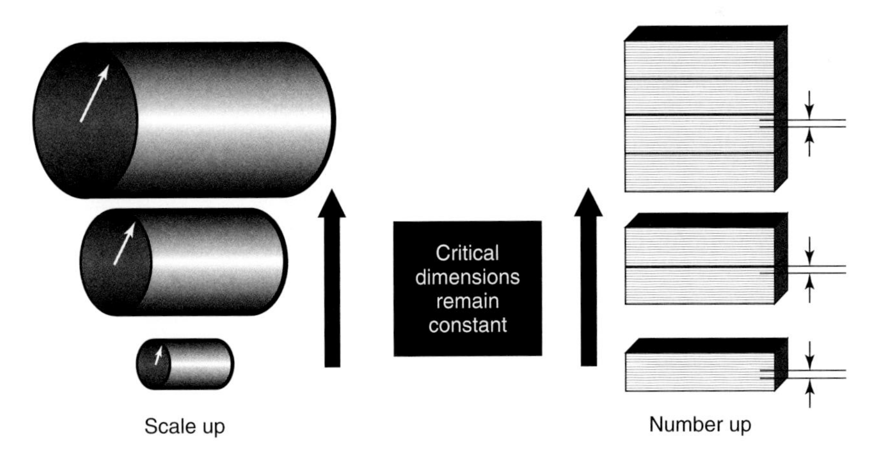

Figure 11.1 Scale-up versus numbering up for microchannels. All critical channel dimensions remain constant in a microchannel system independent of the overall process capacity.

However, many reactions of commercial interest have chemistry, mechanical, or system requirements that preclude the use of cross-flow reactors. Processes cannot use a cross-flow orientation primarily because of high temperatures and the need to internally recuperate heat such as steam methane reforming (SMR) [12, 13] and oxidation reactions [14]. Counter- and coflow devices require a micromanifold to deliver sufficiently uniform flow to each of the many parallel channels.

As the physical scale of a reactor increases by numbering up more channels, the micromanifold challenge increases. Fluid distribution occurs in multiple dimensions: within a layer [15, 16], from one layer to another [17], and from one reactor to another [18]. An external manifold, also known as the *macromanifold* or *tube connection*, as shown in Figure 11.2a, brings the fluids from inlet pipes to the many parallel layers in medium- to large-capacity reactors.

The internal layer flow distribution, or micromanifold, is substantially more challenging; an example is shown in Figure 11.2b. A single flow enters a layer that then feeds tens to thousands of parallel channels. Design variables in the micromanifold region before the connecting process channels are an option to control flow distribution [19–23].

One of the challenges to manifold design for flow distribution is the change in flow regime across the manifold combined with turning losses. Flow within any individual microchannel is typically laminar based on the small hydraulic diameter. As a single manifold channel feeds many microchannels, the flow and the corresponding Reynolds number are substantially larger than for any individual microchannel. The challenge of the micromanifold design is to understand and control flow where the inlet stream is highly turbulent and traverses through transition flow to laminar while concurrently metering a portion of the stream to parallel microchannels.

11.4
Overcoming the Scale-up Challenges

In microchannel reactors, a model-driven design process is essential as the physical size of the process increases from a single channel to a network of thousands of channels. For passive control of flow distribution, pressure losses must be understood for every hydrodynamic region and for various channel geometries. No single model describes the phenomenon. Therefore, models for pressure drop, heat transfer, and reactor performance are combined with mechanical models to predict channel shape and dimension after fabrication, as well as mechanical integrity during operation. Models to evaluate the range of dimensions from manufacturing assess the robustness of a design that integrates all fluid physics.

The scale-up challenges for microchannel reactors are addressed through integrated mathematical models to describe all elements of the physics. Integrated

(a)

(b)

Figure 11.2 (a) An assembly of parallel full-scale reactors, where the macromanifold or tube connections bring fluids into the microchannel reactors. (b) A micromanifold is shown at the bottom of the 0.8 m × 0.5 m sheet. Each of the six manifold passages feeds 12 parallel channels.

models to assist reactor scale-up have been well described in the literature [24–28]. The use of many types of models to design microchannel systems has also been described in the literature [8, 20–33]. This work focuses on the key models for scale-up.

Four elements of microchannel scale-up models will be described: pressure-drop design, heat-transfer design, reactor design, and mechanical and manufacturing designs.

11.4.1 Pressure-Drop Design

Numbering up microchannels to large-scale capacity reactors is driven by a rigorous understanding of pressure drop in every parallel circuit. Passive flow distribution permits sufficient flow to each channel. No serious evaluation of microvalves or actuators has been undertaken for high-capacity systems with thousands to tens of thousands of channels. Beyond the technical issues required to integrate many thousands of sensors and controls, the cost to do so would be prohibitively high for the foreseeable future.

Flow is typically laminar in microchannel devices, although not always rigorously so. Correlations for fully developed laminar flow in perfectly rectangular microchannels have been validated in the literature [33–35]. Transition and turbulent flows in a microchannel have no such consistent treatise, and are highly dependent upon channel shape, aspect ratio, and surface characteristics [36, 37].

Flow in nonideal channels resulting from manufacturing processes has also not been fully explored in the literature. Paul has presented friction coefficient data that suggest both a higher-than-expected value from theory and variation based on the method of manufacturing channels – where the laser-machined channel with more rectangular edges is reported to have lower pressure loss than electrochemically machined channels with trapezoidal edges [38]. The irregularities cited in this work must be carefully understood for reactor scale-up.

For most medium- and large-scale micromanifold structures, where one passage feeds multiple parallel channels, flow traverses through turbulent and transition flows in the micromanifold region. This fluid in turbulent to transition flow also turns in the micromanifold region as it drops flow into parallel microchannels, which are primarily in the laminar flow regime.

The challenge of transition-flow physics in microchannels has been illuminated by the work of Wibel and Ehrland [36, 37] who have presented a video of micro-particle image velocimetry (PIV) experimental results. Visualization of local turbulence effects in a microchannel reveals transient flow starting from a Reynolds number as low as 1000. Increasing levels of surface roughness lead to higher turbulence intensity and corresponding pressure loss, a phenomenon that is further exacerbated by increasing channel aspect ratio. The lack of microchannel flow stability as a function of surface roughness and channel aspect ratio, and at Reynolds numbers beyond 1000 defines the heart of the micromanifold design challenge for medium- to large-scale processes. Manifold design requires a rigorous analysis of pressure losses throughout all flow regimes and of physical geometry of the device.

A detailed design methodology for a complex manifold has been described in the literature [20, 22]. Experimental validation was conducted to verify the design methodology and numerical solution approach. Two 0.6-m-wide microchannel layers were constructed and joined with gaskets and clamps to measure the flow in each of the 72 parallel channels. Figure 11.3 shows a schematic of an experimental validation device, Figure 11.4a–d shows pictures of the actual test device, and Figure 11.5a,b shows experimental data with reasonable model agreement and sufficient flow uniformity across a 0.6 m $\times$ 0.6 m device.

Modeling manifold physics is an essential element for defining all geometry elements required to provide sufficient flow distribution to many thousands of parallel channels. No manifold can ever be perfect, and thus the idealized goal of uniform flow distribution will remain elusive. In some cases, tailored rather

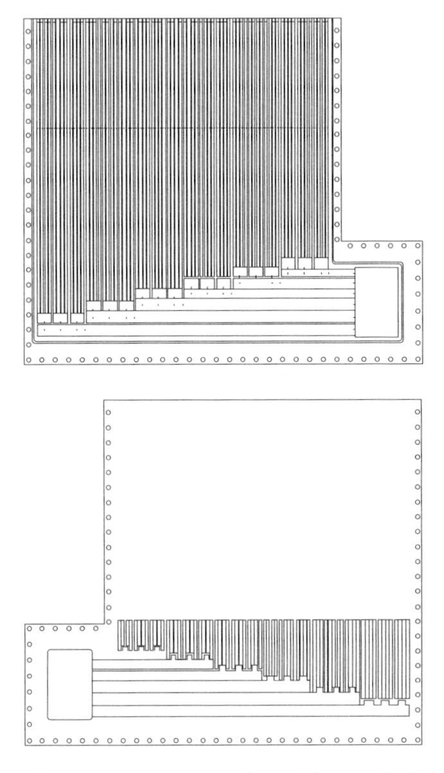

Figure 11.3 Two microchannel layers which form a complete flow circuit including a micromanifold region and connecting channels. The bottom layer is placed on the first to create a common plenum that connects to six submanifolds. Each submanifold connects to 12 parallel microchannels.

than equalized flow distribution has been developed [31]. Through the use of careful modeling, a sufficient flow distribution – which may be toward uniform or tailored – can be achieved to operate a medium- to large-scale microchannel-based plant while achieving commercial objectives.

11.4.2 Heat-Transfer Design

Heat transfer in laminar single-phase, two-stream microchannels has been well described in the literature [33, 39–41]. The challenge for medium- to large-scale processes is an added complexity if more than two fluids are required and also the micromanifold region.

Multiple fluid streams in medium- to large-scale processes become important for many oxidation reactions. Microchannel operation allows safe operation in the explosive regions [42, 43], but this advantage ceases as the dimensions become

Figure 11.4 (a–d) Picture of a large micromanifold test device showing a gasketed 0.6 m × 0.6 m plate, a close-up of the submanifold and connecting channels, a picture of the mating plate that contains flow gates to balance flow, and a final assembled and instrumented device.

conventional. Macropipes connect fluids to the microchannel reactor, thus negating premixed feeds in macromanifolds. Addition of the oxidant stream within the reactor may occur near a manifold region or throughout the device. For both cases, an understanding of the thermal profile determines design dimensions to control flow.

Different types of heat-transfer models are useful depending upon the physics of a problem. Reactive systems require integrated kinetics and heat-transfer models to adequately describe the local temperature profile. Medium- to large-scale chemical process applications push the microchannel productivity to high levels. Designing a nearly isothermal or tailored temperature profile in a high-capacity reactor section, while aided by the inherent microchannel heat transfer advantage, is far from guaranteed.

An example of integrated heat-transfer modeling and reactor design is shown in Figure 11.6. A predicted thermal profile for the reactor section of a combined reactor–heat exchanger is the solid line, while the discrete points are experimentally measured temperatures along the reactor length. The thermal profile is controlled

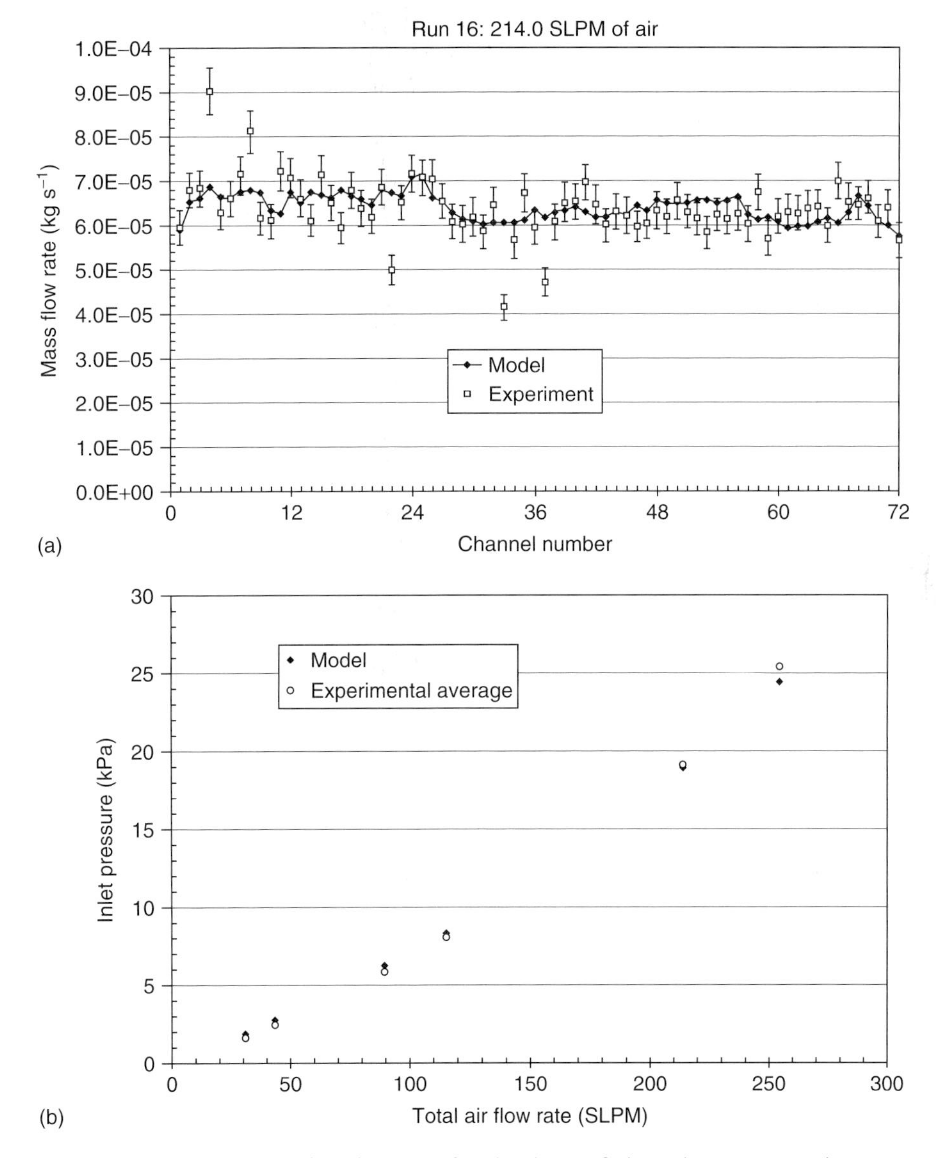

Figure 11.5 (a) Experimental data showing predicted versus experimental mass flow of nitrogen per channel at ambient conditions. Channels 37–72 showed excellent agreement with model predictions. Channels 1–36 were found to have a faulty gasket connection that varied the local channel dimension from the desired value. (b) Agreement between predicted and overall pressure drop for flow in 72 parallel microchannels.

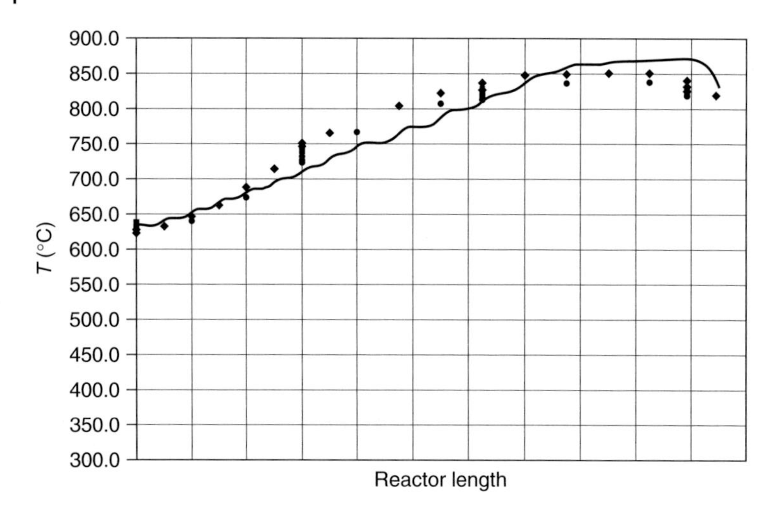

Figure 11.6 Model (solid) versus experiments (points) for the temperature profile of an integrated SMR reactor.

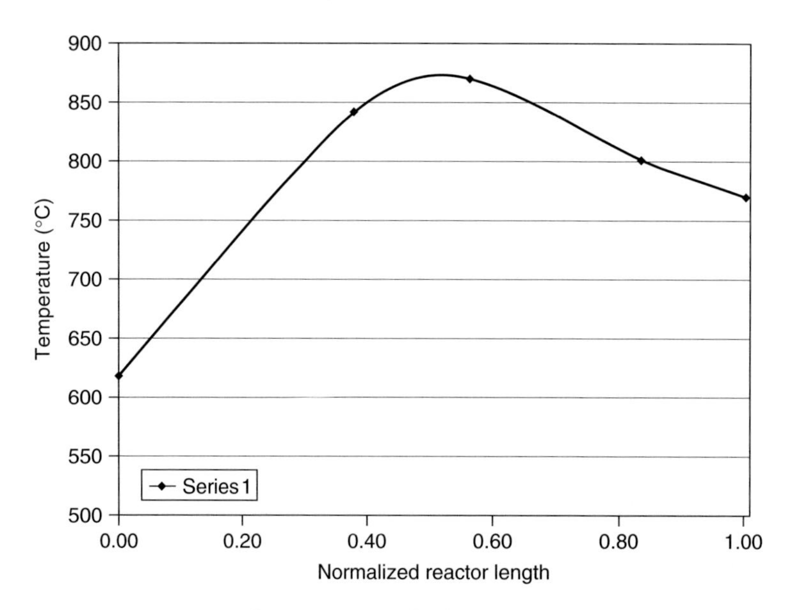

Figure 11.7 Experimental temperature profile for an early SMR reactor with an unoptimized design.

by a carefully designed staged oxidant addition approach to meter air into a combustion channel to drive an endothermic methane steam reforming stream.

The use of integrated reactor and heat-transfer models is essential for scale-up. Figure 11.7 shows an early reactor design for the same chemistry that was developed without the use of integrated models. Other unoptimized designs with temperature spikes have also been reported [12, 44]. Integrated models were used to tune the design to eliminate undesired thermal spikes as shown with experimental performance results in Figure 11.6.

Heat transfer in micromanifolds is a second challenge to the scale-up of microchannel systems. Large micromanifolds are not isothermal. Temperature variations within the distribution zone add to the challenge of predicting pressure drop to control flow [45]. Three-dimensional heat-transfer models are needed to quantify the local metal and fluid temperatures to provide input to local pressure-loss terms. Flow in the micromanifold region of a coflow or counterflow reactor often begins as cross-flow and then moves into counter- or coflow as the fluid turns toward the parallel channels.

A thermal map of a single micromanifold layer is shown in Figure 11.8. The average temperature of the fluid is 60 °C at the inlet and varies significantly along the length of each of six submanifolds. The first submanifold traverses the shortest distance before connecting to 12 parallel channels, while the longest submanifold is close to 0.5 m in length. From the shortest to the longest submanifold, the fluid gains from less than 20 °C to more than 100 °C from the average inlet condition, respectively. The change in density and viscosity affects the local pressure drop and is iteratively factored into the design.

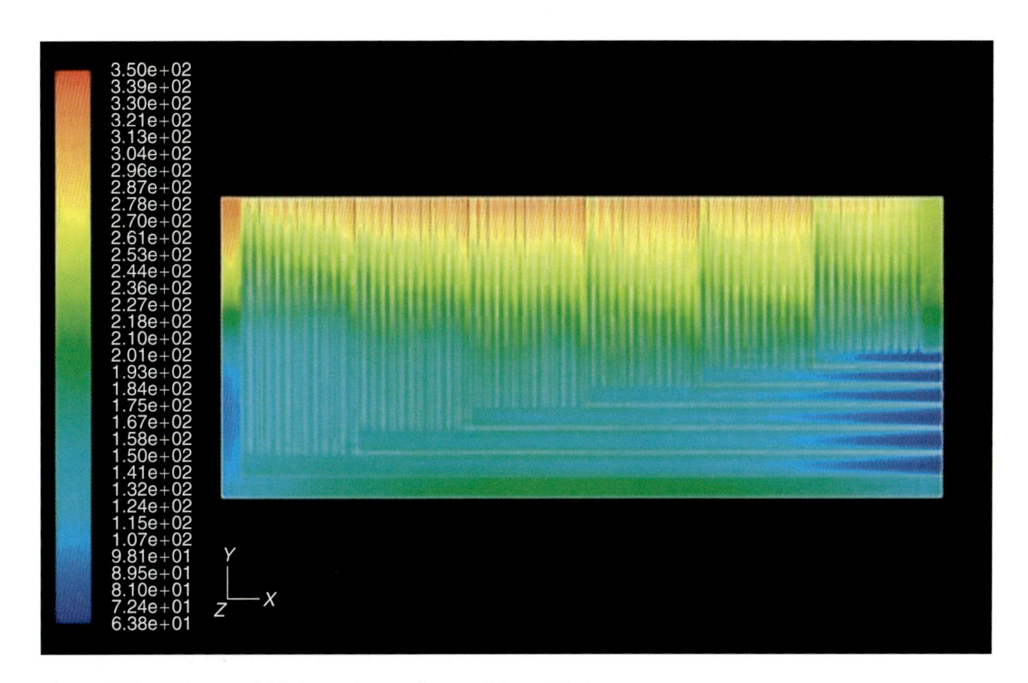

Figure 11.8 Micromanifold thermal map from a 0.6-m wide microchannel plate which contains six submanifolds, where each submanifold feeds 12 parallel channels.

11.4.3
Reactor Design

As the capacity and scale of a microchannel reactor are increased, the design and resulting performance of the reactor section govern the overall product success. The reaction channel dimensions are bound by chemical kinetics, pressure-drop constraints, heat-transfer requirements, and manufacturing limitations. Rigorous reactor models are essential to balance these physics. The reactor length dimension is critical for scale-up and not to be underestimated. As an example, if a chemical reaction is demonstrated with a residence time of 100 ms in a short microchannel reactor 1 cm in length, 1 cm in width, and having a 1-mm gap, where the fluid will have a modest average velocity of $0.1\,ms^{-1}$; then operating at the same average residence time and channel cross-section in a 1-m long microchannel, the fluid will have a velocity of $10\,ms^{-1}$. The pressure drop will be vastly different between the scales, affecting pressure-sensitive reactions. In addition, the local heat release profile may differ between the short and long microchannels. A reactor-modeling strategy that integrates chemical kinetics with heat transfer and hydrodynamic effects is essential to capture the expected performance in a long-length microchannel reactor that is typically used for medium- and large-scale processes.

A process for microchannel reactor scale-up using multiple-scale models for microchannels has been described in the literature [14, 33]. Local models simulate the chemical reaction kinetics for a selected catalyst. These local models are developed by iteratively testing small reactors and comparing results with model predictions. Intermediate models incorporate chemical kinetics with heat transfer and fluid hydrodynamics, and attempt to describe a single full commercial length microchannel. Global models integrate the results from a single channel into a network of tens of thousands of channels to describe the plant operation.

In the example described by Yang *et al.*, local models iterated through hundreds of elementary reactions are described in the literature for the oxidative dehydrogenation of ethane to ethylene. Eight key reactions, including the gas-phase production of soot, were found to control the reaction. The reduced reaction set was incorporated into the intermediate model, which included recuperative heat exchangers integrated with a high-temperature reaction section.

Two reactors were built and put into operation. The design of the first reactor was based on minimal intermediate reactor modeling and subsequently operated for less than 4 h in the laboratory. From the pictures of the reactor after operation, shown in Figure 11.9, and operational pressure-drop data, shown in Figure 11.10, it was readily concluded that carbon deposition was a problem. A second reactor was designed with the full use of an intermediate model to overcome initially unrecognized limitations in the first reactor.

Steady performance data from the second reactor are shown in Figure 11.10, where the pressure drop did not rise exponentially and the conversion and selectivity remained at 75 and 83%, respectively. The reactor was further analyzed after operation, shown in Figure 11.11, to confirm the lack of carbon deposition. Reactor models were pivotal to developing a robust design for this high-temperature and

Figure 11.9 Picture of reaction zone in an original reactor with significant pressure increase in <4 h due to heavy coking.

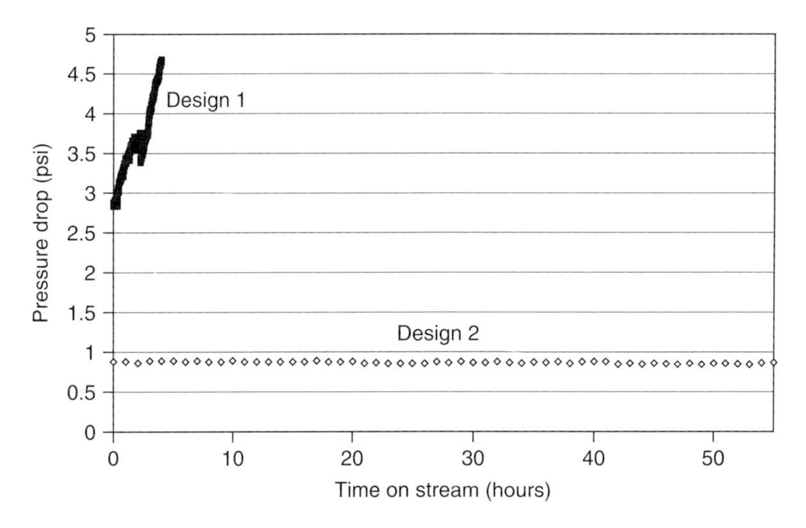

Figure 11.10 Pressure drop during reaction for designs 1 and 2.

Figure 11.11 Second reactor after operation, with no buildup of carbon.

coke-prone chemistry. Scale-up to medium- and large-scale reactor capacity using microchannels is not possible without a consistent and rigorous use of reactor models based on the integration of chemical kinetics, flow physics, and heat transfer.

11.4.4 Mechanical and Manufacturing Designs

Mechanical and manufacturing designs add to the challenge of scaling up microchannel technology to commercial capacity. Large-capacity microchannel devices are made of many sheets or shims, which are joined to form a hermetically sealed reactor. Microchannels are formed in shims either by a partial process or through a feature formation process. Processes for forming microchannels include stamping, machining, etching, water jet, and coining, among others.

The stacking process requires precision alignment to connect like-fluid paths and maintain leak-free communication between different fluids. The stack is joined by diffusion bonding, brazing, welding, or other methods. Diffusion bonding is typically preferred for processes that operate at high temperatures and pressures. The diffusion-bonding process requires high pressure, high temperature, and sufficient time to ensure adequate grain growth across shim lines. The very process that forms a strong joint is concurrent with mechanical creep of the metal, which may act to deform the physical shape and dimension of the microchannel. The degree of deformation is dependent upon the bonding time, temperature, pressure, and the design of internal channel geometry. Figure 11.12 shows a badly deformed microchannel, which includes collapse of some channels. Similar deformation phenomenon has been reported in the literature [38].

Quantifying the final channel shape and dimension is a critical element in understanding the fluid pressure loss through the channel and setting the micromanifold design to allow sufficient flow distribution through a large reactor. Mechanical design models are used to quantify the mechanical environment during bonding to predict the final channel shape and dimension. The shape and dimension of the actual manifold channel and individual microchannel determine the pressure loss that determines the flow distribution.

Figure 11.12 Badly deformed bonded array of microchannels with poorly understood design and intermediate phase diffusion-bonding parameters.

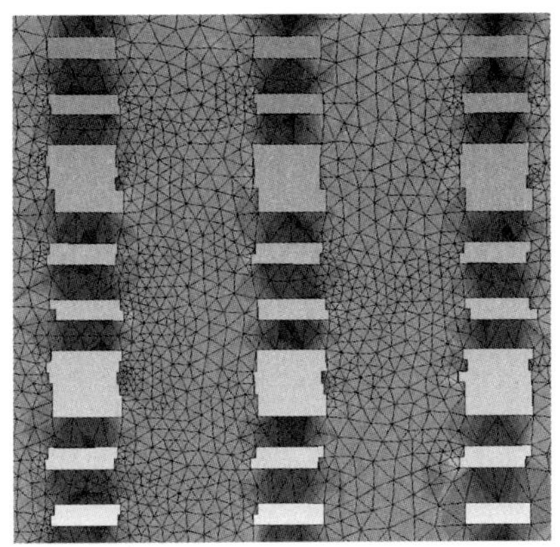

(a)

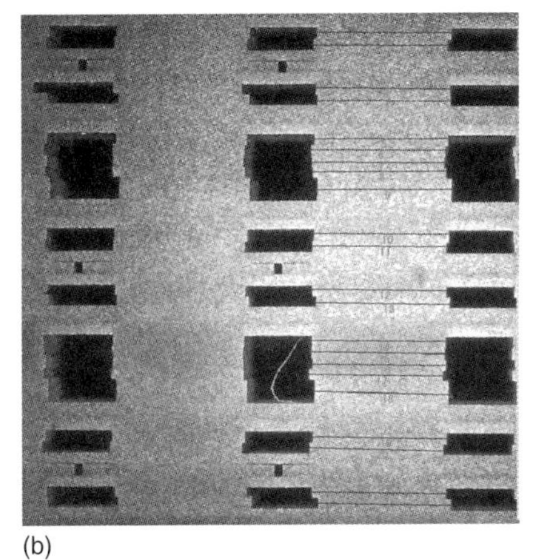

(b)

Figure 11.13 (a) A mechanical model of a diffusion-bonded stack to understand the impact of manufacturing factors on the final reactor dimensions. (b) A reactor based on a mechanical design model, including shim-bonding parameters. The experiments have both successfully avoided channel collapse and quantified the impact of unavoidable manufacturing tolerance and limitations for large-scale systems.

Figure 11.13a,b shows an improved stack with the selection of bonding conditions aided by mechanical models to avoid channel collapse and severe deformation. It is further noted that manufacturing limitations preclude perfect alignment of microchannels, either within a layer or from layer to layer, as seen in Figure 11.13b. It can also be seen from the figure that the dimensional control of the channel width as an example is not perfect, where manufacturing tolerances create a distribution of channel dimensions. The use of integrated design models is crucial in enabling robust designs that can accommodate dimensional variations from manufacturing processes.

Scale-up to medium- and large-scale capacity requires the integration of many parts and many features within an individual part. Each feature and each part may have a nominal design dimension, but in reality nothing can be manufactured perfectly. Each feature will have plus/minus dimensional variation – a tolerance. The scale-up design challenge is to understand manufacturing tolerances and design robust processes based on feasible tolerances.

An intuitive example of the tolerance challenge for microchannels is discussed next. Microchannels enable enhanced heat and mass transfer based on the use of many small passageways. Using a microchannel gap of 0.25 mm as a nominal starting point, the manufacturing process that requires a tolerance of ±0.03 mm, a mere 30 μm, on the gap dimension would allow final channel gaps that range from 0.22 to 0.28 mm. While these dimensions do not appear to be widely different in the lens of conventional technology, they give rise to a pressure drop that varies by a span of more than 20%. This single-dimensional variation would allow flow for high and low flows in parallel channels to vary by 20% – very likely unacceptable for many reactions.

For the same example, if the nominal flow gap were increased to 1 mm, then a variation of ±0.03 mm in flow gap would give a span of 6% for high to low flows in a microchannel. While the latter case appears to provide a much more reasonable tolerance situation, it comes at the price of a single layer replacing four layers. This reduction in layers reduces the surface area for heat transfer and the heat-transfer coefficient, which is inversely proportional to hydraulic diameter in laminar flow. The large-gap microchannel would give rise to an overall physical volume of the reactor that is more than 4 times larger than that of the small-gap microchannel. The true penalty for a large reactor is cost – cost is always a critical factor when competing with conventional technology for medium- to large-scale processes. At the same time, if the design cannot accommodate real manufacturing tolerances, then scale-up is not practical.

If death and taxes are synonymous with unavoidable phenomenon, then so are manufacturing tolerances for medium- and large-scale microchannel reactors. Tolerance impact cannot be avoided, but rather must be managed. A detailed statistical analysis methodology for measuring the impact of manufacturing tolerance is described [46], along with a methodology for measuring the flow variation from dimensional tolerances [47] – this serves as a useful starting point for understanding the impact of every manufacturing variable on the process. The integration of all reactor design models defines the maximum allowable range of flow maldistribution. A rigorous understanding of the tolerance range and how this affects the design defines the allowable types of manufacturing methods for a particular process.

11.5 Example of Scale-up through Concurrent Modeling

Fischer–Tropsch (FT) synthesis is an important medium-to-large-scale application for microchannel reactors. The FT process was first developed by Franz Fischer and Hans Tropsch in Germany in the 1920s and 1930s. The chemistry is based on making longer chain hydrocarbons by passing synthesis gas, a mixture of carbon monoxide (CO) and hydrogen (H_2), over a catalyst at elevated pressure and temperature. The majority of the products from FT synthesis are paraffinic waxes based on the following chemical equation.

$$nCO + (2n + 1)H_2 \rightarrow C_nH_{2n+2} + H_2O \tag{11.1}$$

These products can be fairly easily processed into high-quality diesel and jet fuel; in theory, any source of carbon can be used to generate synthesis gas. These facts along with the growing need for petroleum alternatives have renewed interest in FT synthesis. During the twentieth century, the FT process was used to produce fuels from coal in large and costly reactors. Recently, this megasize approach has been applied to world-scale GTL plants in Qatar. However, to tap abundant biomass resources and stranded natural gas reserves, a smaller scale, yet economically viable, FT process is needed.

The application of microchannel technology is a natural fit for the production of synthetic fuels via the FT process. The primary limitations of conventional FT technology include the removal of process heat that can produce hot spots and severely shorten catalyst life, and effective management of two-phase flow as synthesis gas transforms into liquid hydrocarbons. Both these issues can be addressed with microchannel technology, which greatly improves heat transfer and precisely controls flow through thousands of parallel channels.

Since the efficacy of the FT process was proven in a single microchannel, scale-up has been driven by concurrent modeling and experimental validation. Fortunately, the FT process lends itself to a cross-flow configuration, which simplifies both the design and fabrication of microchannel devices. As shown in Figure 11.14, the process channels flow from top to bottom and the coolant flows horizontally. The downflow process channels are filled with a particulate catalyst, while heat is removed by the use of boiling water to maintain a near-isothermal reactor during operation. Loading and unloading of particulate catalysts in long microchannels are not straightforward and have required a great deal of art to resolve. The flow passage length for each stream ranges between 0.5 and 1 m, which is considered long when compared to that of small-scale microchannels described in the literature but short

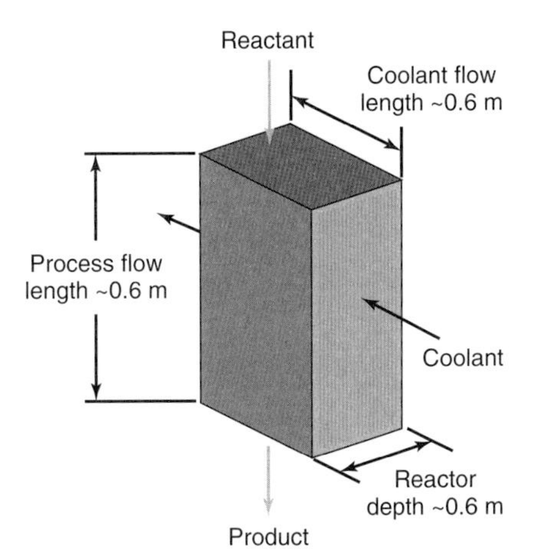

Figure 11.14 Schematic of microchannel FT reactor for large-scale processing.

when compared to that of conventional FT reactors, which have tube lengths of 10 m or more.

As described above, microchannel reactor scale-up requires integrated models, which include the reaction chemistry with heat transfer, pressure drop, flow distribution, and manufacturing tolerances. The culmination of scale-up models is their successful demonstration.

Catalyst kinetics were developed in a small, single-channel microchannel reactor. The kinetics were coupled with heat transfer, pressure drop, and flow distribution models to design, build, and operate a pilot-scale reactor with 450 parallel processes and 250 parallel coolant channels. Integrated reactor modeling identified multiple improper designs before iterations revealed a successful one. Figures 11.15a,b and 11.16a,b show the predicted water quality and metal temperature for both improper and acceptable designs. Subtle design differences would result in large temperature variations, exceeding 50 °C, in the improper design. The proper design

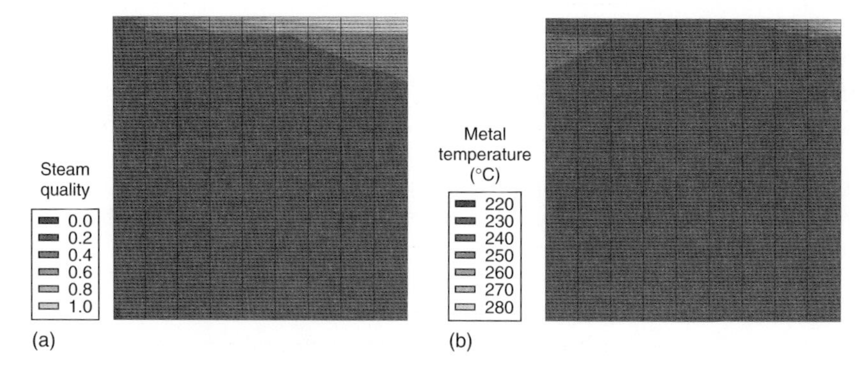

Figure 11.15 (a–b) An FT Reactor (0.6 m × 0.6 m × 0.6 m) with improper design, showing liquid dry out (steam quality equals 1) and a metal hot spot up to 280 °C over the nominal 230 °C.

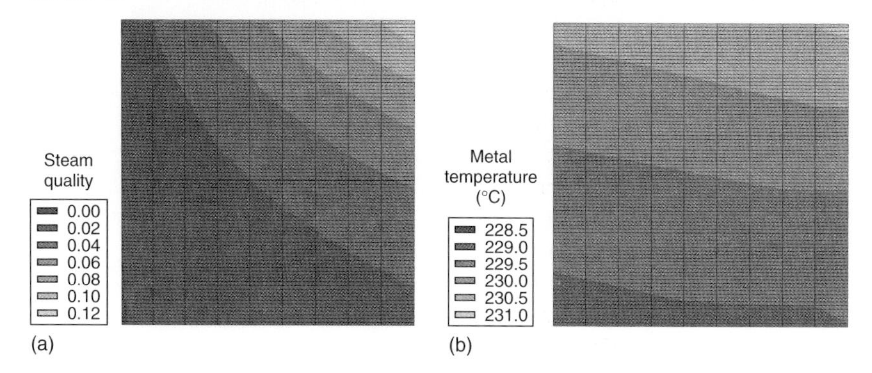

Figure 11.16 (a–b) FT Reactor (0.6 m × 0.6 m × 0.6 m) with a proper design, showing neither liquid dry out nor a metal hot spot. Metal temperature is maintained nearly constant at 230 °C.

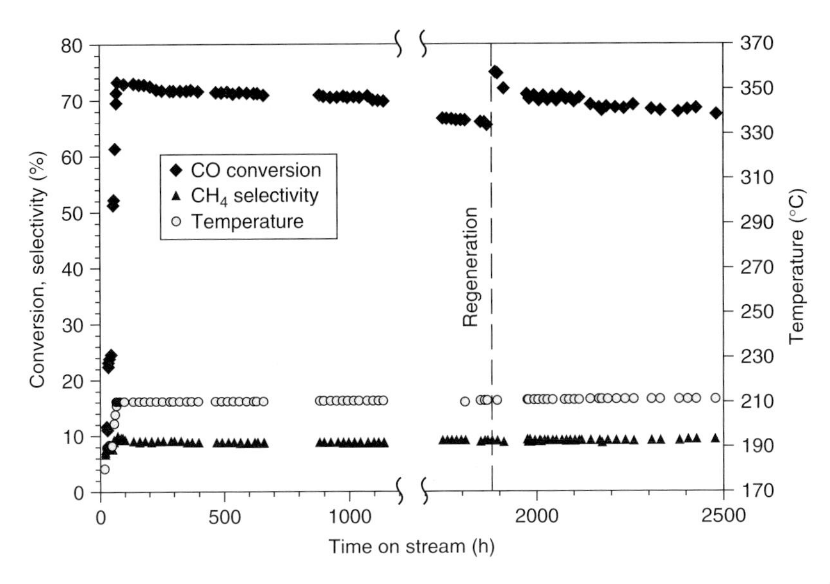

Figure 11.17 Pilot reactor performance data at 290 ms, 20 bar total pressure, 2 : 1 for H2 : CO, 16% nitrogen dilution, and 210 °C average temperature.

maintains a near-isothermal operation within a few degrees celsius without dryout of the partially vaporized liquid coolant.

The proper design was integrated in the pilot-scale reactor, which produced 2 l per day of FT liquids. The operation was smooth, in thermal control, and matched expectations from the small test reactor. The operation ran over 2000 h as shown in Figure 11.17. Commercial interest in FT reactor systems ranges from scales of 5000 to 50 000 tons per annum of synthetic fuels. Numbering up channels beyond the 2-l-per-day pilot demonstration requires an increase in all channel lengths while at the same time managing flow distribution and manufacturing tolerances.

A manufacturing scale-up device is shown in Figure 11.18; the projected capacity of the device as shown is nearly 10 barrels per day, or 450 tons per annum. The scale-up challenge for numbering up FT channels is flow distribution as it relates to manufacturing tolerances. The main challenge in the operation of a cross-flow reactor is to maintain stable boiling in thousands of parallel coolant channels to remove the exothermic reaction heat. The literature has clearly shown the challenge of microchannel boiling, visually revealing flow instabilities [35]. The scale-up challenge for controlling flow for many thousands of partially boiling microchannels should not be underestimated.

The device as shown in Figure 11.18 has more than 10 000 combined coolant and process channels. Cold-flow testing of the flow variability demonstrates less than ±9% variation in channels throughout the device. Figure 11.19 shows a color map of high to low flow as measured in the device and lumped into flow ranges. The measured flow variability across the coolant channels resulting from manufacturing tolerances has been shown to be acceptable for reactor operation.

Figure 11.18 Manufacturing scale-up device for microchannel FT. The device geometry is 0.6 m × 0.6 m × 0.15 m and contains more than 10 000 coolant and process channels.

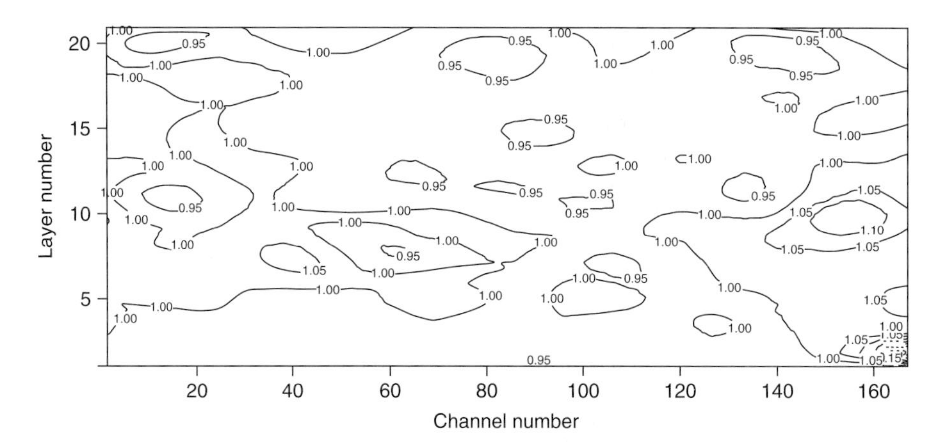

Figure 11.19 Manufacturing scale-up device with submanifolds for passive flow distribution to feed thousands of parallel microchannels. Flow is mapped using ranges to show variations from a normalized value.

Operation of a large-scale reactor with thousands of parallel channels is expected in 2011.

11.6 Conclusions

Medium- to large-scale microchannel processes offer tremendous economically promising advantages in the chemical and energy process industries, and many such demonstrations are underway. Moving from one channel to tens of thousands of parallel channels requires a substantial effort to understand the scale-up principles for microchannel technology. While numbering up channels sounds quite simple, the reality of flow distribution and manufacturing limitations and tolerances necessitate the need for a robust mathematical-modeling effort. Medium- and large-scale processes will succeed by developing integrated models for pressure drop, heat transfer, reaction kinetics, mechanical design, and manufacturing tolerances.

References

1. Seris, E., Abramowitz, G., Johnston, A., and Haynes, B. (2008) *Chem. Eng. J.*, **135S**, S9–S16.
2. Thayer, A. (2005) *Chem. Eng. News*, **83** (22), 43–52.
3. Franke, R., Jucys, M., Lob, P., and Rehfinge, A. (2008) *Evonik Sci. Newsl.*, **22**, 20–24.
4. Klemm, E., Schirrmeister, S., Albrecht, J., Becker, F., Schutte, R., Caspary, K., and Klemm, E. (2005) *Chem. Eng. Technol.*, **28** (4), 459–464.
5. Klemm, E., Doring, H., Geisselmann, A., and Schirrmeister, S. (2008) *Chem Eng. Technol.*, **30** (12), 1615–1621.
6. Cottrill, A. (2008) *Upstream*, **6**, 34–36.
7. Short, P. (2008) *Chem. Eng. News*, **86** (42), 37.
8. Brandner, J., Benzinger, W., Schygulla, U., Zimmermann, S., and Schubert, K. (2007) *ECI Symp. Ser.*, **RP5** (49), 383–393.
9. Brown, D. (2007) *Chem. Aust.*, **74** (7), 14–17.
10. Jarosch, K., Tonkovich, A., Perry, S., Kuhlmann, D., and Wang, Y. (2005) *Microreactor Technology and Process Intensification*, in ACS Symposium Series, vol. 914, American Chemical Society, New York, pp. 258–273.
11. Kratochwill, M., Glatzer, E., and Farrell, L., Nexant, Inc. (2008) Report Technical Expert's Report on Velocys Technology. October 31, 2008.
12. Tonkovich, A., Perry, S., Wang, Y., Qiu, D., LaPlante, T., and Rogers, W.A. (2004) *Chem. Eng. Sci.*, **59**, 4819–4824.
13. Tonkovich, A., Kuhlmann, D., Rogers, A., McDaniel, J., Fitzgerald, S., Arora, R., and Yuschak, T. (2005) *Chem. Eng. Res. Des.*, **83** (A6), 634–639.
14. Yang, B., Yuschak, T., Mazanec, T., Tonkovich, A., and Perry, S. (2008) *Chem. Eng. J.*, **135S**, S147–S152.
15. Ashmead, J., Blaisdell, C., Johnson, M., Nyquist, J., Perrotto, J., and Ryley, J. Jr. (1996) US Patent 5, 534,328.
16. Wegeng, R., Drost, M.K., and McDonald, C. (1997) US Patent 5, 611,214.
17. Jensen, K. (2001) *Chem. Eng. Sci.*, **56**, 293–303.
18. Kiwi-Minsker, L. and Renken, A. (2005) *Catal. Today*, **110**, 2–14.
19. Commenge, J., Falk, L., Corriou, J., and Matlosz, M. (2002) *AIChE J.*, **48** (2), 345–358.
20. Fitzgerald, S., Tonkovich, A., and Arora, R. (2005) Proceedings of the 18th International Conference on Microreaction Technology.
21. Golbig, K., Autze, V., Born, P., and Drescher, C. (2007) US Patent 7, 241,423.
22. Fitzgerald, S., Tonkovich, A., Arora, R., Qiu, D., Yuschak, T., Silva, L., Rogers, W., Jarosch, K., and Schmidt, M. (2008) US Patent 7, 422,910.
23. Nagasawa, H. and Mae, K. (2006) *Ind. Eng. Chem. Res.*, **45** (7), 2179–2186.
24. Harris, C., Roekaerts, D., and Rosendal, F. (1996) *Chem. Eng. Sci.*, **51** (10), 1569–1594.
25. Lerou, J. and Ng, K. (1996) *Chem. Eng. Sci.*, **51** (10), 1595–1614.
26. Donati, G. and Paludetto, R. (1997) *Catal. Today*, **34**, 483–533.
27. Wintermantel, K. (1999) *Chem. Eng. Sci.*, **54**, 1601–1620.
28. Dautzenberg, F. and Mukherjee, M. (2001) *Chem. Eng. Sci.*, **56**, 251–267.

29. Hessel, V., Hardt, S., and Lowe, H. (2004) *Chemical Micro Process Engineering: Fundamentals, Modeling, and Reactions*, Wiley-VCH Verlag GmbH.

30. Rebrov, E., de Croon, M., and Schouten, J. (2001) *Catal. Today*, **69**, 183–192.

31. Rebrov, E., Duinkerke, S., de Croon, M., and Schouten, J. (2003) *Chem. Eng. J.*, **4139**, 1–16.

32. Moulijn, J. and Stankiewicz, A. (2006) Proceedings of the 9th International Symposium on Process Systems Engineering, pp. 29–37.

33. Kockmann, N. (2007) *Transport Phenomenon in Micro Process Engineering*, Springer-Verlag.

34. Obot, N. (2003) *Nanoscale Microscale Thermophys. Eng.*, **6** (3), 155–173.

35. Kandlikar, S. and Grande, W. (2003) *Heat Transfer Eng.*, **24** (1), 3–17.

36. Wibel, W. and Ehrhard, P. (2007) Proceedings of the 5th International Conference on Nanochannels, Microchannels, and Minichannels, pp. 1–8.

37. Wibel, W. and Ehrhard, P. (2008) presented at the 10th International Microreaction Engineering and Technology Conference, New Orleans.

38. Paul, B. (2006) *Micromanufacturing and Nanotechnology*, Springer, pp. 323–355.

39. Sobhan, C. and Garimella, S. (2001) *Microscale Thermophysl. Eng.*, **5**, 293–311.

40. Muwanga, R. and Hassan, I. (2005) Proceedings of the 3rd International Conference on Microchannels and Minichannels, Toronto Ontario.

41. Lee, P. and Garimella, S. (2006) *Int. J. Heat Mass Transf.*, **49**, 3060–3067.

42. Veser, G. (2001) *Chem. Eng. Sci.*, **56**, 1265–1273.

43. Hesse, D. and Jarosch, K. (2006) US20060035182A1.

44. Venkataraman, K., Wanat, E., and Schmidt, L. (2003) *AIChE J.*, **49** (5), 1277–1284.

45. Delsman, E., de Croon, M., Kramer, G., Cobden, P., Hofmann, C., Cominos, V., and Schouten, J. (2004) *Chem. Eng. J.*, **101**, 123–131.

46. Amador, C., Gavriilidis, A., and Angeli, P. (2004) *Chem. Eng. J.*, **101** (1-3), 379–390.

47. Pfeifer, P., Wenka, A., Schubert, K., Liauw, M., and Emig, G. (2004) *AIChE J.*, **50** (2), 418–425.

12
Intensification of Heat Transfer in Chemical Reactors: Heat Exchanger Reactors

Michael Cabassud and Christophe Gourdon

12.1
Introduction

12.1.1
Chemical Reaction Intensification

Nowadays, the chemical industry has to deal with new challenges. In addition to increasing the productivity and decreasing the time to market, inherently safer and cleaner production must be performed. Alternatives have thus emerged to dramatically improve the chemical processes. Over the last few decades, new perspectives have emerged as a result of process intensification (PI) [1]. This trend opens a new way of thinking concerning the evolution and the future design of production units in the chemical industry – the challenge is to decrease their size while increasing their efficiency [2, 3]. In this field, PI can be considered as a method that allows to prevent and to reduce risks related to major industrial accidents. To reach this objective, one solution is to create new types of equipment with innovative features and significantly higher performances in order to attain better heat transfer and subsequently safer conditions compared to traditional batch or semibatch operations. These processes authorize modification of the operating conditions by employing higher concentrations and by using less solvent and reducing reaction holdups. Focusing on the reaction zone, some successful projects have already demonstrated that it is possible to conceive original reactors, miniaturized, multifunctional, and/or continuous. For instance, ICI dramatically improved one of their production plants by combining the qualities of a heat exchanger (HEX) and a chemical reactor in the same apparatus [4]. These encouraging results illustrated how the chemical industry could move toward new directions and how engineering science could be used to replace traditional processes with new smaller, better, and safer promising reactors. Since its initial applications, PI has been growing and at present we can find numerous studies on innovative pilot projects [5, 6].

Finally, the opportunities expected from PI lie primarily in the following three areas [2, 7]:

Novel Concepts in Catalysis and Chemical Reactors: Improving the Efficiency for the Future.
Edited by Andrzej Cybulski, Jacob A. Moulijn, and Andrzej Stankiewicz

ISBN: 978-3-527-32469-9

- **Costs**: PI should lead to substantially cheaper processes, particularly in terms of production cost (much higher production capacity and/or number of products per unit of manufacturing area), investment cost (smaller equipment, reduced piping, etc.), cost of raw materials (higher yields/selectivities), cost of utilities (energy, in particular), and cost of waste-stream processing (less waste, in general).
- **Safety**: PI may drastically increase the safety of chemical processes. It is obvious that smaller is safer. For instance, material inventories will be lower, which is safer in case of hazardous substances. Moreover, keeping processes under control is easier because of PI, for instance, by efficient heat removal from exothermic reactions.
- **Improved chemistry**: PI leads to a better control of the reaction environment (temperature, etc.). Thus, chemical yields, conversions, and product purity are improved. Such improvements may reduce raw material losses, energy consumption, purification requirements, and waste disposal costs as discussed above.

Yet, in spite of these benefits linked to PI, there are still significant barriers [7] that hinder changes in chemical industry. One of these, it is now well admitted, is the transposition from batch to continuous operation, which has to be carefully addressed since this generally allows operating under radically different operating conditions in comparison with the those for batch operations, and sometimes under conditions even nonreachable with a fed-batch mode.

12.1.2
Heat Exchanger Reactor (HEX Reactor)

Among the various technologies, HEX reactors play an important part in PI since many chemical reactions are temperature dependent. The principle of the HEX reactor consists in combining a reactor with a HEX in only one unit. The following are the benefits of using HEX reactors with PI:

- reduction of waste of energy and raw materials;
- high selectivity and yields due to enhanced heat and mass transfer;
- minimized risk of runaway reaction due to enhanced heat transfer;
- smaller and cheaper plant.

The four main parameters that are important to transpose a reaction from batch reactor to continuous HEX reactor are thermal behavior, hydrodynamics, reactor dynamics, and residence time.

12.1.2.1 Thermal Intensification

Most of the chemical reactions in the process industry are temperature dependent. They are either exothermic or endothermic. As a consequence, it is often necessary to remove the heat generated by an exothermic reaction to control the reaction temperature and to avoid thermal runaway reactions or to suppress endothermic by-product reactions, for instance [8].

Therefore, many traditional designs, such as stirred tank reactors, incorporate heat transfer in the process (jacket, external or internal coil, etc.). However, in these devices, there is a significant distance between the heat transfer site and the site of the chemical reaction where heat is released. As a consequence semibatch mode is implemented while batch mode and/or systems are diluted.

The aim of thermal intensification is to reduce this distance by supplying or removing the heat almost as rapidly as it is absorbed or generated by the reaction [9] that is, combining the reaction and heat transfer into a single piece of equipment, using, for instance, a HEX as a chemical reactor, the so-called HEX reactor. As a result, this technology may offer better safety (through a better thermal control of the reaction), improved selectivity (through more isothermal operation), and reduction in by-products.

A comparison of the heat transfer performance of different reactors presented in Table 12.1 shows the main advantages of HEX reactors [10].

12.1.2.2 **Flow Intensification**

Flow intensification is made with the use of apparatuses in which flow follows a perfect plug flow; the internal parts of the reactor have to be designed accordingly. Indeed, dead zones, that is, reactant accumulation, must be avoided not only in order to have better selectivity and yield but also to avoid formation of hot spots, which would generate safety problems.

Moreover, flow geometries should be defined to enhance mixing between reactants. Indeed, mixing, and particularly, micromixing, has an important influence [8, 10] on product quality in industrial reactions such as polymerization or precipitation, especially if the characteristic reaction time is close to the micromixing time (instantaneous or fast reactions). As a consequence, by investigating the combined influences of heat transfer and mixing in the same apparatus, it appears that a small change in the mixing intensity in a HEX reactor has a substantially larger effect than an equivalent change in an adiabatic reactor such as a simple static mixer [11, 12].

12.1.2.3 **Reactor Dynamics**

The volume of HEX reactor is generally small; therefore product holdup is small. It is obviously an advantage from the point of view of safety. However, it is also important to take characteristic times into consideration. Indeed, as these times are of short duration, HEX reactors are relatively flexible and easier to manipulate during the start-up and shutdown procedures.

12.1.2.4 **Residence Time**

The main limitation of HEX reactors is the short residence time, typically from a few seconds to a few minutes. Indeed, the apparatuses are smaller than the traditional ones and fast flow velocities are necessary in order to maintain good level of heat-transfer coefficients. However, as described in the previous paragraph, the highlighted transfer properties of HEX reactors allow us to operate in a few minutes, whereas it takes many hours in batch or semibatch mode.

Table 12.1 Heat-exchange performance of different reactors [10].

Process	Compact multifunctional heat exchanger		Tubular exchanger reactor	Batch reactor with outer heat exchanger	Batch reactor with a double jacket
	Metallic foams *Re* = 1000	Offset strip fins *Re* = 2000			
Schematic diagram					
Specific area, S/V ($m^2\ m^{-3}$)	400	800	400	10	2.5
Global heat-transfer coefficient, U ($W\ m^{-2}\ K^{-1}$)	3500	5000	500	1000	400
US/V ($kW\ m^{-3}\ K^{-1}$)	1400	4000	200	10	1

The literature proposes a relatively large number of HEX reactors that have been designed and built of different materials such as glass, stainless steel, polyether ether ketone (PEEK), and silicon carbide (SiC). A presentation can be found in Anxionnaz *et al.* [13].

In the following part, four reactors that have been extensively studied in our lab are described: the open plate reactor (OPR) – the Alfa-Laval reactor technology (ART®) plate reactor, the Shimtec reactor from Chart Industries, the Corning (glass reactor), and the "DeanHex" reactor which has been constructed with SiC and stainless steel.

12.2 Examples of Heat-Exchanger Reactor Technologies

12.2.1 The Alfa-Laval Reactor

The OPR developed by Alfa-Laval Vicarb is based on a modified HEX design [14] and consists of a modular structure built by stacking different plates (Figure 12.1) related to the circulation of reaction medium and utility fluid (UF) .

Each section consists of a reaction plate where the reaction mixture flows, surrounded by two cooling plates containing the UF. The reactants and catalyst are stored separately and put into contact at the opening of the first reaction plate. The pilot holdup is typically 1.5 l. The successive plates of the reactor can be represented as shown in Figure 12.1. Inside the reactive plate (RP), the environment of the reaction mixture is composed of PEEK. The UF flows between two stainless steel plates, the sandwich plate (SP) and the transition plate (TP).

Some specific flow inserts introduced inside the plates (both reactant and utility plates) enhance the mixing and residence time and guarantee a good heat transfer capacity [15–17].

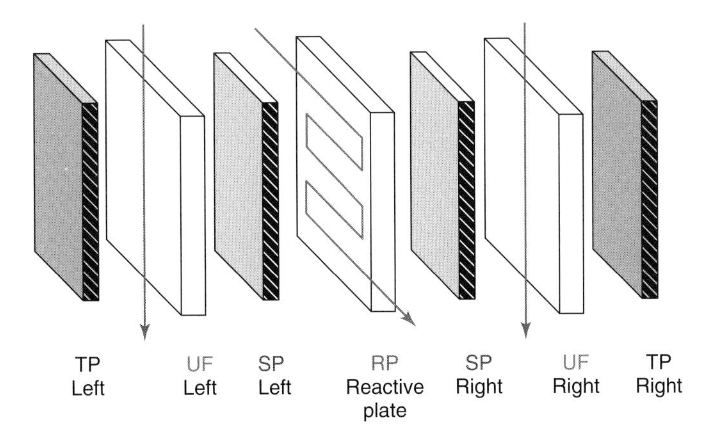

Figure 12.1 Successive plates contained in a block of the "heat-exchanger/reactor" OPR.

12.2.2 The ShimTec Reactor (Chart)

The ShimTec reactor from Chart Industries [18, 19] is an example of a HEX reactor with diffusion-bonded plates (Figure 12.2). The reactor is composed of three plates: one process plate sandwiched between two utility plates. All the units are made of 316 stainless steel, unless stated otherwise. Swagelok fittings are used for all the connections.

The geometrical data and characteristic sizes of the HEX reactor studied in our lab are described in Table 12.2.

However, several other geometries are also available [20].

Figure 12.2 A typical design of Shimtec plate. (Courtesy of Chart Industries.)

Table 12.2 Details of the Shimtec reactor.

Details	Process stream	Utility stream
Number of parallel channels	2	70
Number of layers of each streams	1	2
Individual channel width (mm)	2.0	1.05
Individual channel depth (mm)	2.0	3.0
Average flow area per channel (mm^2)	4.0	2.61
Individual channel length, L (mm)	1855	125
Hydraulic diameter, d_h (mm)	2.0	1.141
Total fluid volume, V (mm^3)	14 836	45 712
Total surface area, A (mm^2)	29 673	160 209
Thickness of the metal between streams (upper bound), e (mm)	1.2	

Figure 12.3 Corning reactor. (Courtesy of Corning.)

12.2.3 The Corning Reactor

The Corning reactors [21] are based on glass technology; this material offers chemical compatibility with a wide range of chemicals and offers one of the highest corrosion resistances to almost all chemicals. The HEX reactor integrates thermal management capability in the form of one or more high-flow buffer fluid passages or layers. The device includes a unitary mixer, that is, a mixing passage through which all of at least one reactant to be mixed is made to pass, the mixing passage being structured so as to promote efficient mixing (Figure 12.3). The reaction and heat-exchange volumes are, for instance, 5 and 16 ml, respectively. Typical dimensions for the reactor channels are in the millimeter and submillimeter range.

Many other plate designs are now available with specific functionalities, and because of modular conception, it is possible to perform all the steps required in a chemical reaction in various zones such mixing zone, residence time zone, and quench zone.

12.2.4 The DeanHex Reactor

This new prototype of the HEX reactor made of ceramic (SiC) has been developed in under a collaboration between BOOSTEC and our laboratory [22]. BOOSTEC is a French innovative small and medium enterprise (SME), specializing in a high-performance all-SiC components and systems. The sintered SiC provides a unique combination of key advantages for the production of high-performance components or systems:

- nearly pure SiC, no secondary phase (to avoid interaction between the material and the reactive medium);
- isotropic physical properties (that guarantee a good homogeneity of the reactor channels);
- high mechanical strength and stiffness, insensitivity to mechanical fatigue;
- high thermal conductivity;
- high stability with time and in aggressive environments.

From a PI point of view, the use of ceramic allows strong corrosion resistance, high heat-transfer capacity, and temperature stability. Consequently, such a reactor appears particularly suited to PI of highly exothermic and/or strongly corrosive applications that are frequently carried out in pharmaceutical and fine chemical industries. In fact, intensification allows reactions to be improved (increase of reactants concentration, of catalyst amount, etc.) and then leads to the use of more aggressive and more corrosive products in an environment where thermal exchanges are enhanced. Moreover, the reactor offers promising perspectives in the fields of safety (reduction of reactive medium amount, minimization of thermal runaway risk), energy efficiency (high thermal transfer performances), productivity (possibility to increase reactant concentration), and environmental impact (reduction of solvent consumption and, therefore, of separation steps).

The concept of the reactor is based on a modular structure built by the alternative stacking of reaction plates and of utility plates containing the thermal fluid (Figure 12.4). The reaction and utility plates are made of SiC; the end plates are

Figure 12.4 Silicon carbide plate heat exchanger.

Table 12.3 SiC reactor characteristics according to flow rate.

Flow rate (L/h)	Reynolds number (–)	Residence time (s)	Pressure drop (bar)	Compacity factor (m^2 m^{-3})	Thermal performances (kW m^{-3} K^{-1})
2	300	60	0.4	2000	9600
5	760	25	0.7		20 000

made of stainless steel to facilitate the connection with the feed lines and with the outlets.

The reactor has been designed for a nominal range of flow rates from 1 to 10 l h^{-1}. In such a range, the reactor offers interesting performances as shown in Table 12.3.

It is important to note that the *compacity factor* is defined by the ratio of the surface area offered to heat transfer over the volume of the reactive medium. The thermal performances are estimated from the product between this compacity factor and the global heat-transfer coefficient. Consequently, owing to the large value of this factor combined with the conductivity performances of the SiC material, the heat-exchange performances are expected to be very high, which can be noticed from the last column of this table.

12.2.5
Influence of the Compacity Factor and of the Material of HEX Reactors

One of the major interests of the HEX reactor is to offer a large ratio surface to reaction volume. Therefore, even if most of the time the laminar flow regime is not suitable to enhance transport phenomena with a moderate overall coefficient, the heat performances are expected to be high, since the compacity factor is always large. This fact is clearly exhibited in Table 12.4, where the results relative to the various HEX reactors studied in our laboratory have been plotted.

Table 12.4 Comparison of different HEX reactors according to the heat transfer performances.

Device	Corning reactor	Alfa-Laval OPR	Chart reactor	SiC reactor
Overall heat transfer coefficient U (W m^{-2} K^{-1})	700	4500	4000	5000
Residence time	min	min	s	min
Compacity factor S/V	2500	400	2000	2000
Intensification factor US/V (kW m^{-3} K^{-1})	1750	1800	8000	10 000

Table 12.5 Effusivity values according to the reactor material.

	SiC	Steel	Glass
$\lambda(20\,^\circ\mathrm{C})$ ($\mathrm{W\,m^{-1}\,K^{-1}}$)	180	16	1
C_p ($20\,^\circ\mathrm{C}$) ($\mathrm{J\,kg^{-1}\,K^{-1}}$)	680	500	800
ρ($\mathrm{kg\,m^{-3}}$)	3210	7900	2600
Effusivity b	20 000	8000	1500
$b = (\lambda C_p \rho)^{1/2}$			

There is also another key parameter linked to the choice of the material for the reactor. First, the choice is obviously determined by the reactive medium in terms of corrosion resistance. However, it also has an influence on the heat transfer abilities. In fact, the heat transport depends on the effusivity relative to the material, defined by $b = (\lambda \rho C_p)^{1/2}$; the effusivity b appears in the unsteady-state conduction equation.

For a given system, this classical equation has the following form:

$$b \cdot \sqrt{t} \approx C$$

From the point of view of dynamic response, the larger that b is, the higher is the reactor dynamics and the surface temperature is preferentially imposed by the material rather than by the reactive medium. Some values of material effusivities are given in Table 12.5.

Consequently, for a given design, the material properties are expected to widely contribute to the heat transport phenomena, with SiC appearing to be the most attractive option under this criterion.

12.3 Methodology for the Characterization of the HEX Reactor

This section presents the different methods used to characterize the performances of the HEX reactors related to

- plug flow behavior
- pressure drop
- mixing
- heat transfer.

12.3.1 Residence Time Distribution Experiments (RTD)

The first characteristic that has to be addressed is the plug-flow behavior of the reactor. This result is important in view of implementing chemical syntheses.

Indeed, the fact that there are no dead zones indicates that there will be no accumulation of reactants, which increases the safety of the process by avoiding the development of hot spots. Moreover, a good mixing and the absence of stagnant zones will have a positive impact on yields and selectivity.

The characterization is performed by means of residence time distribution (RTD) investigation [23]. Typically, holdup is low, and therefore the mean residence time is expected to be relatively short. Consequently, it is required to shorten the distance between the pulse injection and the reactor inlet. Besides, it is necessary to use specific experimental techniques with fast time response. Since it is rather difficult, in practice, to perfectly perform a Dirac pulse, a signal deconvolution between inlet and outlet signals is always required.

According to what has been stated above, good results have been obtained as a result of a spectrophotometric technique that entails a colored tracer. Two measuring probes are set up one at the inlet and the other at the outlet of the device. The acquisition time is set to 0.12 s. The operating protocol adopted during RTD experiments is as follows:

- the reactor is fed at a given flow rate;
- when steady state is reached, a known amount of tracer is instantaneously injected (close to a Dirac pulse) with a syringe through a septum at the inlet of the reactor;
- the tracer concentration is recorded at the reactor outlet.

For example, Figure 12.5 shows the RTDs obtained with two different flow rates in the Shimtec reactor. Each experiment has been performed at room temperature with a total process flow rate varying from 5 to $20\,\mathrm{l\,h^{-1}}$.

The curves have been represented using a reduced time variable.

The aspect of the RTDs shows that the behavior of the Shimtec reactor can be compared to the plug-flow reactor. Curve (b) presents a shorter tail than curve (a); this is due to the influence of the flow rate. Indeed, the higher the flow rate is, the better is the plug-flow behavior. Moreover, Figure 12.5 shows that there is no tail beyond the value 2. The intensity of absorbance is equal to the one before the injection. Thus, it seems that the specific design of the Shimtec reactor offers good mixing performances and has no dead zones, recirculations, or stagnant volumes.

12.3.2 Pressure Drops

Pressure drop is a key parameter from the energy point of view, and it is important for the choice of the auxiliaries and pumping equipment that will be implemented with these reactors.

Pressure drops are referred to in the literature data, assuming that they result from two main contributions: friction between the liquid and the channel wall,

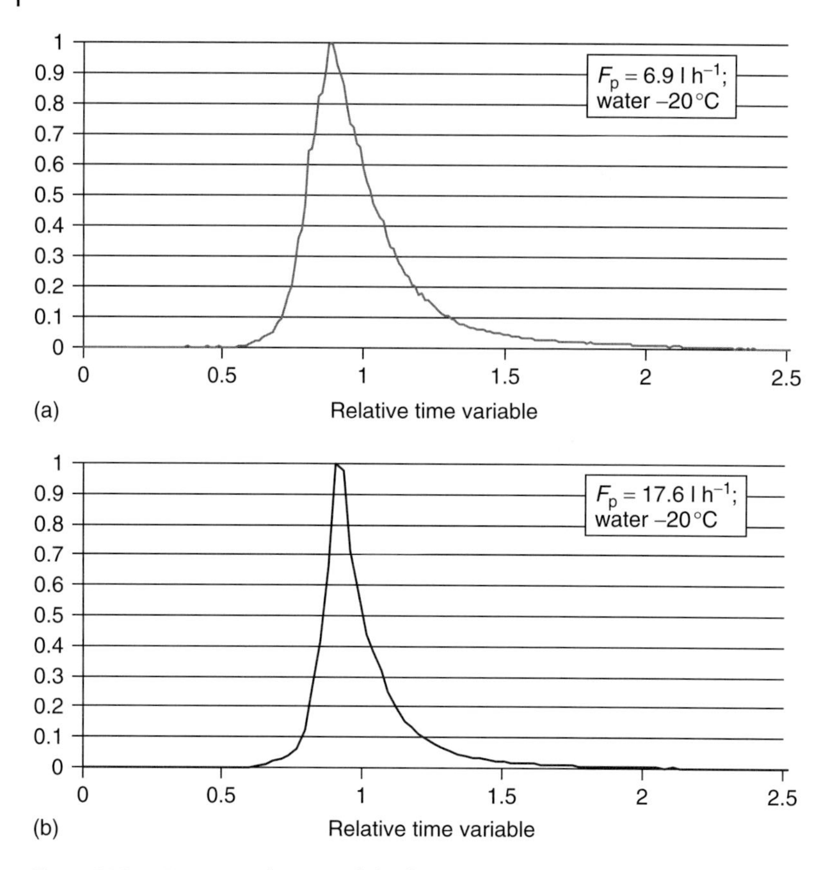

Figure 12.5 RTD curves, function of the flow rate.

called *regular pressure drops* [24, 25] and the pressure drops related to the channel singularities, called *singular pressure drops.*

Usually, the general expression of the regular pressure drop is written as

$$\Delta P = \Lambda \cdot \frac{L}{d_h} \cdot \rho \cdot \frac{u^2}{2} \tag{12.1}$$

with

$$\Lambda = a \cdot Re^b \quad \text{(Darcy coefficient)} \tag{12.2}$$

and

$$Re = \frac{\rho \cdot u \cdot d_h}{\mu} \tag{12.3}$$

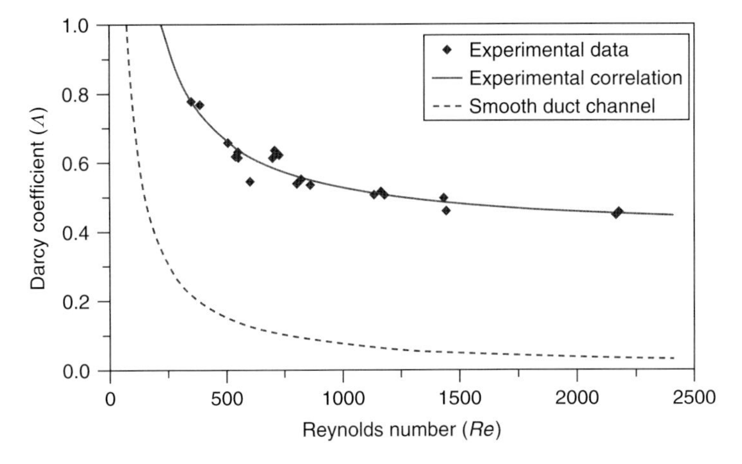

Figure 12.6 Darcy coefficient of pressure drop versus the Reynolds number in a Corning reactor.

Finally, the general expression of the Darcy coefficient including singularities is as follows:

$$\Lambda_{\text{sing}} = \left(\Lambda_{\text{lam}} + \frac{K \cdot d_h}{L}\right) \tag{12.4}$$

Coefficient of friction – "Regular pressure drop" (Λ_{lam})

Coefficient of singularities – "Singular pressure drop" ($\frac{K \cdot d_h}{L}$)

with

$$\Lambda_{\text{lam}} = 0.9 \cdot \frac{64}{Re} = \frac{57.6}{Re} \quad \text{in the case of square channels} \tag{12.5}$$

and K: coefficient of singularities.

This correlation shows that the coefficient of friction, Λ_{lam}, depends on the Reynolds number, whereas the coefficient of singularities, K, is a constant depending on the geometry of each apparatus.

Pressure drops have been measured in a Corning glass HEX reactor with pressure sensors located on reactive and utility lines and estimated for different fluids (water, glucose solutions) at various flow rates, from 2 to $10\,\text{l}\,\text{h}^{-1}$, and various temperature levels (from 20 to 50 °C). The results are presented in Figure 12.6.

12.3.3 Mixing

There is an extensive bibliography related to the mixing notions in chemical reaction engineering. Though important, this topic is not dealt with here. Our aim is only to mention some of the methods that could be easily taken into consideration in view of globally characterizing the mixing performance.

For instance, in order to characterize the mixing performance of any transparent reactor, the reaction of discoloration of an iodine solution with sodium thiosulfate could be used according to the following reaction scheme:

$$I_2(\text{brown}) + 2Na_2S_2O_3(\text{colorless}) \rightarrow 2NaI(\text{colorless}) + Na_2S_4O_6(\text{colorless}) \quad (12.6)$$

This homogeneous reaction is instantaneous and mixing is limited. As a consequence, at steady state, by measuring the required volume to complete the discoloring of iodine, it is possible to determine the global mixing time very easily.

For instance, the mixing performances of different channel designs for the DeanHex reactor have been investigated in our lab, and the respective operating conditions are given in Table 12.6, while Figure 12.7 shows the mixing time versus the Reynolds number.

We do not intend to discuss here in detail the influence of the channel designs (corrugation angle, channel size, straight length between two successive bends, etc.) on the resulting mixing time [26, 27]. It is noticed that the mixing time is generally decreasing with the Reynolds number, that is, with the mechanical energy dissipation. The results are not very sensitive to the channel design provided the Reynolds number is large enough resulting in 1 s or less for the mixing characteristic time. Otherwise, at very low *Re* values, the mixing times can exceed 10 s.

Table 12.6 Operating conditions for the characterization of mixing (the ratio between the flow rates being equal to 1).

C_i (mol l^{-1})	C_t (mol l^{-1})	$F_i + F_t$(l h^{-1})
5×10^{-3}	1.4×10^{-2}	1 1.5 2.25

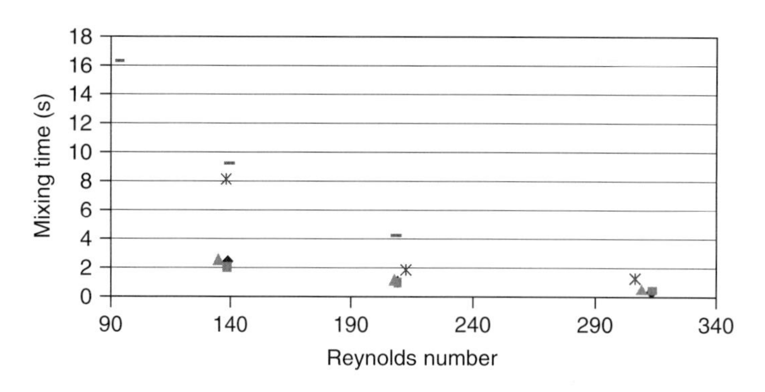

Figure 12.7 Mixing time versus Reynolds number for different channel designs.

12.3.4
Heat Transfer

Industrial chemical syntheses are exo- or endothermic. Therefore, the thermal study aims at characterizing its performances according to the following two steps:

- study and check the thermal performances;
- estimate the heat-transfer coefficients.

Thermal study is based on experiments which aim at cooling the process fluid. The process fluid temperature can, for instance, easily vary from 20 to 60 °C. Process fluid can be water at room temperature or heated with thermostat, whereas UF is water at 15 °C. Experiments with water are necessary since they allow the study of thermal performances of the reactor regardless of any other phenomenon (reaction, mass transfer, etc.). As a consequence, thermal study is an important preliminary step before performing chemical reactions. For each experiment, the operating protocol is as follows:

- first, only process channels are fed, and once steady state is reached, thermal losses can be determined;
- then, utility plates are fed with raw water until steady state is reached, and heat exchanged between the two fluids is calculated;
- finally, utility is shut down when steady state is reached, and the experiment is stopped.

During the last two steps, starting and stopping UF will allow the determination of the reactor dynamics.

The heat-exchange during the process is computed using data collected from the process fluid (flow rate and temperature variation).

After this preliminary step, which gives an outline of the heat transfer abilities of the investigated technology, it is possible to consider a chemical reaction and to carry out some energy balances.

On the process fluid, the global heat exchanged (W) is defined according to the following equation:

$$Q_{\text{global}} = F_{\text{p}}^{\text{in}} \cdot C_{\text{p}} \cdot \left(T_{\text{p}}^{\text{out}} - T_{\text{p}}^{\text{in}} \right) + Q_{\text{generated}} \quad \text{(heat balance on reactor)} \tag{12.7}$$

and

$$Q_{\text{generated}} = w_{\text{i}} \cdot F_{\text{p}}^{\text{in}} \cdot \Delta H_{\text{Ri}} \quad \text{(reaction heat balance)} \tag{12.8}$$

where w_{i} and F_{p}^{in} (mass concentration and flow rate) are related to the limiting reactant.

In the case of an instantaneous exothermic reaction, we can assume that all the reaction heat is generated at the first part of the reactor. Considering a position where the process fluid temperature has not yet reached the corresponding utility temperature in order to avoid the temperature pinch, it is possible to derive a global heat-transfer coefficient U with Eq. (12.9):

$$U = \frac{Q_{\text{position}}}{A \cdot \Delta T_{\text{lm}}} \quad \text{(global heat transfer coefficient)} \tag{12.9}$$

with

$$Q_{\text{position}} = F_{\text{p}}^{\text{in}} \cdot C_{\text{p}} \cdot \left(T_{\text{p}}^{\text{position}} - T_{\text{p}}^{\text{in}}\right) + Q_{\text{generated}} \quad \text{(heat balance at the position)} \tag{12.10}$$

and

$$\Delta T_{\text{lm}} = \frac{\left(T_{\text{p}}^{\text{in}'} - T_{\text{u}}^{\text{in}}\right) - \left(T_{\text{p}}^{\text{position}} - T_{\text{u}}^{\text{position}}\right)}{\ln \dfrac{\left(T_{\text{p}}^{\text{in}'} - T_{\text{u}}^{\text{in}}\right)}{\left(T_{\text{p}}^{\text{position}} - T_{\text{u}}^{\text{position}}\right)}} \quad \text{(logarithmic-mean temperature difference)} \tag{12.11}$$

As the whole reaction occurs at the beginning of the reactor, the process fluid inlet temperature has been modified with the adiabatic increase of temperature:

$$T_{\text{p}}^{\text{in}'} = T_{\text{p}}^{\text{in}} + \Delta T_{\text{ad}} \quad \text{(correction on the inlet temperature)} \tag{12.12}$$

In case of a noninstantaneous reaction, the previous calculation has to be performed between two successive experimental points (and not from a global point of view).

Owing to the high values of the utility flow rates, there is no significant difference between utility inlet and outlet temperature values. Thus, it is not possible to use this information to present heat balances on the UF.

In order to illustrate the approach, we present some results obtained with the OPR Alfa-Laval reactor in the case of an instantaneous exothermic reaction.

To evaluate the heat exchange/productivity performances of the device and its environment, an acid–base neutralization involving sulfuric acid and soda has been performed. It is an instantaneous and exothermic reaction with $\Delta H = -92.4\ \text{kJ mol}^{-1}$ (NaOH). Each experiment is characterized by the initial concentration of the reactants (from 10 to 30% in mass of soda and from 5 to 12% in mass of sulfuric acid). These concentrations are varied in order to evaluate the behavior of the reactor with respect to different amounts of heat generated (from 0.4 to 1.3 kW). Each run is performed with a variable utility flow rate (from 1 to 3 $\text{m}^3\ \text{h}^{-1}$).

Figure 12.8 shows the recording of the process fluid temperature at steady state along the reactor for the 1.5 $\text{m}^3\ \text{h}^{-1}$ utility flow rate and for different T_{p}^{in}.

It is possible to calculate U through Eq. (12.3), the global heat-exchange coefficient. Table 12.7 presents the experimental results. U varies from 3900 to 5000 $\text{W m}^{-2}\ \text{K}^{-1}$ and has a mean value of 4500 $\text{W m}^{-2}\ \text{K}^{-1}$. These values are in the same order of magnitude of the coefficients obtained in plate exchangers and are higher than the ones obtained in tubular reactors, and far away from values measured in batch reactors.

It is possible to observe that U increases with the utility flow rate and initial process fluid temperature. This corresponds to a classical evolution of the Nusselt number function of Reynolds and Prandtl numbers.

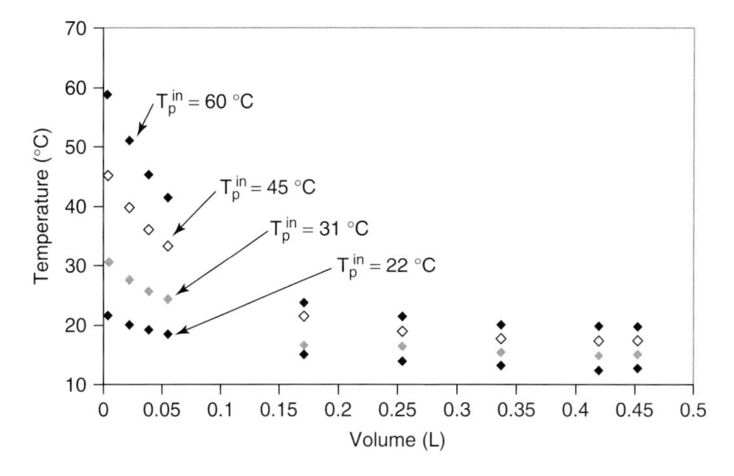

Figure 12.8 Temperature profile along the reactor at steady state for different T_p^{in}.

Table 12.7 Overall heat transfer coefficient for all experiments (with reaction).

wt% NaOH	wt% H_2SO_4	T_p^{in}(°C)	$\Delta T_{adiabatic}$ (°C)	$Q_{generated}$ (kW)	F_u^{in} ($m^3\ h^{-1}$)	U ($W\ m^{-2}\ K^{-1}$)
10.5	5.9	19.4	7.0	0.43	1	3979
10.5	5.9	19.6	6.7	0.41	3	4190
14.6	6.1	18.4	10.1	0.62	1	3938
14.6	6.1	17.5	10.8	0.67	2.8	4389
17.0	6.9	19.7	12.3	0.79	1	4446
27.0	11.7	18.1	19.9	1.28	1	4414
27.0	11.7	18.2	19.6	1.26	3	4939

12.4
Feasibility of HEX Reactors

A number of reactions presenting different characteristics have been carried out in HEX reactors [28]. In the following, some of those that exhibit different intensification opportunities are discussed:

- It is possible to strongly accelerate reactions assumed to be slow by increasing the reaction temperature (for example, by acting on the pressure), increasing the amount of catalyst, reducing the amount of solvent, and therefore increasing the concentration of the reactants.
- In case of exothermic reactions, the heat-exchange capacities of the reactor allow to rapidly evacuate the heat generated by the reaction and therefore to perform a transposition of a pure batch operating mode into a continuous one. The main point is the ability to avoid, as far as possible, an initial increase of the temperature as soon as the reactants are mixed.

As widely mentioned in the literature [29, 30], a batch operation with a constant temperature is, most of the time, a better operating mode than a fed batch, especially for selectivity reasons. Therefore, the transposition from a batch to a plug-flow continuous mode should be a promising operation if the operating temperature is strictly controlled.

12.4.1 Oxidation Reaction

Oxidation reactions are generally problematic because of their large heat release. For instance, the oxidation reaction of sodium thiosulfate, $Na_2S_2O_3$, by hydrogen peroxide, H_2O_2, for which the stoichiometric scheme is

$$2Na_2S_2O_3 + 4H_2O_2 \rightarrow Na_2S_3O_6 + Na_2SO_4 + 4H_2O \quad (12.13)$$

is an irreversible homogeneous reaction, fast and highly exothermical [31, 32], and the heat generated by the reaction is $\Delta H_r = -586.2\ \text{kJ mol}^{-1}$ of $Na_2S_2O_3$. The reaction is temperature sensitive. As a consequence, the conversion rate will depend on the operating temperature of reactants and on the cooling power of the utility stream. That is the reason this oxidation reaction is well adapted to test thermal performances of continuous intensified reactors [14, 33].

Some results obtained in our lab with the Shimtec reactor of Chart are described. Mixing of the two reactants is made at the process stream inlet and product storage is cooled with ice and diluted to prevent thermal runaway. The solubility limits of H_2O_2 and $Na_2S_2O_3$ in water are respectively 30 and 35% by weight. In order to avoid very high temperatures in case of an adiabatic rise, process streams of H_2O_2 and $Na_2S_2O_3$ were both diluted at 10% weight. Finally, a Dewar vessel was used in order to evaluate the conversion by calorimetry [34].

Two experiments have been performed by varying the reactant flow rates and subsequently the residence time in the reactor. Details of each experiment are shown in Table 12.8.

The utility stream gets started at operating temperature and flow rate. In the following experiments, the utility stream is heated so as to initiate the reaction. The main and secondary process lines are fed with water at room temperature and with the same flow rate as one of the experiments. Once steady state is reached, operating parameters are recorded. Process lines are then fed with the reactants, hydrogen peroxide and sodium thiosulfate. At steady state, operating parameters are recorded, and a sample of a known mass of reactor products is introduced in the Dewar vessel. Temperature in the Dewar vessel is recorded until equilibrium is reached, that is, until the reaction ends. This calorimetric method is aimed at calculating the conversion rate at the product outlet and thus the conversion rate in the reactor. The latter is also determined by thermal balances between process inlet and outlet of the reactor. Finally, the reactor is rinsed with water. This procedure is repeated for each experiment.

Table 12.8 Operating conditions of oxidation reaction experiments.

		Experiment A	Experiment B
H_2O_2	wt (%)	11.5	
	H_2O_2 flow rate ($l\,h^{-1}$)	3.3	1.3
	T_{p1}^{in} (°C)	22	
$Na_2S_2O_3$	wt (%)	10.8	
	$Na_2S_2O_3$ flow rate ($l\,h^{-1}$)	6.7	4
	T_{p2}^{in} (°C)	22	
Process outlet	T_p^{out} (°C)	49	
Utility (water)	F_u^{in} ($l\,h^{-1}$)	50	
	T_u^{in} (°C)	47	
	T_u^{out} (°C)	50	

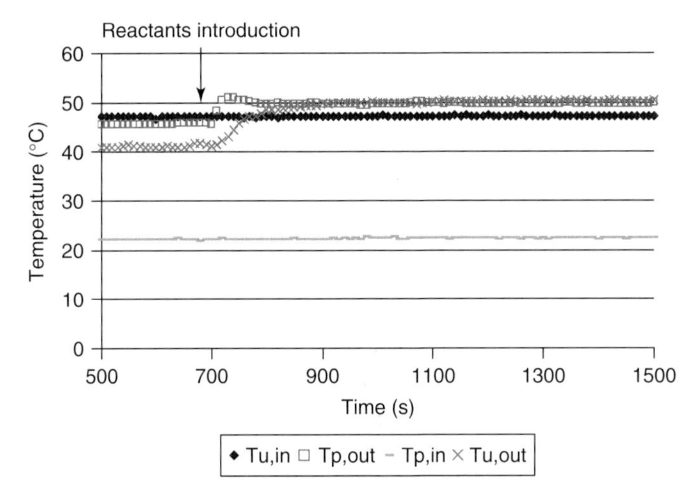

Figure 12.9 Temperature recording for experiment A.

Figure 12.9 shows the temperature recording during the experiment A at the inlet and outlet of the reactor for process and utility streams. At $t = 700$ s, the reactor is fed with reactants instead of water.

From these results, it is possible to evaluate the heat generated by the reaction and thus the conversion rate.

Two methods have been used to calculate the conversion rate in the reactor. They are based on thermal balances: first between inlet and outlet of process and utility streams in the reactor and then between sampling and thermal equilibrium in the Dewar vessel. The former leads to the conversion rate obtained in the reactor, χ; and the latter gives the conversion rate downstream from the reactor outlet, $1 - \chi$.

12.4.1.1 Thermal Balance in the Reactor

This approach consists in estimating heat exchanged by each stream in order to determine the heat provided to the system by the reaction. Actually, at steady state, the heat of reaction will lead to a temperature rise of process and utility streams, while the utility stream aims at limiting this increase (cooling effect). The conversion rate, χ, is easily calculated by

$$\chi = \frac{Q_{\mathrm{l,th}} + F_{\mathrm{p}} \cdot Cp_{\mathrm{p}} \cdot (T_{\mathrm{p}}^{\mathrm{out}} - T_{\mathrm{p}}^{\mathrm{in}}) + F_{\mathrm{u}} \cdot Cp_{\mathrm{u}} \cdot (T_{\mathrm{u}}^{\mathrm{out}} - T_{\mathrm{u}}^{\mathrm{in}})}{Q_{\mathrm{r}}^{\mathrm{total}}} \tag{12.14}$$

with

$$Q_{\mathrm{r}}^{\mathrm{total}} = \dot{N}_{\mathrm{i}}^{0} \cdot \Delta H_{\mathrm{r}} \tag{12.15}$$

12.4.1.2 Thermal Balance in the Dewar Vessel

This approach consists in measuring the adiabatic temperature increase of a sample taken at the outlet of the reactor. Sampling is made in an adiabatic vessel (Dewar vessel) and temperature is recorded until the reaction ends, that is, until an equilibrium temperature is reached. The conversion rate is thus written as

$$\chi = 1 - \frac{(m_{\mathrm{p}} \cdot Cp_{\mathrm{p}} + m_{\mathrm{dw}} \cdot Cp_{\mathrm{water}}) \cdot (T_{\mathrm{dw}}^{\mathrm{eq}} - T_{\mathrm{p}}^{\mathrm{out}})}{n_{\mathrm{i}}^{0} \cdot \Delta H_{\mathrm{r}}} \tag{12.16}$$

Table 12.9 gives the experimental results.

The estimated error is proportional to

$$\text{error} \approx k \cdot \frac{2 \cdot T}{\Delta T} (\%) \tag{12.17}$$

Temperature difference in the reactor was less important than in the Dewar vessel because of the efficient exchange with the utility stream. That is the reason error estimation in the reactor is higher than in the Dewar vessel.

Conversion is not total because of the very short residence time in the reactor (<15 s). It can be noticed that in spite of the decrease of global process flow rate between experiment A and experiment B, the conversion is not improved and even decreases. This phenomenon is explained by the $[H_2O_2]/[Na_2S_2O_3]$ ratio. Indeed, in experiment A, this ratio is equal to 2.5, whereas in experiment B it is equal to 1.6. As a consequence, in experiment A, sodium thiosulfate is the limiting reactant whereas in experiment B, this is hydrogen peroxide. As a consequence, two phenomena interfere: increasing residence time, which tends to increase conversion rate, and, in contrast, decreasing concentration, which tends to do the opposite. Besides,

Table 12.9 Conversion rates of oxidation reaction experiments.

		Experiment A	**Experiment B**
Conversion – reactor	χ (%)	68 (±13)	61 (±26)
Conversion – Dewar		72 (±5)	74 (±9)

the experiments show that conversion was not complete because of a very short residence time. This arbitrary barrier can be removed by simply adding plates in the reactor and thus increasing the volume and therefore the residence time.

Finally, the oxidation reaction has to been run under strict conditions of temperature, which are impossible to be operated in a batch reactor. Indeed, utility stream in the Shimtec reactor was heated to 47 °C, which first initiates the reaction, accelerates its kinetics, and then controls the temperature when the heat of the reaction is too important. In a batch reactor, working with such UF temperature is impossible because of security constraints. It would certainly lead to a reaction runaway. We now consider this question in the next section.

12.4.2
Comparison with Semibatch Operation

Comparison between the heat exchanged per unit of volume during oxidation experiment in the Shimtec reactor and the maximal heat exchanged in a classical batch reactor (with a double jacket) highlights the effectiveness of the former. Indeed, in oxidation reaction experiments, a mean value of the heat exchanged per unit of volume in the HEX reactor is estimated with utility stream temperature of 47 °C:

$$\frac{Q}{V} = 20 \times 10^3\ \mathrm{kW \cdot m^{-3}} \tag{12.18}$$

Then, the quantity of heat that could be removed in batch reactors whose volume varies from 1 l to 1 m^3 is calculated. In order to compare with experimental results, the temperature gradient is fixed at 45 °C (beyond which water in the utility stream would freeze and another cooling fluid should be used). The maximum global heat-transfer coefficient is estimated at an optimistic value of 500 W m^{-2} K^{-1}. The calculated value of the global heat transfer area of each batch reactor, A, is in the same range as the one given by the Schweich relation [35]:

$$\frac{A}{V} = (4.9 \pm 0.6) \cdot V^{-\frac{1}{3}} \tag{12.19}$$

Table 12.10 summarizes the geometrical parameters and the heat exchanged per unit of volume of the batch reactors in the same reaction conditions as the HEX ones.

Table 12.10 Characteristics and heat release for batch reactors with a double jacket.

Volume (m^3)	1×10^{-3}	1×10^{-2}	0.1	1
Diameter (m)	0.08	0.2	0.4	0.9
Height (m)	0.2	0.3	0.8	1.4
Area (m^2)	0.055	0.22	1.13	4.6
Q_{max}/V (kW m^{-3})	1200	500	250	100
t_c (s)	230	560	1120	2800

It also details the feeding times (t_c) of a semibatch reactor in which oxidation reaction would be operated. Feeding times are calculated with the same operating conditions as in the Shimtec reactor: around 10% weight of sodium thiosulfate and hydrogen peroxide and a stoichiometric ratio between sodium thiosulfate and hydrogen peroxide equal to 2. The main hypotheses are that the final volume of reactant mixture is equal to the volume of the semibatch reactor and that the reactor is filled with hydrogen peroxide before starting, while sodium thiosulfate is in the feeding device. Feeding rates are calculated so that the heat generated by the reaction and relative to the feeding rate of sodium thiosulfate should not exceed the maximum heat that can be removed by the double jacket, Q_{max}:

$$t_c = \frac{m_{thio}}{M_{thio} \cdot \dot{N}_{thio}^{max}} \tag{12.20}$$

with

$$\dot{N}_{thio}^{max} = \frac{Q_{max}}{\Delta H_r} \tag{12.21}$$

Since the final volume of the reactant mixture (equal to the volume of the reactor) and that of the mass fraction of sodium thiosulfate are known, the total mass of sodium thiosulfate that has to be added is also known.

Clearly, the oxidation reaction could not have been implemented in a pure batch operating reactor. Indeed, heat removal capacity would not have been sufficient (100–1200 kW m^{-3} removed versus 20×10^3 kW m^{-3} generated). As a consequence, a semibatch mode is necessarily required. Besides, Table 12.10 shows that the feeding times are much higher than the residence time of the Shimtec reactor (around 15 s).

As expected, heat exchanged per unit of volume in the Shimtec reactor is better than the one in batch reactors (15–200 times higher) and operation periods are much smaller than in a semibatch reactor. These characteristics allow the implementation of exo- or endothermic reactions at extreme operating temperatures or concentrations while reducing needs in purifying and separating processes and thus in raw materials. Indeed, since supply or removal of heat is enhanced, semibatch mode or dilutions become useless and therefore, there is an increase in selectivity and yield.

12.4.3 Inherently Safer Characteristics of HEX Reactors

The feasibility of operating highly exothermic reactions in a HEX reactor has been demonstrated, some considerations can also be given concerning the inherently safer characteristics of an intensified continuous HEX reactor. This type of evaluation has been conducted on the OPR, using the esterification of propionic anhydride by 2-butanol as test reaction [36, 37].

This study shows that in case of a major deviation corresponding to the stop of both the utility and the reactant lines, part of the energy released by the reaction would be immediately dissipated in the plates closer to the reaction mixture.

Therefore, consideration of this "mass" in heat-transfer phenomenon leads to a significant temperature decrease reached after deviation and increases the time to maximum rate. This reveals an intrinsically safer behavior of this apparatus compared to that of batch reactors.

12.5 Conclusions

In the field of PI, the transposition of the chemical reaction operation from batch to continuous mode represents a critical point. In this chapter, some examples of HEX reactors, allowing this transposition, have been presented.

The main characteristics of these apparatuses are their plug-flow behavior allowing a continuous production, their ability to exchange rapidly a large amount of heat authorizing intensified new chemical synthesis operating conditions, and the low holdup inside the reactor.

New chemical synthesis routes leading to a better productivity and increased selectivity could be defined with regard to the new opportunities offered by HEX reactors. For example, they can lead to solvent-free operation or operations with at least dramatically reduced amount of solvent, to increase the reaction temperature or to engage in more efficient catalysis.

The geometry of the reactor is of great importance owing to the fact that it is necessary to obtain good heat and mass transfer characteristics while working with laminar hydrodynamic conditions in order to assume a sufficient residence time compared to the reaction time.

The choice of the material is also strategic for two reasons: to obtain an apparatus resisting to the corrosion of the different chemical products, like glass for glass-lined batch reactors, but also to keep high heat transfer performances, like aluminum in the case of HEXs. Steel and more so SiC appear to be very interesting compromises for both aspects.

Before putting a HEX in operation, it is necessary to characterize its properties in terms of heat and mass transfer, pressure drop, and hydrodynamics. A number of experimental methods have been presented and exemplified. These data are important to perform simulations and to define optimal operating conditions.

Examples of chemical reactions have been presented and carried out in different types of HEX reactors. Applications such as oxidation, nitration, hydrosilylation, are still in progress in the industrial companies

From the process safety point of view, the evaluation of the intrinsic safe character of HEX compared to batch or semibatch reactors has been investigated [33, 37]. Two points clearly show the interest in the HEX reactors:

- a comparison with semibatch reactors in the case of a very highly exothermic reaction;
- an evaluation of the role of the mass and the material properties of the reactor in case of an accident linked to the stoppage of both utility and process fluids.

A number of HEX reactors are presented in the literature, some of which have been designed for a specific chemistry. It is necessary to remember the necessary flexibility and polyvalence which are required in fine chemicals or pharmaceutical industries. Nevertheless, owing to the low holdup and the rapidity of putting HEX reactors in operation, one can imagine types of HEX reactors devoted to a portfolio of reactions linked to the exothermicity, product characteristics (such as corrosion) or to the nature of the phases (liquid–liquid, gas–liquid, solid, etc.).

Some questions could arise concerning process control and especially concerning the problem of deviation detection. Solutions can be found by using software able to monitor and supervise the process in order to guarantee production quality and safety of the plant and operators [38].

Acknowledgment

The authors gratefully thank their colleagues who contributed to most of the results presented in this chapter: Dr P. Cognet, Dr S. Elgue, Pr. N. Gabas, Dr L.E. Prat, the PhD and postdoc students Z. Anxionnaz, Dr W. Benaïssa, Dr A. Devatine, and also the technical staff of the LGC who participated in the pilot design and experimentation.

List of Symbols

A	Global heat transfer area (m^2)
C_p	Specific heat capacity ($J\ kg^{-1}\ K^{-1}$)
d_h	Hydraulic diameter (m)
e	Thickness of the wall between process and utility streams (m)
F	Flow rate ($kg\ s^{-1}$)
h	Local heat-transfer coefficient ($W\ m^{-2}\ K^{-1}$)
K	Coefficient of singularities
L	Channel length (m)
m	mass (kg)
m_{dw}	Mass of water equivalent to the mass of the Dewar vessel (kg)
m_p	Mass of the sample (kg)
m_{thio}	Mass of sodium thiosulfate in the feeding device (kg)
M_{thio}	Molecular weight of sodium thiosulfate ($kg\ mol^{-1}$)
$n_i{}^0$	Initial mole number of the limiting reactant (mol)
$\dot{N}_i^0$	Initial molar flow rate of the limiting reactant ($mol\ s^{-1}$)
$\dot{N}_{thio}^{max}$	Maximum feeding rate of sodium thiosulfate ($mol\ s^{-1}$)
P	Pressure (Pa)
$Q_{l,th}$	Thermal losses (W)
Q	Heat exchanged (W)
Q_r^{total}	Total heat of reaction (W)
Re	Reynolds number

t_c	Feeding time (s)
T_{dw}^{eq}	Equilibrium temperature in the Dewar vessel (K)
T	Temperature (K)
U	Global heat-transfer coefficient (W.m^{-2}.K^{-1})
u	Flow velocity (m.s^{-1})
V	Volume of fluid (m^3)
X	Steady state conversion reached during normal operation.

Greek Symbols

ΔH_r	Heat generated by the reaction (J.mol^{-1})
ΔP	Pressure drop (Pa)
ΔT_{ad}	Adiabatic temperature rise (K)
Λ	Darcy coefficient
λ	Thermal conductivity of the material (W m^{-1} K^{-1})
ρ	Density of the fluid (kg m^{-3})
μ	Viscosity of the fluid (Pa s)
χ	Conversion rate.

Subscripts

dw	Dewar vessel
lam	Laminar
sing	Singularities
p	Total process side
r	Reaction
u	Utility

Superscripts

in	Inlet
out	Outlet
position	Position of the thermocouple

References

1. Becht, S., Franke, R., Geißelmann, A., and Hahn, H. (2009) An industrial view of process intensification. *Chem. Eng. Process.*, **48** (1), 329–332.
2. Stankiewicz, A. and Moulijn, J.A. (2002) Process intensification. *Ind. Eng. Chem. Res.*, **41** (8), 1920–1924.
3. Lomel, S., Falk, L., Commenge, J.M., Houzelot, J.L., and Ramdani, K. (2006) The Microreactor: a systematic and efficient tool for the transition from Batch to Continuous Process? *Chem. Eng. Res. Des.*, **84** (5), 363–369.
4. Phillips, C.H., Lausche, G., and Peerhossaini, H. (1997) Intensification of batch chemical processes by using integrated chemical reactor

heat-exchangers. *Appl. Therm. Eng.*, **17**, 809–824.

5. Stankiewicz, A. (2003) *Re-Engineering the Chemical Processing Plant, Process Intensification*, Marshal Dekker, New York.
6. Jachuk, R.J. (2002) Process intensification for responsive processing. *Trans. IChemE*, **80** (A), 233–238.
7. Tsouris, C. and Porcelli, J.V. (2003) Process intensification – Has its time finally come? *Chem. Eng. Prog.*, **99** (10), 50–55.
8. Nilsson, J. and Sveider, F. (2000) Characterising Mixing in a HEX Reactor Using a Model Chemical Reaction. Available on *www.chemeng.lth.se/exjobb/002.pdf*, Department of Chemical Engineering II, Lund (Last accessed May 2010).
9. Ferrouillat, S., Tochon, P., Della Valle, D., and Peerhossaini, H. (2006a) Open loop thermal control of exothermal chemical reactions in multifunctional heat exchangers. *Int. J. Heat Mass Transfer.*, **49** (15-16), 2479–2490.
10. Ferrouillat, S., Tochon, P., and Peerhossaini, H. (2006) Micromixing enhancement by turbulence: application to multifunctional heat exchangers. *Chem. Eng. Process.*, **45** (8), 633–640.
11. Edge, A.M., Pearce, I., and Phillips, C.H. (1997) Compact heat exchangers as chemical reactors for process intensification (PI). Process Intensification in Practice, BHR Group Conference Series, Publication No. 28, pp. 175–189.
12. Thonon, B. and Tochon, P. (2004) *Re-engineering the Chemical Processing Plant, Process Intensification*, (eds A. Stankiewicz and J.A. Moulijn) Chem. Ind., Marcel Dekker, pp. 147–153.
13. Anxionnaz, Z., Cabassud, M., Gourdon, C., and Tochon, P. (2008) Heat exchanger/reactors (HEX Reactors): concepts, technologies: State-of-the-art. *Chem. Eng. Process.*, **47**, 2029–2050.
14. Prat, L., Devatine, A., Cognet, P., Cabassud, M., Gourdon, C., Elgue, S., and Chopard, F. (2005) Performance evaluation of a novel concept "open plate reactor" applied to highly exothermic reactions. *Chem. Eng. Technol.*, **28** (9), 1028–1034.
15. Chopard, F. (2001) Improved device for exchange and/or reaction between fluids, Patent No. FR 0105578, PCT WO 02/085511.
16. Chopard, F. (2002) Flow directing insert for a reaction chamber and a reactor, Patent No. SE 0203395.9, PCT WO 20/04045761.
17. Bouaifi, M., Mortensen, M., Anderson, R., Orcuich, W., Anderson, B., Chopard, F., and Noren, T. (2004) Experimental and numerical cfd investigations of a jet mixing in a multifunctional channel reactor: passive and reactive systems. *Chem. Eng. Res. Des.*, **82**, 274.
18. Enache, D.I., Thiam, W., Dumas, D., Ellwood, S., Hutchings, G.J., Taylor, S.H., and Stitt, E.H. (2007) Intensification of the solvent-free catalytic hydroformylation of cyclododecatriene: comparison of a stirred batch reactor and a heat-exchange reactor. *Catal. Today*, **128**, 18–25.
19. Plucinski, P., Bavykin, D., Kolaczkowski, S., and Lapkin, A. (2005) Liquid phase oxidation of organic feedstock in a compact multichannel reactor. *Ind. Eng. Chem. Res.*, **44**, 9683–9690.
20. CHART Industries (2009) Compact Heat Exchange Reactors. Available on *www.chart-ind.com/app_ec_reactortech.cfm* (accessed March 2009).
21. CORNING, Reactor Technologies (2009) Corning® Advanced-Flow™ Glass Reactors. Available on *www.corning.com/r_d/emerging_technologies/reactors.aspx* (accessed March 2009).
22. Elgue, S., Ferrato, M., Momas, M., Chereau, P., Prat, L., Gourdon, C., and Cabassud, M. (2007) Silicon carbide equipments for process intensification: first steps of an innovation. 1st Congress on Green Process Engineering, April 24-26, 2007, Toulouse.
23. Villermaux, J. (1994) Réacteurs Chimiques – Principes in Techniques de l' ingénieur, traité Génie des procédés, J 4 010, Vol. JB5, pp. 4–10; pp. 37–42.
24. Bird, R.B. and Stewart, W.E. (2002) *Lightfoot Transport Phenomena*, 2nd edn, John Wiley & Sons, Inc., New York, pp. 178–184.
25. Bird, R.B. and Stewart, W.E. (2002) *Lightfoot Transport Phenomena*, 2nd edn,

John Wiley & Sons, Inc., New York, pp. 205–209.

26. Anxionnaz, Z., Cabassud, M., Gourdon, C., and Tochon, P. (2009) Hydrodynamic study and optimisation of the geometry of a heat exchanger/reactor. 2nd European Process Intensification Conference 2009, Venice.
27. Anxionnaz, Z., Cabassud, M., Gourdon, C., and Tochon, P. (2009) Thermal and parametric study of a wavy channel in an intensified heat exchanger/reactor. 8th World Congress of Chemical Engineering 2009, Montréal.
28. Elgue, S., Devatine, A., Prat, L., Cognet, P., Cabassud, M., Gourdon, C., and Chopard, F. (2009) Intensification of ester production in a continuous reactor. *Int. J. Chem. React. Eng.*, **7**, Article 24.
29. Rippin, D.W.T. (1983) Simulation of single and multiproduct batch chemical plants for optimal design and operation. *Comput. Chem. Eng.*, **7** (3), 137–156.
30. Garcia, V., Cabassud, M., Le Lann, M.V., Pibouleau, L., and Casamatta, G. (1995) Constrained optimization for fine chemical productions in batch reactors. *Chem. Eng. J.*, **59**, 229–241.
31. Lo, S.N. and Cholette, A. (1972) Effect of channelling on the performance of adiabatic flow tank reactors for exothermic reactions. *Can. J. Chem. Eng.*, **50** (1), 66–70.
32. Lo, S.N. and Cholette, A. (1972) Experimental study on the optimum performance of an adiabatic MT reactor. *Can. J. Chem. Eng.*, **50** (1), 71–80.
33. Benaïssa, W., Elgue, S., Gabas, N., Cabassud, M., Douglas, C., and Demissy, M. (2008) Dynamic behaviour of a continuous heat exchanger/reactor after flow failure. *Int. J. Chem. React. Eng.*, **6** (A23), Available at: *http://www.bepress.com/ijcre/vol6/A23*.
34. Anxionnaz, Z., Cabassud, M., Gourdon, C., and Tochon, P. (2010) Transposition of an exothermic reaction from a batch reactor to an intensified continuous one. *Heat Transfer Eng.*, **31** (9), 788–797.
35. Schweich, D. (2001) *Génie de La Réaction Chimique*, Tec & Doc, Lavoisier.
36. Benaïssa, W. (2006) Développement d'une méthodologie pour la conduite en sécurité d'un réacteur continu intensifié. PhD Thesis, Institut National Polytechnique de Toulouse, France.
37. Benaïssa, W., Gabas, N., Cabassud, M., Carson, D., Elgue, S., and Demissy, M. (2008) Evaluation of an intensified continuous heat-exchanger reactor for inherently safer characteristics. *J. Loss Prev. Process Ind.* **21** (5), 528–536.
38. Orantes, A., Kempowsky, T., Le Lann, M.-V., Prat, L., Elgue, S., Gourdon, C., and Cabassud, M. (2007) Selection of sensors by a new methodology coupling a classification technique and entropy criteria. *Chem. Eng. Res. Des.*, **85** (A6), 825–838.

13
Reactors Using Alternative Energy Forms for Green Synthetic Routes and New Functional Products

Guido Mul, Tom Van Gerven, and Andrzej Stankiewicz

13.1
Introduction

Alternative energy forms are increasingly being investigated and applied to activate and facilitate both long-known and novel chemical processes. Several reasons explain the booming interest from chemical process industry. First, some energy forms are selective in nature – such as microwaves coupling with materials having a high dielectric loss, or magnetic energy preferentially interacting with ferromagnetic materials – and can thus be used to activate the molecules, particles, or phases of interest. Secondly, traditional bottlenecks in chemical processes can possibly be bypassed using these new forms of energy. An example is the safety issue when performing chemical reactions at high temperature or pressure. Being able to attain high temperature or pressure locally within a bulk volume that remains at atmospheric conditions, such as is the case with ultrasound, offers new perspectives. Thirdly, new technologies also bring forth new product properties. High-gravity technology, for instance, is able to produce particles with a smaller particle size and a smaller particle-size distribution than attainable with any other technology. In some cases, reactions are even not possible except when using e.g., ultrasound. Finally, as (bio)chemical reactions are improved by optimizing catalytic and enzymatic activity, process conditions become the weakest link in the overall chemical process. Improvement of heat transfer, mass transfer, mixing, and so on, should follow the pace of chemical innovation in order to achieve overall intensification.

This chapter reviews the reported effects of different types of energy on chemical processing. Many of them are already known for a long time, but were, until recently, mostly used in nonreactive systems such as separation or drying. The focus here is on the (assumed) mechanism, reported effects, and known industrial applications of reactive chemical systems.

Novel Concepts in Catalysis and Chemical Reactors: Improving the Efficiency for the Future.
Edited by Andrzej Cybulski, Jacob A. Moulijn, and Andrzej Stankiewicz

ISBN: 978-3-527-32469-9

13.2 Energy of Electromagnetic Field

13.2.1 Microwave Reactors

Microwave energy has been investigated in drying, sintering, polymer, and food processing since the 1950s [1]. It has been successfully scaled up to industrial applications in these sectors, taking advantage of the selective heating and subsequent evaporation of molecules (e.g., water) sensitive to microwave coupling. Microwave-assisted chemical reactions, however, have only started to attract attention since the mid-1980s.

Microwaves are electromagnetic waves with frequencies between those of infrared radiation and radio waves, ranging from 0.3 to 300 GHz, which corresponds to wavelengths between approximately 1 mm and 1 m. The most often used frequency is 2.45 GHz, with a wavelength of about 12 cm. Microwaves transfer their energy to materials not primarily by conduction or convection, but by dielectric loss. The potential of a material to undergo microwave heating ("to couple with microwaves") is therefore highly dependent on its dielectric properties. Materials couple with microwave energy mainly by two mechanisms: dipole rotation and ionic conduction. Dipole rotation refers to the alignment by rotation with the electric field component of the electromagnetic microwave radiation of molecules that have permanent or induced dipoles. Ionic conduction is the migration of dissolved ions with the oscillating electric field. Both mechanisms generate movement, and therefore molecular friction and collisions, thus producing heat. Besides this undisputed effect of molecular agitation, other effects such as improved transport properties of molecules, changes to the transition states, and decrease of the activation energy in Arrhenius law were also postulated. Recently, increasingly more authors, however, agree that the effect of microwaves has a purely thermal character. The enhancement effect of microwaves on homogeneous liquid-phase reactions is well documented. In recent years several reviews have appeared [2–4], showing the ability of microwave heating to accelerate organic reactions with acceleration factors from several to more than a thousand. Besides the process rate, sometimes also the product yield increases as a result of microwave heating. In heterogeneously catalyzed systems and solvent-free synthesis reactions, microwave effects have also been observed [5–8]. It is suggested that selective heating of the catalyst results in an increased reaction rate at the hot spot. Also the occurrence of arcing and subsequent plasma formation has been postulated. These effects are currently still under investigation.

Microwave technology has now matured into an established technique in laboratory-scale organic synthesis. In addition, the application of microwave heating in microreactors is currently being investigated in organic synthesis reactions [9–11] and heterogeneous catalysis [12, 13]. However, most examples of microwave-assisted chemistry published until now have been performed on a

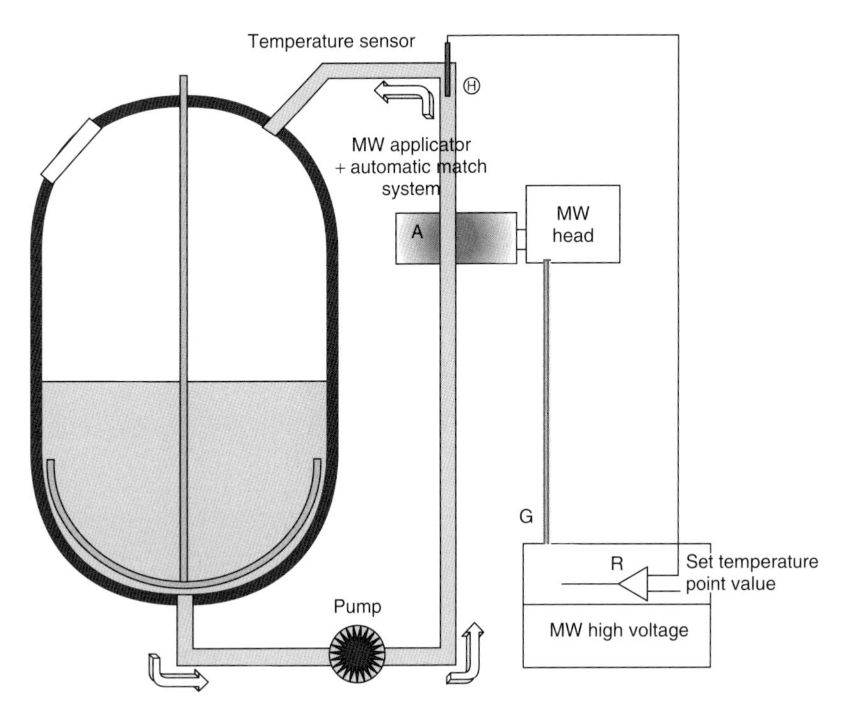

Figure 13.1 Process scheme of the industrial microwave-assisted production of laurydone. (Courtesy: SAIREM.)

scale less than 1 g and 25 ml [1]. In order to become a fully accepted and implemented industrial technology in the future, it is clear that larger scale techniques should be developed that can routinely provide products on a multikilogram and even multiton scale. Several (interrelated) barriers to large-scale implementation exist and the most important ones from an engineering perspective are limited penetration depth of microwaves, (un)wanted occurrence of hot spots and the related phenomenon of arcing, lack of appropriate models/equipment to predict and control the system. The most prominent examples on the larger scale are the continuous microwave dry-media reactor for the production of waxy esters using montmorillonite clay as a catalyst delivering quantities of 100 kg per day of the corresponding product in high purity [14, 15], and the tubular reactor for the esterification of linalool on a 100-tons-per-year scale [16, 17]. So far the largest reported microwave reactor for organic synthetic applications has been operated by SAIREM in France since 2003 for the production of laurydone [1]. The reactor was developed and designed in collaboration with BioEurope and De Dietrich. The stop-flow system consists of a 1 m^3 batch reactor with an external recycling loop which is irradiated by a 6000-W microwave generator (Figure 13.1). The claimed features include 40% less energy requirement, five times shorter processing time, and elimination of catalyst compared to the conventional thermal approach.

13.2.2
Photochemical Reactors

To activate chemical bonds, light is an attractive form of energy. In the field of synthetic organic photochemistry, light of high enough energy (deep UV) is used to directly activate molecules [18]. While in the laboratory synthetic photochemistry is often applied, a relatively limited amount of photon-induced industrial processes exist. One of the factors responsible is the often suboptimal reactor design. Incorporation of annular Hg-lamps typically applied in industrial photoreactors limits design flexibility, while Hg-lamps are costly, have a limited lifetime, and produce a significant amount of heat requiring extensive cooling [19, 20]. Recent developments in reactor design on the laboratory scale include thin (liquid) film reactors, or microbatch reactors [20]. The most innovative design is based on the microreactor concept. LED light sources are available to illuminate microreactors with well-defined wavelengths, including in the UV. Without being exhaustive, microreactors have been successfully applied for various photochemical conversions, including cyclo-additions, cyanation of aromatic compounds, photooxygenation of alkenes (dienes), and photochlorination. These examples show that, in particular, in gas/liquid configurations microreactors with their well-defined control of the gas–liquid distribution are feasible photoreactors, which lead to larger photon efficiency and space time yields. Also, effects on selectivity have been observed [20], both as a result of the use of monochromatic light (the advantage of LEDs vs e.g., Hg-lamps) and excellent control over the residence times of the reactants. Application of microreactors in combination with photochemistry appears quite feasible for the pharmaceutical industry.

Photochemical conversion can also be induced with the aid of a catalyst, that is, by photocatalysis [21]. Here light is used to activate a catalyst, typically a semiconductor, rather than the reactant. Photocatalysis is applicable for a large variety of reactions [22–24]. By far, the dominant research area of heterogeneous photocatalysis is the photodegradation of organic compounds in either air or water [25]. TiO_2 is the most investigated photocatalyst in these applications, which converts contaminants to CO_2 and H_2O, and, if applicable, harmless inorganic ions. The superior performance of TiO_2 is attributed to the ability to form a relatively large concentration of oxidizing holes and hydroxyl radicals [23]. Hydroxyl radicals are considered as the reactive oxidant converting the contaminants [26].

Heterogeneous photocatalysis is less explored in the field of organic synthesis [27]. However, the possibility to induce selective, synthetically useful redox transformations has become increasingly more attractive and promising. Finally, the combination of a photocatalyst and light allows the conversion of solar energy into chemical energy, by, for example, water splitting and/or CO_2 activation. An increasing intensity in research efforts can be observed in this field [24].

Besides improvements in catalyst characteristics [28], the low productivity of a photocatalytic process can also be improved by reactor design. In photocatalytic research on a laboratory scale, the most widely applied reactors are the top illumination or annular reactors containing a suspended catalyst [29]. This type of reactor provides ease of construction and high catalyst loading but there are several drawbacks. These are difficulties of downstream separation and recycle of the catalyst from the reaction mixture, and low light-utilization efficiencies due to light scattering and shielding by the slurry mixture [29]. Therefore, reactors containing immobilized photocatalysts are preferred. There are basically two configurations containing immobilized photocatalysts. The first configuration uses lamps external to the reactor with the catalyst coated on the reactor wall (similar to the solar reactor used for water decontamination), while the second configuration is based on lamps placed within the reactor with the catalyst coated on the lamp housing or reactor wall. In these two configurations, the total illuminated surface area is limited by the geometry of standard light sources. An improvement in light distribution can be obtained by using an optical fiber bundle (optical fiber reactor), on which the catalyst is coated [30]. A novel concept in photocatalytic reactor design based on optical fibers is the realization of the so-called internally illuminated monolith reactor (IIMR). In the concept of the IIMR, side-light emitting fibers are placed inside the channels of a ceramic monolith, in which a TiO_2 photocatalyst is coated on the wall of each individual channel (Figure 13.2) [31].

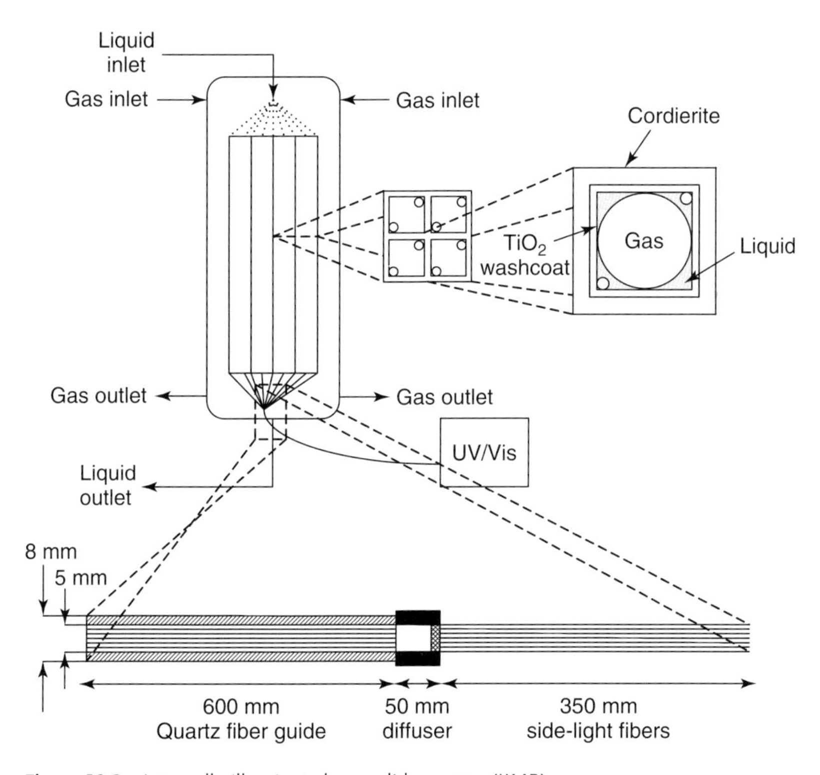

Figure 13.2 Internally illuminated monolith reactor (IIMR) scheme with a detail of the cross section of the monolith channels and the fiber-optic bundle.

The advantages of microreactors, for example, well-defined control of the gas–liquid distributions, also hold for photocatalytic conversions. Furthermore, the distance between the light source and the catalyst is small, with the catalyst immobilized on the walls of the microchannels. It was demonstrated for the photodegradation of 4-chlorophenol in a microreactor that the reaction was truly kinetically controlled, and performed with high efficiency [32]. The latter was explained by the illuminated area, which exceeds conventional reactor types by a factor of 4–400, depending on the reactor type. Even further reduction of the distance between the light source and the catalytically active site might be possible by the use of electroluminescent materials [19]. The benefits of this concept have still to be proven.

13.3 Energy of Electric Field

13.3.1 Electrochemical Reactors

Electrochemical reactions have been performed for a long time in various reactor configurations, either as electrolytic or as galvanic cells. With respect to the latter, much research is dedicated to fuel cells as a cleaner substitute for combustion engines. This also illustrates the general trend of applying electrochemistry in miniaturized systems. Other examples of electrochemical microprocessing can be found in the domains of combinatorial (bio)organic synthesis and chemical analysis [33].

An interesting application of electrocatalysis is the direct electrochemical reduction of CO_2 in aqueous solutions over Cu-electrodes. Various reviews on this topic have appeared in the open literature. Recently, it has been demonstrated that certain Cu-electrodes produce products in CO_2 electroreduction with a distribution as typically obtained in the Fischer–Tropsch reaction of syn-gas over heterogeneous Co- or Fe-based catalysts. This is an important scientific discovery, since this is the first tangible evidence that chain propagation occurs during CO_2 electroreduction over Cu-electrodes [34]. Furthermore, this work shows that a gas-to-liquid process starting directly from CO_2 is potentially feasible over Cu-electrodes.

Electric fields are also investigated in chemical processes to improve mass transfer, heat transfer, and, more recently, mixing [35]. In liquid–liquid extraction, enhancement factors between 2 and 10 have been reported. Via electric-field-induced emulsification, the surface area per unit volume can be increased up to 500 times compared to conventional extraction processes. Improvement of heat transfer has been reported in different systems such as in boiling liquids (increase by a factor of 4–10), in falling film evaporators (sevenfold enhancement) and in silicone oil rising (two- to threefold increase). When electric fields are applied to microstructured systems, mixing has been shown to intensify by a factor of 35 (shortening of mixing length) to 70 (increase of mixing index).

13.3.2 Plasma Reactors

A specific form of electrical energy is plasma, which can be classified into the thermal (hot) and the nonthermal (cold) variant. In particular, the nonthermal plasma technology is investigated for chemical processes and destruction of pollutants. Nonthermal plasma technology can generate cold plasma through various types of excitation such as silent, glow, corona, microwave, or radio-frequency electrical discharges, operated at atmospheric or low pressure [36, 37]. These discharges ionize the gas near the electrodes in such a way that energy is taken up only by the electrons, gaining typical temperatures in the range of 10 000–250 000 K (1–25 eV), which are not in thermal equilibrium with the heavier components of the atoms, hence "cold" plasma [37]. In the dielectric barrier discharge, an example of the silent discharge, two electrodes are placed in (miniaturized) planar or annular arrangement and separated from each other by at least one dielectric layer and a discharge gap [38]. The discharge gap is the volume where process streams can be treated. Another type of discharging is the glow discharge, where no dielectric barrier is used. An example is the GlidArc technology, where glow discharge is combined with arcing [36]. Gliding discharges are produced between diverging electrodes and across the flow. The discharges start at the spot where the distance between the electrodes is the shortest and glide progressively along the electrodes in the direction of the flow until they break up and disappear at a certain distance. The electrical discharges immediately reform at the initial spot. The continuous displacement of the discharge roots on the uncooled electrodes prevents their chemical corrosion or thermal erosion, which is common in other high current arc configurations.

Numerous chemical processes have been investigated using nonthermal plasma reactors [35]. An important area is the destruction or abatement of pollutants in air streams for emission control. Successful results have been found for the reduction of emitted NO_x, SO_2, H_2S, and diesel particulate matter, as well as the decomposition of VOC (e.g., toluene, trichloroethylene, and benzene). Also destruction of organic dyes and of microbiological organisms in aqueous solutions has been reported. Besides applications such as environmental technology, plasma reactors are being researched for chemical synthesis reactions, such as the production of methanol and ethylene from methane, the production of syn-gas from the partial oxidation of methane or propane, and ethanol steam reforming. The best-known industrial application of plasma technology in the field of chemical reactions is the generation of ozone from oxygen, air, or other N_2/O_2 mixtures [38]. O_2 molecules dissociate in dielectric barrier discharges by electron impact and/or by reactions with N atoms or excited N_2 molecules. Ozone is then formed in a three-body reaction involving O and O_2. The capacity of a large ozone generator can be up to $100\,kg\,h^{-1}$. The main applications are in wastewater treatment (Figure 13.3), pulp bleaching, and organic synthesis.

Figure 13.3 A large ozone generator for water treatment producing 60 kg h^{-1}. (Courtesy: Ozonia Ltd.)

13.4 Energy of Magnetic Field

Magnetic fields have been used for separation processes since the nineteenth century. It is now being used to separate materials ranging from coarse to colloidal and from strongly magnetic to diamagnetic [39]. Magnetically assisted chemical separation disperses magnetic microspheres in liquid waste streams where they sorb pollutants and then remove the particles from the solution by magnetic force [40, 41]. Press filtration has also been shown to improve in a magnetic field by slowing down and structuring cake formation [42]. A large area of research is devoted to magnetic aggregation/deaggregation of particles in suspensions or aerosols. While most results show that magnetic fields can induce aggregation, the group of Martens has recently observed that a permanent magnetic field can also assist the breakup of nanoparticle aggregates suspended in a turbulent liquid flow [43].

The application of magnetic fields in (bio)chemical processing is a much younger discipline. Ueno and Harada [44] describe the effects of magnetic fields on combustion, biological processes, and enzymatic activities. During catalyzed combustion of hydrocarbons and alcohol, it appeared that gradient magnetic fields influenced the spatial pattern of combustion temperature and that this redistributed pattern varied with the reactants. More recently, Digilov and Sheintug [45] report trapping the reaction zone of ethylene glycol oxidation with a colloidal suspension of iron oxide nanoparticles under a gradient magnetic field. They were able to keep the reaction zone steady or rotate it by moving the magnetic field relative to the reaction dish. Another type of stable colloidal suspensions – ferrofluids – were oscillated near the gas–liquid interface by a magnetic field, achieving 40–50% increase in mass transfer rates [46]. Most work regarding magnetic-field-assisted chemical processing is focusing on magnetohydrodynamics. Larachi and coworkers have been applying gradient magnetic fields to enhance the hydrodynamics in trickle-bed reactors [47–49]. They take advantage of the magnetization force on liquid and gas flows to control pressure drop, liquid hold-up (11% improvement), and wetting efficiency of the catalyst particle (18% improvement). Fluidization properties have also been investigated for particles in gas streams (e.g., [50]) and for solid–gas reactions (e.g., [51]).

13.5 Energy of Acoustic Field

Ultrasound is another energy form that is gaining importance for intensifying chemical processing [52–55]. The range of ultrasonic frequencies used for sonochemical applications is from 16 to 100 kHz (sometimes extended to 1 MHz). When applying ultrasound in this frequency range to a liquid, the acoustic wave generates alternate compression and expansion (i.e., rarefaction) zones. When during the rarefaction phase the large negative pressure exceeds the intermolecular van der Waals forces locally, small cavities or gas-filled microbubbles are formed. Once these microbubbles are formed, they will either oscillate around a mean radius and exist for many acoustic cycles (stable cavitation), or grow more during the expansion phase than they shrink during the compression phase (transient cavitation) due to the unequal diffusion of gases and vapor from the bulk liquid into the bubble. The occurrence of stable or transient cavitation depends on ultrasound intensity. In the case of transient cavitation, growth of the bubbles continues over several oscillation cycles until the pressure of the surrounding bulk liquid on the bubble exceeds a critical value at which moment the bubble implodes, creating extremely high local temperature (up to 5000 K) and pressure (up to 50 000 atm has been suggested).

Cavitations generate several effects. On one hand, both stable and transient cavitations generate turbulence and liquid circulation – acoustic streaming – in the proximity of the microbubble. This phenomenon enhances mass and heat transfer and improves (micro)mixing as well. In membrane systems, increase of flux through the membrane and reduction of fouling has been observed [56].

In solid–liquid and biological systems, acoustic streaming also thins the boundary layer at the solid or cell surface [57, 58]. On the other hand, transient cavitation induces other more extreme effects. First, the high local temperature and pressure occurring in the collapsing microbubble induce pyrolysis reactions of the water vapor and/or chemical reactants leading to the release of free radicals. Secondly, the sudden collapse of the bubble also results in an inrush of the liquid to fill the void, producing shear forces in the surrounding bulk liquid. These shear forces are capable of forming liquid jets targeted at solid surfaces in solid–liquid systems. This causes fragmentation of solids, removal of passivating surface layers, and exposure of new reaction surfaces [59, 60]. In immiscible liquid–liquid systems, the shear forces will cause disruption of the interface and mixing [60]. In biological systems, induced forces will be able to cause cell destruction [58], although less-intensive ultrasound has been shown to enhance metabolism and subsequent growth rate and biomass yield. Meticulously controlled ultrasonication is also able to extract extracellular polymer substances while avoiding cell lysis [61].

Ultrasound can thus be used to enhance kinetics, flow, and mass and heat transfer. The overall results are that organic synthetic reactions show increased rate (sometimes even from hours to minutes, up to 25 times faster), and/or increased yield (tens of percentages, sometimes even starting from 0% yield in nonsonicated conditions). In multiphase systems, gas–liquid and solid–liquid mass transfer has been observed to increase by 5- and 20-fold, respectively [35]. Membrane fluxes have been enhanced by up to a factor of 8 [56]. Despite these results, use of acoustics, and ultrasound in particular, in chemical industry is mainly limited to the fields of cleaning and decontamination [55]. One of the main barriers to industrial application of sonochemical processes is control and scale-up of ultrasound concepts into operable processes. Therefore, a better understanding is required of the relation between a cavitation collapse and chemical reactivity, as well as a better understanding and reproducibility of the influence of various design and operational parameters on the cavitation process. Also, reliable mathematical models and scale-up procedures need to be developed [35, 54, 55].

13.6 Energy of Flow

The energy of fast fluid flow can be utilized to intensify processes in chemical reactors and there are two basic ways of doing it: by purposefully creating the cavitation conditions in the reacting liquid or by using a supersonic shockwave for fine phase dispersion.

13.6.1 Hydrodynamic Cavitation Reactors

Purposeful hydrodynamic cavitation in a chemical reactor can be created in two ways. One alternative is to let the liquid pass through a throttling valve, orifice plate, or any other mechanical constriction. If the pressure in vena contracta falls below the cavitation pressure (usually the vapor pressure of the medium), microcavities will be generated. These cavities will subsequently collapse as the liquid jet expands and pressure recovers. Another possibility to create cavitation is to use the so-called liquid whistle, already applied in food industry for homogenization and emulsification [62]. Here, the liquid is accelerated in a jet and then flows across a steel blade, which vibrates as liquid passes over it at high velocity. The frequency of these vibrations can be adjusted in such a way that cavitation is created. This way large liquid volumes could, in principle, be processed. Liquid whistle, however, suffers from several important shortcomings, such as very low vibrational power, low intensity of cavitation generated, high pumping costs, and erosion of the blade in the presence of particulate matter [63].

Hydrodynamic cavitation reactors have been investigated for more than a decade now in the UDCT Department of Bombay University [63–66]. When applied to some industrially relevant reactions, the hydrodynamically created cavitation appeared to deliver on average an order of magnitude higher cavitation yields than the acoustic cavitation. In addition, the processing volumes could be up to about 100 times larger than in the conventional sonochemical reactors. So far, there is no information about the industrial applications of the hydrodynamic cavitation reactors, although some concepts have already been patented [67].

13.6.2
Supersonic Shockwave-Based Reactors

The energy of the supersonic shockwave is used in another promising alternative method for intensification of the phase contacting and transport processes. In 1995, Mattick *et al.* [68] described a supersonic shockwave reactor for pyrolysis of hydrocarbons, in which a gas cooled to subpyrolysis temperature by expansion to supersonic speed was mixed with a supersonic flow of feedstock. The ethylene yield in such a reactor increased by 20% as compared to conventional technology, while the energy consumption dropped by 15%. Interestingly, most research in the field of supersonic chemical processing has been carried out in industrial environment and has led to a number of commercial applications. For example, Praxair investigated a supersonic gas–liquid reactor for carrying out fast processes [69, 70]. The energy of the supersonic shockwave is used here to disperse gas into tiny microbubbles and thus create an enhanced interfacial area for mass transfer. Results of experiments carried out with oxygen–water system show that the oxygen transfer rate in a supersonic reactor is up to 10 times higher than in a tee-mixer. Mass transfer coefficients exceeding $2.0\ s^{-1}$ are reported [69]. The German company Messer Griesheim GmbH has patented and commercialized a supersonic nozzle for fluidized-bed applications [71, 72]. The concept was subsequently applied on the industrial scale in a fluidized-bed reactor for iron sulfate decomposition at Bayer AG. Supersonic injection of oxygen has increased the capacity of the reactor by 124% [73]. The same technology has also been applied to the sludge combustion reactors increasing the throughput by approximately 40% [74]. Recently, DSM

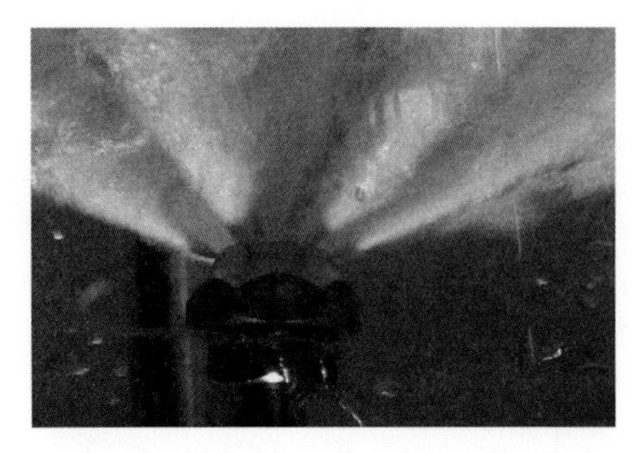

Figure 13.4 DSM transonic oxygen injection nozzle. (Courtesy: DSM.)

reported development and application of a transonic oxygen injection technology in one of their large-scale fermentation processes. This new type of injection system (Figure 13.4) has doubled the yeast productivity of a large-scale fermenter [75].

The several industrial applications reported in the literature prove that the energy of supersonic flow can be successfully used as a tool to enhance the interfacial contacting and intensify mass transfer processes in multiphase reactor systems. However, more interest from academia and more generic research activities are needed in this field, in order to gain a deeper understanding of the interface creation under the supersonic wave conditions, to create reliable mathematical models of this phenomenon and to develop scale-up methodology for industrial devices.

13.7
Energy of Centrifugal Fields–High-Gravity Systems

The use of centrifugal fields in chemical processing is quite well established and includes physical operations such as pumping, compression, extraction, and solid–liquid separations. The use of these fields in chemical reaction engineering is much less mature. Two basic families of reactors employing the centrifugal fields are reported in scientific literature: rotating packed bed reactors (RPBRs) and spinning disc reactors (SDRs). Both reactor families are characterized by substantial intensification of transport processes compared to the conventional systems, either mass transfer or heat transfer, or both.

13.7.1
Rotating Packed Bed Reactors

Since the pioneering work by Ramshaw [76], the high-gravity (HiGee) technology employing rotating packed beds (RPBs) has been primarily developed for separation processes, such as absorption, distillation, or stripping. However, various concepts of using these systems for carrying chemical reactions are also reported in the literature, along with first commercial-scale applications.

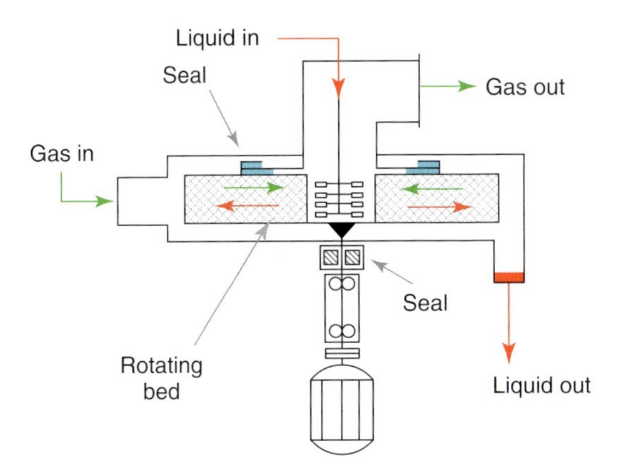

Figure 13.5 Scheme of a rotating packed bed reactor.

The scheme of a typical RPB for countercurrent gas–liquid contacting is shown in Figure 13.5. The rotor of the apparatus consists of a continuous packing. The liquid enters the center of the rotor and moves toward its perimeter, while the gas enters the stationary housing and passes through the rotor from outside to inside. Two mechanisms are responsible for the substantially increased mass transfer rates in the RPBs. First, the liquid forms a thin laminar film on the packing surface, with the average thickness between 10 and 80 μm ([77] – data for water). Secondly, atomization of the liquid inside the packing occurs, which creates high-surface-area liquid droplets in addition to the film wetting of the packing. This results in significant additional gas–liquid mass transfer. The second mechanism prevails and allows using low-surface-area packings that result in lower cost packings, reduced pressure drop, and higher throughputs. The mass transfer improvements in RPBs are significant. The patent by Ramshaw and Mallison [78] describes a distillation process in an RPB and reports a 27- to 44-fold increase of the liquid-side mass transfer coefficient and 4- to 9-fold increase of the gas-side mass transfer coefficient, with respect to the stationary bed of 0.5 in. Intalox saddles. Zheng and coworkers investigated hydrodynamics and heat transfer in a cross-flow RPB and found the height of the mass transfer unit for the liquid-side controlling processes to be as low as 2.5–4 cm [79]. Lin and coworkers investigated the methanol–ethanol distillation process in an RPB and reported height equivalent of theoretical plate (HETP) values of 3–9 cm, compared to 30–40 cm for the conventional structured packings [80]. Also, the data provided in the review paper by Rao and coworkers indicate that up to 200 times increase of the $k_{L}a$ in the RPBs with respect to the conventional packed columns is possible [81].

On the other hand, RPBs suffer from poor heat transfer possibilities. Heat input could theoretically be achieved by use of eddy currents, microwaves, or sonic energy, and thus endothermic reactions are, in principle, possible. The heat removal is more problematic and exothermic reactions must be conducted adiabatically within the rotor. Alternating packing and heat transfer plates could perhaps be an option, although it would greatly increase the complexity and the price of the reactor.

In processes where evaporation is compatible with the chemical reaction, evaporation could be used for heat removal. An additional downside of RPBs is the short residence time in these devices, usually in the range of 0.2–2 s, which makes them applicable to fast and very fast processes only.

Nevertheless, first commercial-scale applications of RPBs as chemical reactors are reported, perhaps the most impressive being the reactive stripping in Dow's hypochlorous acid process [77]. In this process, the hypochlorous acid has to be very quickly removed from the reaction environment due to its decomposition reaction with NaCl. The conventional process using stripping and absorption columns to remove the gaseous product has been replaced by three RPBRs (Figure 13.6), in which fast reactive stripping process takes place. The effects reported by Dow are impressive: about 15% increase in yield, about 40 times reduction of the equipment size, and about one-third reduction in wastewater and chlorinated by-products. Other interesting examples of applications of RPB technology to reacting systems include production of nanoparticles via reactive precipitation. The high-gravity technology allows for manufacturing of nanoproducts with required crystal forms and very narrow particle-size distribution [82]. Various types of mostly pharmaceutical nanoparticles have been synthesized so far, including barium titanate [83], zinc sulfide [84], calcium carbonate [85], or barium carbonate [86]. In their review paper, Chen and Shao [87] describe the commercial-scale facility for production of calcium carbonate nanoparticles in RPBRs. Five such

Figure 13.6 HiGee technology for hypochlorous acid manufacturing at Dow Chemical. Three rotating packed beds shown in front offer the same production capacity as the conventional plant behind them. (Courtesy: Dow Chemical.)

Table 13.1 Comparison between the reactive precipitation of $CaCO_3$ in conventional equipment and in rotating packed bed reactor.

Items	Conventional technology	HiGee technology
Reactor types	Stirred tank or tower, bubbling reactor	RPB
Rate of micromixing	Slow, 5–50 ms	Fast, 0.01–0.1 ms
Rate of mass transfer, HTU	Slow, 1–2 m	Fast, 1–3 cm
Product particle sizes	>100 nm without crystal growth inhibitor; <100 nm with crystal growth inhibitor. Morphology is not easy to control	15–30 nm without crystal growth inhibitor. Morphology is controllable
Batch reaction time	Long, 60–75 min, additional three days aging	Short, 15–25 min
Reactor volumes (3000 $t\,a^{-1}$)	30–50 m^3, three reactors. Investment is expensive	About 4 m^3. Investment is inexpensive
Controllability of production and production cost	Difficult, repeatability of product quality is poor, cost is high	Easy, product quality is stable, cost is low
Scale-up effects	Large, hard to scale up	No scale-up effects

[a]After [87].

facilities are in operation, with a total production capacity of about 36 000 tons per annum. The HiGee technology has clear advantages over the conventional one, as shown in Table 13.1. In recent years, several research papers appeared dealing with possible environmental applications of the RPBRs in the ozonation reactions [88–90]. In all the cases, significant improvement of the ozone transfer rates was observed.

13.7.2 Spinning Disc Reactors

Contrary to RPBRs, in SDRs, intensified heat transfer presents the most important advantage. Liquid reactant(s) are fed on the surface of a fast rotating disk near its center and flow outward. Temperature control takes place via a cooling medium fed under the reaction surface. The rotating surface of the disc enables to generate a highly sheared liquid film. The film flow over the surface is intrinsically unstable and an array of spiral ripples is formed. This provides an additional improvement in the mass and heat transfer performance of the device.

The mass and heat transfer performance of the SDR is indeed impressive. Aoune and Ramshaw [91] found local heat transfer coefficients ranging from about 10 000 to about 30 000 $W\,m^{-2}\,K$ and local mass transfer coefficients between about 4E-04 and 1E-03 ms^{-1}. The enhancement of the local mass transfer coefficients appears to be associated with the passage of the ripple. Excellent heat transfer and short residence times (usually between 1 and 5 s) make the SDR particularly effective when fast processes with high heat fluxes or viscous liquids are involved [92]. Consequently, highly exothermic fine-chemical and pharmaceutical processes as well as polymerization reactions constitute the vast majority of systems investigated in SDRs.

On the fine-chemical/pharmaceutical side, the SDR is reported to deliver a substantial improvement in a phase-transfer-catalyzed Darzen's reaction for preparing a drug intermediate. The SDR allows here 99.9% (i.e., 1000 times) reduction in the reaction time, 99% reduction of the equipment inventory, and 93% reduction of the impurities level [93]. In the rearrangement of α-pinene oxide to campholenic aldehyde, a continuously operated SDR with catalytically activated surface delivers a substantial reduction of the reaction time, from 3600 down to 1 s, and increase of the conversion, from 50 to 85%. At the same time, catalyst separation from the product stream is eliminated [94]. Among the large chemical companies in this sector, SmithKline Beecham has claimed an SDR-based method for epoxidizing substituted cyclohexanones [95], while Procter and Gamble has patented a process for making esters and amides using the SDR [96]. In polymerization processes, the SDR is shown to deliver substantial savings in the reaction times. In the prepolymerization of styrene and in the polyesterification reaction between maleic anhydride and ethylene glycol time savings of 15–58 and 40–50 min, respectively, were reported [97, 98]. Also, in the cationic polymerization of styrene on the silica-supported BF_3 catalyst higher rates and higher molecular weights of the product are observed [99]. Finally, some photochemical processes were also studied in the SDR, including the UV-initiated polymerization of butyl acrylate [100, 101] and TiO_2-based photocatalytic oxidations [102, 103].

The SDRs reported so far are mainly applicable to liquid reaction systems and their main limitations are short residence times and low production capacities. These problems are addressed by the so-called rotor–stator SDR, which allows a countercurrent gas–liquid contacting at higher throughputs and longer residence times [104].

13.7.3
Centrifugal Fields in Microreactors

In the microprocessing systems, the high-gravity fields are primarily used to achieve an intensified mixing at high throughputs as well as to realize specific contacting patterns. In the Co-mix Coriolis Microreactor developed at the University of Freiburg [105], liquid is pumped through radial microchannels of a microreactor rotating with frequencies between 20 and 120 Hz and reaching high-gravity conditions of up to 10 000 g. This allows fast mixing at high throughputs up to 1 ml/s/channel. Ducrée *et al.* [106] report shortening of the mixing times by up to 2 orders of magnitude in cocurrent centrifugal flow through straight microchannels.

13.8 Conclusion

This chapter has described a wide variety of alternative energy forms for chemical processing, with focus on reactive systems. Many promising results have been reported, sometimes with spectacular improvements in heat transfer rate, mass transfer rate, mixing time and subsequent conversion, selectivity, and product quality. It is clear that the number of commercial applications is still lagging behind, compared to the processes investigated in scientific literature. One of the reasons is that many of these processes enhanced by alternative energy forms are investigated from a mostly chemical point of view. The engineering approach is less investigated until now, although this part is equally essential to finally reach a process that is feasible on the industrial scale. Both academia and industry are needed to explore this domain more systematically, and industrial partners should include not only the end users but also the equipment producers and contracting companies. Only with an integrated approach, success can be achieved.

It will have occurred to the reader that all of these energy forms need to be produced from electricity. Magnetic and light energy can be partly exceptions to this rule. In assessing the overall economics of a process, electricity cost and the efficiency of converting electricity to the desired energy form will play a vital role. As an example, the overall efficiency of combustion of gas to produce electricity and subsequent conversion to microwaves is in the order of 25%, whereas the conversion of gas to steam reaches an efficiency of 85%. This might hamper many industrial applications for the time being. It can be expected, however, that fossil fuel consumption will become less interesting in the future due to increasing prices and decreasing availability, whereas electricity generation based on wind or solar energy will become more cost effective. Also the efficiency of alternative energy production from electricity is expected to increase in future.

References

1. Kremsner, J.M., Stadler, A., and Kappe, C.O. (2006) *Top. Curr. Chem.*, **266**, 233–278.
2. Lidström, P., Tierney, J., Wathey, B., and Westman, J. (2001) *Tetrahedron*, **57**, 9225–9283.
3. Larhed, M., Moberg, C., and Hallberg, A. (2002) *Acc. Chem. Res.*, **35** (9), 717–727.
4. Kappe, C.O. (2004) *Angew. Chem. Int. Ed.*, **43**, 6250–6284.
5. Chemat, F., Esveld, D.C., Poux, M., and DiMartino, J.L. (1998) *J. Microwave Power Electromagn. Eng.*, **33** (2), 88–94.
6. Loupy, A., Petit, A., Hamelin, J., Texier-Boullet, F., Jacquault, P., and Mathé, D. (1998) *Synthesis*, **9**, 1213–1234.
7. Zhang, X., Lee, C.S.M., Mingos, M.P., and Hayward, D.O. (2003) *Appl. Catal., A*, **249**, 151–164.
8. Will, H., Scholz, P., and Ondrushka, B. (2004) *Top. Catal.*, **29** (3-4), 175–182.
9. He, P., Haswell, S.J., and Fletcher, P.D.I. (2004) *Sens. Actuators, B*, **105**, 516–520.
10. Comer, E. and Organ, M.G. (2005) *J. Am. Chem. Soc.*, **127**, 8160–8167.
11. Jachuck, R.J.J., Selvaraj, D.K., and Varma, R.S. (2006) *Green Chem.*, **8**, 29–33.

12. Cecilia, R., Kunz, U., and Turek, T. (2007) *Chem. Eng. Process.*, **46** (9), 870–881.
13. Sturm, G.S.J., Van Gerven, T., Verweij, M., and Stankiewicz, A. (2009) International Conference on Process Intensification, GPE-EPIC, June 14–17, 2009, Venice.
14. Esveld, E., Chemat, F., and van Haveren, J. (2000) *Chem. Eng. Technol.*, **23** (3), 279–283.
15. Esveld, E., Chemat, F., and van Haveren, J. (2000) *Chem. Eng. Technol.*, **23** (5), 429–435.
16. Nüchter, M., Ondrushka, B., Bonrath, W., and Gum, A. (2004) *Green Chem.*, **6**, 128–141.
17. Bierbaum, R., Nüchter, M., and Ondrushka, B. (2005) *Chem. Eng. Technol.*, **28** (4), 427–431.
18. Hoffmann, N. (2008) *Chem. Rev.*, **108**, 1052–1103.
19. Van Gerven, T., Mul, G., Moulijn, J.A., and Stankiewicz, A. (2007) *Chem. Eng. Process.*, **46**, 781–789.
20. Coyle, E. and Oelgemöller, M. (2008) *Photochem. Photobiol. Sci.*, **7**, 1313–1323.
21. Schiavello, M. (1997) *Heterogeneous Photocatalysis*, John Wiley & Sons, Ltd, Chichester.
22. Mills, A. and LeHunte, S. (1997) *J. Photochem. Photobiol., A*, **108**, 1–35.
23. Herrmann, J.M. (2005) *Top. Catal.*, **34**, 49–65.
24. Carp, O., Huisman, C.L., and Reller, A. (2004) *Progr. Solid State Chem.*, **32**, 33–177.
25. Hoffmann, M.R., Martin, S.T., Choi, W.Y., and Bahnemann, D.W. (1995) *Chem. Rev.*, **95**, 69–96.
26. de Lasa, H., Serrano, M., and Salaices, M. (2005) *Photocatalytic Reaction Engineering*, Springer, New York.
27. Maldotti, A., Molinari, A., and Amadelli, R. (2002) *Chem. Rev.*, **102**, 3811–3836.
28. Almeida, A.R., Moulijn, J.A., and Mul, G. (2008) *J. Phys. Chem. C*, **112**, 1552–1561.
29. Du, P., Moulijn, J.A., and Mul, G. (2006) *J. Catal.*, **238**, 342–352.
30. Lin, H.F. and Valsaraj, K.T. (2005) *J. Appl. Electrochem.*, **35**, 699–708.
31. Du, P., Carneiro, J.T., Moulijn, J.A., and Mul, G. (2008) *Appl. Catal., A*, **334**, 119–128.
32. Gorges, R., Meyer, S., and Kreisel, G. (2004) *J. Photochem. Photobiol. A*, **167**, 95–99.
33. Stankiewicz, A. (2007) *Ind. Eng. Chem. Res.*, **46** (12), 4232–4235.
34. Shibata, H., Moulijn, J.A., and Mul, G. (2008) *Catal. Lett.*, **123**, 186–192.
35. Stankiewicz, A. (2006) *Chem. Eng. Res. Des.*, **84** (A7), 511–521.
36. Czernichowski, A. (1994) *Pure Appl. Chem.*, **66** (6), 1301–1310.
37. Van Durme, J., Dewulf, J., Leys, C., and Van Langernhove, H. (2008) *Appl. Catal., B*, **78**, 324–333.
38. Kogelschatz, U., Eliasson, B., and Egli, W. (1999) *Pure Appl. Chem.*, **71** (10), 1819–1828.
39. Svoboda, J. and Fujita, T. (2003) *Min. Eng.*, **16**, 785–792.
40. Nunez, L. and Kaminsky, M.D. (1998) *Filtr. Sep.*, **5**, 349–351.
41. Yavuz, C.T., Prakash, A., Mayo, J.T., and Colvin, V.L. (2009) *Chem. Eng. Sci.*, **64**, 2510–2521.
42. Stolarski, M., Fuchs, B., Kassa, S.B., Eichholz, C., and Nirschl, H. (2006) *Chem. Eng. Sci.*, **61**, 6395–6403.
43. Stuyven, B., Chen, Q., Van de Moortel, W., Lipkens, H., Caerts, B., Aerts, A., Giebeler, L., Van Eerdenbrugh, B., Augustijns, P., Van den Mooter, G., Van Humbeeck, J., Vanacken, J., Moschalkov, V.V., Vermant, J., and Martens, J.A. (2009) *Chem. Commun.*, **1**, 47–49.
44. Ueno, S. and Harada, K. (1986) *IEEE Trans. Magn.*, **22** (5), 868–873.
45. Digilov, R.M. and Sheintug, M. (2005) *Appl. Phys. Lett.*, **86**, 082507, 1–3.
46. Suresh, A.K. and Bhalerao, S. (2001) *Indian J. Pure Appl. Phys.*, **40**, 172–184.
47. Iliuta, I. and Larachi, F. (2003) *AIChE J.*, **49** (6), 1525–1532.
48. Munteanu, M.C., Iliuta, I., and Larachi, F. (2005) *Ind. Eng. Chem. Res.*, **44**, 9384–9390.
49. Munteanu, M.C. and Larachi, F. (2009) *Chem. Eng. Sci.*, **64** (2), 391–402.
50. Rosensweig, R.E. (1979) *Science*, **204**, 4388, 57–60.

51. Dahikar, S.K. and Sonolikar, R.L. (2006) *Chem. Eng. J.*, **117**, 223–229.
52. Keil, F.J. and Swamy, K.M. (1999) *Rev. Chem. Eng.*, **15** (2), 85–155.
53. Thompson, L.H. and Doraiswamy, L.K. (1999) *Ind. Eng. Chem. Res.*, **38**, 1215–1249.
54. Cravotto, G. and Cintas, P. (2006) *Chem. Soc. Rev.*, **35**, 180–196.
55. Gogate, P.R. (2008) *Chem. Eng. Process.*, **47**, 515–527.
56. Muthukumaran, S., Kentish, S.E., Stevens, G.W., and Ashokkumar, M. (2006) *Rev. Chem. Eng.*, **22** (3), 155–194.
57. Neis, U. (2002) *TU Hamb. Harb. Rep. Sanit. Eng.*, **35**, 79–90.
58. Chisti, Y. (2003) *Trends Biotechnol.*, **25** (2), 89–93.
59. Horst, C., Kunz, U., Rosenplänter, A., and Hoffman, U. (1999) *Chem. Eng. Sci.*, **54**, 2849–2858.
60. Gogate, P.R., Mujumdar, S., and Pandit, A.B. (2003) *J. Chem. Technol. Biotechnol.*, **78**, 685–693.
61. Yu, G.-H., He, P.-J., Shao, L.-M., and Lee, D.-J. (2007) *Appl. Microbiol. Biotechnol.*, **77**, 605–612.
62. Mason, T.J., Paniwnyk, L., and Lorimer, J.P. (1996) *Ultrason. Sonochem.*, **3**, S253–S260.
63. Gogate, P.R. and Pandit, A.B. (2001) *Rev. Chem. Eng.*, **17** (1), 1–85.
64. Pandit, A.B. and Moholkar, V.S. (1996) *Chem. Eng. Prog.*, **92** (7), 57–69.
65. Moholkar, V.S., Kumar, P.S., and Pandit, A.B. (1999) *Ultrason. Sonochem.*, **6**, 53–65.
66. Gogate, P.R. and Pandit, A.B. (2005) *Ultrason. Sonochem.*, **12**, 21–27.
67. Kozyuk, O.V. (1998) Patent WO98/50146.
68. Mattick, A.T., Russell, D.A., Hertzberg, A., and Knowlen, C. (1995) Shock controlled chemical processing, in *Shockwaves at Marseille*, Proceedings of the International Symposium (eds R. Brun and L.Z. Dumitrescu), Springer-Verlag, Berlin.
69. Cheng, A.T.Y. (1997) A high-intensity gas-liquid tubular reactor under supersonic two phase flow conditions, in *Process Intensification in Practice – Applications and Opportunities* (ed. J. Semel), Mechanical Engineering Publications Limited, Bury St Edmunds, pp. 205–219.
70. Cheng, A.T.Y. (1999) Patent EP0995489.
71. Gross, G. (1998) Patent DE19718261.
72. Gross, G. (1998) Patent DE19722382.
73. Gross, G. (2000) Supersonic oxygen injection doubles the capacity of fluidized-bed reactors, in *ACHEMA 2000, International Meeting on Chemical Engineering, Environmental Protection and Biotechnology, Abstracts of the Lecture Groups Chemical Engineering and Reaction Technology*, Dechema, Frankfurt am Main, pp. 161–162.
74. Gross, G. and Ludwig, P. (2003) *Chem. Anlagen Verfahren*, **36** (3), 84–86.
75. Groen, D.J., Noorman, H.J., and Stankiewicz, A. (2005) Improved method for aerobic fermentation intensification, in *Proceedings of the International Conference Sustainable (Bio)Chemical Process Technology, Delft 27–29 September 2005* (eds P.J. Jansens, A. Green, and A. Stankiewicz), BHR Group Ltd, Cranfield, UK, pp. 105–112.
76. Ramshaw, C. (1983) *Chem. Eng.*, **389**, 13–14.
77. Trent, D. (2004) Chemical processing in high-gravity fields, in *Re-Engineering the Chemical Processing Plant: Process Intensification* (eds A. Stankiewicz and J.A. Moulijn), Marcel Dekker Inc., New York, pp. 33–67.
78. Ramshaw, C. and Mallison, R.H. (1981) Patent US4,282,255.
79. Guo, F., Zheng, C., Guo, K., Feng, Y., and Gardner, N.C. (1997) *Chem. Eng. Sci.*, **52** (21/22), 3853–3859.
80. Lin, C.-C., Ho, T.-J., and Liu, W.-T. (2002) *J. Chem. Eng. Japan*, **35** (12), 1298–1304.
81. Rao, D.P., Bhowal, D.A., and Goswami, P.S. (2004) *Ind. Eng. Chem. Res.*, **43** (4), 1150–1162.
82. Chen, J., Wang, Y., Jia, Z., and Zheng, C. (1997) Synthesis of nano-particles of $CaCO_3$ in a novel reactor, in *Process Intensification in Practice: Applications and Opportunities* (ed. J. Semel), Mechanical

Engineering Publications, Ltd, Bury St Edmunds, pp. 157–164.
83. Chen, J.-F., Shen, Z.-G., Liu, F.-T., Liu, X.-L., and Yun, J. (2003) *Scr. Mater.*, **49** (6), 509–514.
84. Chen, J.-F., Li, Y., Wang, Y., Yun, J., and Cao, D. (2004) *Mater. Res. Bull.*, **39** (2), 185–194.
85. Wang, M., Zhou, H.-K., Shao, L., and Chen, J.-F. (2004) *Powder Technol.*, **142** (2-3), 166–174.
86. Tai, C.Y., Tai, C., and Liu, H. (2006) *Chem. Eng. Sci.*, **61** (22), 7479–7486.
87. Chen, J.-F. and Shao, L. (2007) *J. Chem. Eng. Jpn.*, **40** (11), 896–904.
88. Lin, C.-C. and Liu, W.-T. (2003) *J. Chem. Technol. Biotechnol.*, **78** (2-3), 138–141.
89. Ku, Y., Ji, Y.-S., and Chen, H.-W. (2007) *Water Environ. Res.*, **80** (1), 41–46.
90. Chiu, C.-Y., Chen, Y.-H., and Huang, Y.-H. (2007) *J. Hazard. Mater.*, **147** (3), 732–737.
91. Aoune, A. and Ramshaw, C. (1999) *Int. J. Heat Mass Transfer*, **42**, 2543–2556.
92. Ramshaw, C. (2004) The spinning disc reactor, in *Re-Engineering the Chemical Processing Plant: Process Intensification* (eds A. Stankiewicz and J.A. Moulijn), Marcel Dekker Inc., New York, pp. 69–119.
93. Oxley, P., Brechtelsbauer, C., Ricard, F., Lewis, N., and Ramshaw, C. (2000) *Ind. Eng. Chem. Res.*, **39** (7), 2175–2182.
94. Vicevic, M., Scott, R.J.J., Clark, J.H., and Wilson, K. (2004) *Green Chem.*, **6** (10), 533–537.
95. Brechtels-Bauer, C.M.H. and Oxley, P. (2001) Patent WO 01/14357.
96. Burns, M.E., Gibson, and York, D.W. (2002) Patent WO 02/18328.
97. Boodhoo, K.V.K., Jachuck, R.J., and Ramshaw, C. (1997) Spinning disc reactor for the intensification of styrene polymerization, in *Process Intensification in Practice: Applications and Opportunities* (ed. J. Semel) Mechanical Engineering Publications, Ltd, Bury St. Edmunds, pp. 125–133.
98. Boodhoo, K.V.K. and Jachuck, R.J. (2000) *Green Chem.* **2** (5), 235–244.
99. Boodhoo, K.V.K., Dunk, W.A.E., Vicevic, M., Jachuck, R.J., Sage, V., Duncan, J., and Clark, J.H. (2006) *J. Appl. Polymer Sci.*, **101** (1), 8–19.
100. Dalglish, J., Jachuck, R.J., and Ramshaw, C. (1999) Photo-initiated polymerization using spinning disc reactor, in *Process Intensification for the Chemical Industry* (ed. A. Green), Professional Engineering Publishing, Ltd, Bury St. Edmunds, pp. 209–215.
101. Boodhoo, K.V.K., Dunk, W.A.E., and Jachuck, R.J. (2001) *Polym. Prepr.*, **42** (2), 813–814.
102. Yatmaz, H.C., Wallis, C., and Howarth, C.R. (2001) *Chemosphere*, **42**, 397–403.
103. Dionysiou, D.D., Burbano, A.A., Suidan, M.T., Baudin, I., and Laîné, J.-M. (2002) *Environ. Sci. Technol.*, **36**, 3834–3843.
104. Meeuwse, M., van der Schaaf, J., Kuster, B.F.M., and Schouten, J.C. (2010) *Chem. Eng. Sci.*, **65**, 466–471.
105. Haeberle, S., Schlosser, H.-P., Zengerle, R., and Ducrée, J. (2005) Proceedings of the 8th International Conference Microreaction Technology, (IMRET 8), AIChE, New York, p. 129f.
106. Ducrée, J., Haeberle, S., Brenner, T., Glatzel, T., and Zengerle, R. (2006) *Microfluid. Nanofluid.*, **2**, 97–105.

14
Switching from Batch to Continuous Processing for Fine and Intermediate-Scale Chemicals Manufacture

E. Hugh Stitt and David W. Rooney

14.1
Introduction

"Fuelled by a need to reduce costs and improve efficiencies, continuous processing may be the next paradigm shift in pharmaceutical manufacturing" [1]. There has been significant discussion on switching the production of "fine chemicals" from batch to continuous (B2C) processes over a number of years. Epithets such as that above are legion. But how much progress has there really been? What are the drivers and blockers in the "fine chemicals" industry? How can a given process be assessed for its potential to switch to continuous process? What is the role of "new" process technology and process intensification (PI) in enabling the switch from B2C?

14.1.1
Fine and Intermediate-Scale Chemicals

The term *fine chemicals* is widely used (abused?) as a descriptor for an enormous array of chemicals produced at "small scale" and is frequently assumed to infer a significant added value of the product derived from the degree of complexity (number of functional groups, geometric isomers, and enantiomers) and precision in their manufacture. Whether the term *fine chemicals* refers to the finesse of the chemistry or to the small scale of manufacture is far from clear. However, in order to assist our discussion the following division can be adopted [2]:

1) **Custom chemical for a specific customer**: produced on a "toll" basis, or the result of collaboration into developing the manufacturing route or, indeed, the specific chemistry. The product value for the manufacturer is dependent on the degree of ownership of intellectual property (IP), including both patents and know-how, relating to the molecule itself or its manufacture. Product pricing will tend toward a value-minus basis – where the magnitude of the minus will depend on the strength of the manufacturer's IP position.

Novel Concepts in Catalysis and Chemical Reactors: Improving the Efficiency for the Future.
Edited by Andrzej Cybulski, Jacob A. Moulijn, and Andrzej Stankiewicz

ISBN: 978-3-527-32469-9

2) **Generic chemical for multiple customers:** the chemistry and manufacturing technology are commonly available and chemical product will be available from other suppliers, and thus, manufacturing costs impact on competitiveness and operating margins. Product pricing tends toward a cost-plus basis and thus value to the manufacturer is dependent on capability and cost base. Marketing strategies such as customer and applications support may mitigate the "cost-plus" pricing basis somewhat but those services need to provide real value in order to generate a net gain to the supplier [3].

The former may be considered the true "fine chemical" while the latter may conveniently be termed *intermediate-scale* chemicals; commodity chemicals produced at a relatively small scale. This crude division starts to indicate the complexity of the economic drivers for B2C that is discussed later. First, however, progress in making the switch is reviewed and the essential elements of plant design structures and costs are briefly introduced.

14.2 Progress in Switching from Batch to Continuous

While the top 300 organic chemicals are made continuously, of the next 2700, 90% are made batch wise, rising to 97% for numbers 3001–30 000 [4], indicating that batch operations dominate the chemical industry as a whole. Of these, fermentation serves as a useful starting point for B2C discussion, given both its long history and the increasing usage in larger fine chemical industries [5]. In line with the bulk of the general academic literature, it has been shown that significant improvements can be gained by making the B2C switch. For example, Banat *et al.* identified that continuous fermentation using a biocatalyst [6] significantly increased ethanol productivity ($5\,^\circ C \approx 22–25\,^\circ C$) and, owing to the lower operating temperatures, subsequently improved the quality of the final product.

The advantages of continuous fermentation processes had been reported for some time. In 1956, DM breweries (New Zealand) patented a method for continuous brewing, although it was their competitor who began operating the first exclusively continuous fermenting process in 1957 [7]. They later reverted back to batch processing in order to increase their product portfolio, indicating that the gains obtained were offset by the increase in lead times to generate new products. DB Breweries, however, still employ continuous beer production using a series of stirred vessels with a holding time of 40–120 h [8]. Hygiene control is, however, essential as highlighted by Godoy *et al.* who conducted studies of 51 batch and 11 continuous Brazilian distilleries over a nine-year period to examine the advantages and disadvantages of each. Here they verified the commonly held belief that the continuous systems were cheaper to install. However, poor productivity levels associated with bacterial and wild yeast contamination necessitated more frequent cleaning and higher antibiotic costs [9], and thus, in contrast to popular belief, overall productivity was higher in the batch system.

So while chemical engineering principles would strongly support operating the process plant continuously in order to reduce capital and operating costs, the reality is much more complex and thus requires a broader viewpoint. Toyota's post WW2 strategy designed to address the general need for fast flexible processes which give customers, both internal and external, high-quality affordable products when they want it, has developed into current lean manufacturing philosophies. These require a way of thinking that focuses on making the product flow through value-added processes without interruption replenishing each subsequent operation at short time intervals [10]. For the intermediate and fine chemicals industry this is often translated to B2C strategies facilitated by PI technology and modular engineering; however, the philosophy in no way requires that every stage of the process must be made to operate continuously. For example, optimized penicillin production is achieved by operating two or more batch fermentation vessels out of phase and employing buffer vessels [11]. This represents one example of the potential synergy between batch and continuous processing in order to improve overall process economics.

As an analogy, the human body performs a mixture of batch, semicontinuous and continuous processes in order to operate effectively. Solids-handling operations are conducted batchwise for transfer and semicontinuously for processing, whereas heat exchange (HEx), filtration, pumping, and so on, are performed continuously. It should therefore not be surprising that for relatively small-scale operations involving solids handling within the fine and intermediate chemicals industry, batch operation is preferred. Similarly, continuous processes that involve precipitation or crystallization, a common unit operation in fine chemicals, are rare. Small-scale examples are known, for instance, a continuous crystallization process was used by Bristol-Myres Squibb in order to improve dissolution rates and bioavailability of the product [12]. The above does indicate that not all process or parts thereof are suited for conversion from B2C, given the current technology.

So where has the success of B2C been over the last several years? Unsurprisingly, efforts have focused on reactor conversion using a range of microfabricated devices. The advantages of continuous reactors are well lauded and some of the reasons are discussed in greater detail in Section 14.7.2. Currently, there are a number of commercial devices available for continuous reactions, for example, AIMS Fine Chemicals market their proprietary Process Intensifying Continuous Flow-Through Reaction (PICFTR) reactor system where reductions in operating expenses by 90% and capital expenses by 50% [13] are claimed. Recently, an agitated cell reactor was shown to give significant yield and selectivity enhancements [14] using a series of 10 reaction cells connected by short channels to prevent back-mixing – a design that apparently can be used for slower reactions [15]. Such microfabricated devices have progressed significantly over the last 20 years and excellent reviews on the topic can be found elsewhere [16].

However, we can summarize their main benefits (including microreactors, micromixers, and micro heat exchangers) as having significantly improved mixing and heat transfer performance with associated higher yields and selectivities, lower overall energy costs, and lower hold-up of potentially dangerous intermediates. These advantages are particularly effective for a wide variety of process chemistries, more generally, those that require careful control of heat and mass transfer. La-Mesta's Raptor® unit has been demonstrated for alcohol carbonylations (gas–liquid), isomerizations (liquid–liquid), and hydrogenation of polyunsaturated organic compounds (gas–liquid–solid) [17]; currently three units are installed at their Gilette facility [18]. Using such units, a batch process using 4% catalyst, four volumes of solvent, and a reaction time of 4–5 h yielding a conversion of 95% was upgraded to 99.9% yield, no solvent, and 0.4% catalyst. Alternative designs have shown success in heterogeneously catalyzed reactions [19], although swapping out catalyst in microfabricated devices is significantly more difficult than traditional reactors. Alternative designs such as the ThalesNano H-cube® (who are working with Sanofi-Aventis R&D in the development of continuous-process technologies [20]) circumvent this problem by utilizing catalyst cartridges, thus allowing for easy catalyst exchange. The use of typically large-scale heterogeneous reactors has also shown success. For example, LaPorte *et al.*, (Bristol-Myres Squib) have described the development of a multistep continuous process for the production of an active pharmaceutical ingredient (API) involving an in-line static mixer followed by continuous oxidation in a trickle-bed reactor. This process was scaled using both volumetric and numbering up methods and shown to be successful in pilot trials [21]. In the United Kingdom, Thomas Swann Ltd installed a continuous hydrogenation unit based on a traditional continuous stirred tank reactor (CSTR) using a supercritical CO_2 solvent in 2002 with about 1000 te per year product capacity. Plant data show significant gains in product purity and obviation of the product purification step [22]. United Kingdom-based James Robinson Ltd (a subsidiary of Yule Cato) also ventured into novel reactor technology, installing a Continuous Oscillatory Baffled Reactor (COBR) in 2003 for the manufacture of a dyestuff with hazardous and corrosive intermediates [23]. The 270-l COBR replaced a 16 000-l batch reactor and achieved similar product yield and quality [24].

Of course, the benefits of microfabricated and structured reactors are also applicable to even larger scale processes. For example, Evonik's development of a production-scale microstructured reactor for vinyl acetate manufacture (150 000 t per annum) claims depreciation and operating cost savings of €3 million per year [25].

In contrast to the significant literature publicizing the yield and selectivity improvements of continuous over batch operation, there are only a few published accounts of real fine and intermediate-scale processes employing microstructured reactors. By 2006, it was estimated that between 30 and 40 plants using such reactors were installed worldwide, with about half of these published in the literature [26]. Merck developed and implemented (in 1998) the first industrial example for an organometallic reaction, reporting an increase in yield by 25% [27]. Other examples include DSM Fine Chemicals GmbH in Linz, Austria, which produced 300 t of a high-value polymer product within 10 weeks using a 65-cm-long microstructured reactor, and Clariant who announced an annual production of 1000 t of pigments in 2005 [28]. It is worth noting that in the Clariant example the preparation of the coupling and diazo compounds are still conducted batchwise. In trials of model compounds, it was observed that the quality of the pigment was also higher in laboratory-scale batch process when compared to the microreactor but that this could not be replicated at production scale (volume factor of 40 000) due to mixing issues [29].

Similarly, Kirschneck and Tekautz demonstrated that the installation of a StarLam 3000 microreactor (Institute of Microtechnique, Mainz, Germany) into an existing production plant was able to double the plant capacity of a 10 m^3 batch plant used to perform a strongly exothermic process [30]. At the time of publication, the plant had been running for 2 years without problem at a throughput of 3 t h^{-1}. Both capital (an equivalent batch reactor costed 10 times) and operating savings were reported. More recent examples include Ampac Fine Chemicals (AFC) in California, who implemented a multipurpose continuous small-scale plant for the production of several hundred tons of API per year [31], and Sigma-Aldrich's Fine Chemicals, which has been adding continuous-process technology within its Buch facility [32]. SK Life Science of Korea have also used continuous fixed-bed hydrogenation at high pressure (304 bar) to produce a pharmaceutical intermediate at more than 99% enantiomeric excess at a volume of 120 t per year [31]. It was noted that operation at moderate temperatures using the continuous hydrogenation process produced much higher purity product as well as inherent safety improvements in the handling of hexyl lithium when compared to the batch process. One should not forget the numerous downstream processing operations which are also candidates for making the B2C transition. As discussed earlier, some examples in the areas of continuous crystallization have been reported; another notable one is AFC who have successfully employed an FDA-approved simulated moving bed chromatography unit for the production of APIs [33] for several years.

The question of "is the revolution underway" was addressed by Roberge *et al.* [34], who concluded that "the transformation of the fine and speciality chemicals industry cannot rely solely on the few numbers of industrial players" and thus, "the adoption needs to be generalised in the pharmaceutical industry, and the big players need to be involved." This has begun to happen as in 2007 Novartis announced a $65 million investment to establish the Novartis-MIT Centre for Continuous Manufacturing. Within Europe, research into continuous pharmaceutical manufacturing is also underway. The recent industry-led IMPULSE project included such partners as GlaxoSmithKline, Degussa, and Procter & Gamble. Therefore key companies from the pharmaceutical to consumer chemicals sectors are investing in this area.

14.3 Structure of Batch Processes

Overviews of batch process design are available within the literature [35, 36]; however, it is useful to summarize the main features when considering the arguments for B2C. A batch process is one where the individual steps of the manufacturing cycle are carried out in discrete intervals, resulting in frequent start-up and shutdown of the unit operations involved. Thus, a scheme that requires a reaction followed by a crystallization, drying, and so on, may carry out each of these separately using different main plant items (MPIs); intermediate storage may also be employed to improve the efficiency of the downstream processes. If the vessel is suitably equipped a number of unit operations (e.g., the reaction and crystallization steps) may, however, be performed in the same vessel. For cases where multiple reaction steps are required these could again be conducted in the same reactor, assuming that it is capable of meeting the requirements. This description suggests that all other downstream equipment remains idle until the process completes its current operation and, depending on the size of the equipment and the production target, this may be true. However, in general, the items are scheduled to allow for the individual unit operations to be carried out in parallel using successive batches. These subsequent batches may not necessarily yield the same product; however, the loading and unloading of each equipment item inevitably leads to downtime. Significant effort is expended to minimize this and ensure efficient scheduling of the process equipment, and here computer-based optimization methods are increasingly employed [37]. Taking this effort to the extreme, that is zero downtime and 100% equipment utilization leads, as discussed later, to the design of continuous processes.

Despite their flaws, batch processes have stood the test of time for a number of reasons, the most important of which is the flexibility it brings to the manufacturer in terms of the range of products that the plant can produce, the feedstocks used to produce them, and the speed at which they can be brought to market with very limited information on physical properties, reaction kinetics, and so on (very few, if any, Michelin-starred chefs have ever measured the rheology or kinetics of their latest culinary creation). This flexibility, however, has a price which comes in the form of lower efficiencies in terms of production, energy, labor, and so on, and ultimately efficiency equates to cost. However, one should never underestimate the pull of flexibility particularly, as discussed earlier in the examples of fermentation, where control of important parameters is difficult to achieve.

At the heart of such processes are the reactor stages and built around these are the required workup operations, utilities, and so on. The batch reactor, typically a simple stirred vessel, is itself extremely versatile although once installed modifying the available heat transfer area, impeller type and so on, is difficult, and thus the reactor should be designed at the outset to handle the entire range of expected conditions. However, given the broad set of process conditions used, it is often impractical to have one single reactor and thus many batch plants will have a number of different units available. As discussed later, other factors such as heat transfer are important. Therefore when developing new products, the choice, quantities and delivery of solvents/reagents, and so on, are often adjusted to suit the range of reactors and downstream equipment available. Overall, the chemistry is in general so well defined at the laboratory scale that the design of the process is simply down to the decision of which available equipment is required and for how long.

The product synthesis steps also dictate the type and number of downstream equipment necessary. Depending on the scale of operation this workup may be conducted batchwise, or continuously, and again there are advantages and disadvantages to each. Continuous distillation columns, for example, have an efficient range set by the hydraulic limitations of flooding and weeping, and operating the column outside these limits is not feasible. Batch distillation is thus often encountered when the capacity is too small to permit continuous operation at a practical rate and, in particular, when the feed contains solids or materials that form solids, tars, or resin that plug or foul a continuous distillation column [38].

14.4 Structure of Continuous Processes

As stated above, continuous processes allow the product to move through the entire plant without interruption. Therefore, unlike batch processes different MPIs are required for each of the process steps. As a result, the number of MPIs on the plant will typically be larger than the equivalent batch plant but this does not equate to the plant footprint for a given throughput. The range of equipment designs increases dramatically, with the choice focused on the options necessary to minimize any mass and heat transfer resistances. Therefore, it could be said that engineering is used to force the limiting resistances to the chemistry (including materials chemistry), whereas in batch processes the chemistry is often changed to overcome the limits of the installed engineering. The design of continuous processes therefore relies on the knowledge of the necessary design parameters – kinetics for reaction; activity coefficients for distillation, and so on, and the combination of these with accurate models for the physical properties of the streams involved and how these change with temperature, pressure, and composition. Such models can be used to theoretically test different designs and maximize efficiency of a single unit or the whole plant. Obtaining this information is time consuming and often unnecessary in batch processes where the modeling is further complicated by the inherent nature of the non-steady-state processes, leading to challenging design problems. The steady-state behavior of continuous processes can also improve energy recycle (pinch analysis) on the plant and increase overall efficiency. This is difficult on batch plant unless energy storage is employed, which adds additional capital expense. As the efficiency of continuous processes also relies heavily on the properties of the streams involved, monitoring and control is essential, and thus the level of instrumentation and control on a typical continuous plant is significantly greater than that of a batch facility where off-line analysis and direct operator control are more common.

In terms of downstream processes, the flow-rates, compositions, and so on, dictate the size and number of each unit operation; for example, while a batch distillation may be used to separate a single feed into a number of different product streams, a continuous distillation train would in general require N columns for N different product streams. The fact that a high degree of modeling is used in the design of each MPI, results in the generally held belief that continuous processes are specific to one product and costly to modify. Flexibility can be introduced into the plant, for example, recycle streams can be added across entire sections or on a specific item of equipment, for example, a reactor. Designing flexibility and robustness into the required unit operations is key to the success of B2C.

The costs, both in terms of time and finance, associated with continuous plant modification are currently decreasing due to the implementation of modern modular engineering concepts. By decentralizing production, this concept allows for plant manufacturers to dramatically reduce the overall project length necessary in the construction of factories. In 2002, this technique allowed Novo Nordisk Engineering to complete the production of a pharmaceutical plant within 18 months after basic design was approved. Since then, they have aimed at reducing this time to within 1 year [39]. Hence significantly reduced lead times in the modification of process plant in order to deal with issues such as increased capacity requirements, equipment redesign, and so on, are feasible.

14.5 Capital Cost Considerations

Traditionally, in the bulk chemical industries the capital cost of a plant may be scaled using the exponential costing approach:

$$C_1 = C_2 \left(\frac{Q_1}{Q_1} \right)^n \tag{14.1}$$

where C_i is the plant's capital cost and Q_j its production rate or process throughput. The exponent, n, typically has a value of approximately 0.7. This simple approach works well for bulk, that is, large scale, chemical plants but fails when applied to smaller scale plants. To understand why, it is necessary to briefly describe the structure of the capital costs by considering techniques for capital cost estimation.

14.5.1 Factorial Costing Technique

The basis of the "Factorial Methods" is to use quotes or estimates of the "delivered costs" of the MPIs and to multiply these up by so-called "installation" to obtain a total installed cost [40, 41].

$$C_{\mathrm{Inst}} = F_L \cdot C_{\mathrm{Del}} \tag{14.2}$$

where C_{Inst} and C_{Del} are the installed and delivered costs of the MPI, respectively, and F_L is the installation factor. While there are critical aspects of how best to estimate the MPI costs, discussed in all of the cited texts, what is more interesting in the present context are rough guides for variation of delivered costs with scale and of the composition and scale effect of installation factors. A single factor for the whole plant may be used (referred to as a *Lang factor*) for rapid cost estimates. In the current context, this is less interesting than the item-specific installation factors because it gives little insight into the real capital cost structure of a plant.

14.5.2
Main Plant Item Costs

To a first approximation, the cost of a single MPI is assumed to vary with scale (vessel volume or process throughput) on an exponent of 0.7. The value of this exponent does vary from one plant item type to another and while it typically lies in the range 0.5–0.9 [40, 42] for some equipment types (e.g., centrifuges) it may be at or above unity. This indicates that the purchased cost of equipment per unit production rate, say €/(tons per year), generally increases as manufacturing scale decreases; well known as *economies of scale*, related to large bulk chemical plants.

14.5.3
Installation Factors

The overall installation factor for each MPI is built up of a number of components. These include erection, piping, instruments, electrical, civil, structures/buildings, and lagging. Detailed tabulations of installation factors can be found in most texts on costing techniques [40, 41, 43, 44] and are not reproduced here. The values of the various subfactors are a function of the plant size, with the values always increasing with smaller scale – another factor in economies of scale. Figure 14.1 shows how the (overall) Lang factor varies with the average MPI cost for a plant, showing clearly the relative increase in installation and site costs with reducing scale. How the individual installation factors vary with scale is not, however, uniform.

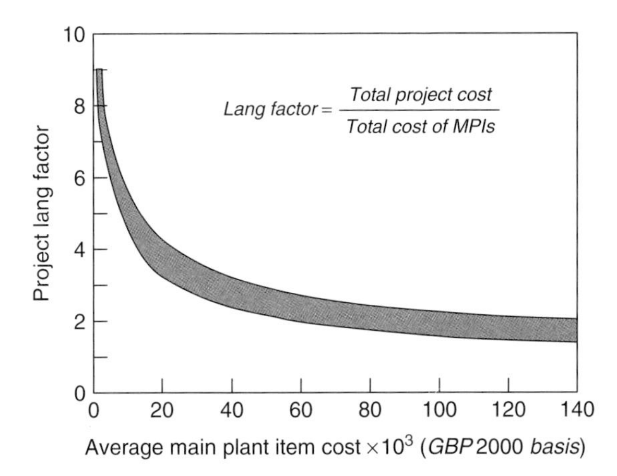

Figure 14.1 Approximate relationship of Lang factor to average main plant item cost. (Redrawn and updated from de la Mare [45].)

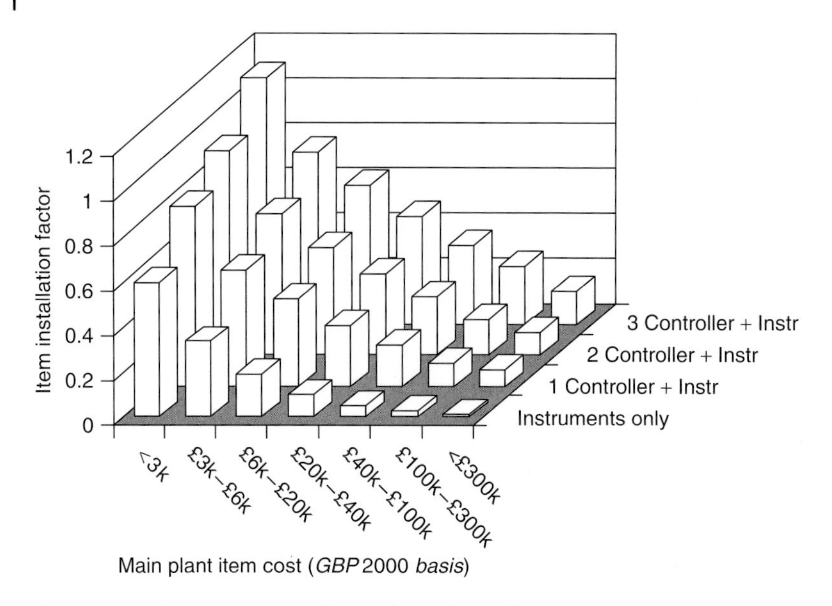

Figure 14.2 Instrumentation and control installation factors as a function of scale and complexity.

It is important to note that installation factors may vary significantly from B2C plant. For example, a typical batch processing plant has relatively little installed instrumentation when compared with a continuous process. This is especially true if the continuous plant exploits any PI; faster processing inevitably needs better measurement and control. The potential impact of this is shown in Figure 14.2, where the installation factor for four different levels of instrumentation and control are shown. While batch plants tend toward the "instrumentation only" classification, small continuous plants for high-value chemicals will tend toward the "controllers and instruments" categories, and as noted above PI is exploited toward the highest of these. This means that although B2C may facilitate smaller equipment, and for PI even more so, each of these incremental savings is offset at least in part by increased costs associated with its installation.

Similar observations can be made on other installation factors – for example, "structures and building." While traditionally (large) continuous plants are built outdoors and (small) batch plants indoors, it may not necessarily imply that small and intermediate-scale continuous plants should follow this rule. Product containment and contamination requirements may dictate the need for an enclosed manufacturing environment, meaning that little saving in capital may be *realized.*

The critical observation from the above is that installation factors are complex for small plants, and particularly when trying to relate B2C plant costs. Attempts to scale and interrelate costs using simple exponential methods of single installation factors for the plant are unlikely to be sufficiently accurate, even for preliminary assessment.

14.5.4
Cost of Fine and Intermediate-Scale Chemical Plants

Notwithstanding the above caveats on oversimplification of the economics of batch versus small continuous plant, it is useful to consider this in general terms, even while reaffirming the caveat that this should not be done quantitatively. The key observations from the preceding sections may be summarized as follows:

- Batch plants will have fewer MPIs due to multipurpose utilization of equipment.
- Specific MPIs will be significantly smaller in continuous plant, and options for PI may reduce process volumes and costs even further.
- The cost of an MPI per unit process volume increases as scale decreases, with an approximate exponent of 0.7, although this may not hold at small process volumes and certainly not for complex designs such as intensified reactors.
- The relative cost of installing an MPI increases with decreasing scale.
- The complexity of installation and measurement and control must be considered in comparing batch and continuous plants.
- The civil structures required for the plant may add significantly to cost. If the new plant, with its relatively small footprint, can be sited within an existing infrastructure, this will lead to significant savings.

14.6
Revenue/Operating Cost Considerations

Operating costs are conveniently considered in two groups, fixed and variable costs: those dependent on production rate and those that are independent of production rate. Operating costs can be further broken down into a number of key elements as indicated in Table 14.1, where estimation guidelines for continuous plant are also given [42]. Of critical interest in the present context is how these values vary between a batch and a continuous plant of nominally the same production rate.

14.6.1
Variable Costs

For the variable cost element, raw materials costs will depend critically on whether improvements in product yield are anticipated (and confidently expected); in many cases this may be a significant driver for the switch. Zhang [47], however, notes that "while the most obvious contributor is the cost of materials (substrates, reagents, catalysts, solvents, filtering media, and their transportation, etc.), it usually accounts for only a small (typically 20–45%) portion of the overall expense in a production setting" with the majority of costs arising from "the cost of labour (operators, analysts, quality control, and other supporting personnel), capital (equipment, instruments, and facility depreciation), utilities (water, steam, electricity, nitrogen,

Table 14.1 Summary of production costs [46].

Variable costs	Typical values
1. Raw materials	From flowsheet
2. Miscellaneous materials	10% of item (5)
3. Utilities/services	From flowsheet
4. Shipping and packaging	Usually negligible
Fixed costs	
5. Maintenance	5–10% of fixed capital
6. Operating labor	From manning estimates
7. Laboratory costs	20–25% of item (6)
8. Supervision	20% of item (6)
9. Plant overheads	50% of item (6)
10. Capital charges	15% of fixed capital
11. Insurance	1% of fixed capital
12. Local taxes	2% of fixed capital
13. Royalties	1% of fixed capital

compressed air, etc.), maintenance, waste treatment, taxes, insurance, and various overhead charges." This indicates that step changes in process costs for a given product are not obtainable by incremental changes in the process chemistry. Even significant improvements in a single step may not significantly impact on the overall process economics and production costs. Roberge [46], however, notes that the relative importance of fixed and variable costs changes through the production chain – with variable costs increasing in importance with each synthesis step. To fundamentally change the economics of a given product step changes in either or both of the chemistry and the production methodology may be required.

14.6.2 Fixed Costs

It is generally accepted that the operating labor needs of batch plants are higher than for continuous plant – but by how much? There is relatively little reliable assistance in the public domain literature. Wessel's work from the 1950s [48, 49], still frequently cited, gives insight into the key factors determining operating labor requirements and their relative importance [49]:

$$L = K \cdot N \cdot Q^{-0.76} \tag{14.3}$$

where L is operating labor in man-hours per ton; N is the number of process steps; Q is the plant capacity in tons per day, and K is a constant, equal to 10 for continuous processes and 23 for batch processes. That is, the operating labor costs for an equivalent batch process are 2.3 times those of the equivalent continuous process. There is some discrepancy in more recent texts but there is an overall consensus of operating labor cost reduction by 50–75% for continuous process relative to batch process [41, 43, 44].

14.7 Key Considerations for B2C Viability

The typical factors to consider when pondering B2C will include size and scale of operation, use of existing equipment, flexibility for multiproduct operation, start-up and shutdown, cleaning frequency, inventory, validation, traceability of product with respect to raw materials, equipment design, and availability as well as operational experience [50].

14.7.1 Process Complexity

As discussed in Sections 14.2 and 14.3, a critical difference between batch and continuous processes lies in equipment utilization. The complexity (or simplicity) of synthesis and isolation is a critical factor in determining whether a whole process is viable for switching from B2C. Given that it takes an average of eight synthetic steps to produce an API from raw materials [51], it is clear that the "average" API manufacturing process is probably too complex in its current form. Reduction in the number of process steps for a continuous process will, to a first approximation, reduce the plant costs *pro rata*.

Switching to continuous production in such complicated processes is therefore not to be treated lightly. Rather, specific process sections are targeted – based on their impact on the overall process economics [34]. While it has been noted that potentially in the order of $1/2 - 3/4$ of fine chemical reactions may be candidates [45], the complexity of many of these processes will preclude wholesale redesign [52]. For those with limited reaction stages, opportunities exist for switching the whole process. As noted in Section 14.4, a second critical factor to consider is the complexity of the product isolation and workup. Here simple one- or two-step product workup, particularly those such as filtration, evaporation, and drying, favor whole process switching and indeed can drive the switch from B2C rather than the reaction step [53].

14.7.2 Reaction Classification

The selection of reactor type in the traditionally continuous bulk chemicals industry has always been dominated by considering the number and type of phases present, the relative importance of transport processes (both heat and mass transfer) and reaction kinetics plus the reaction network relating to required and undesired reactions and any aspects of catalyst deactivation. The opportunity for economic benefit relies on specifically targeting reactor designs that capitalize best on these criteria.

Reactions in fine chemicals can be divided into three categories based on their tendency to accumulate heat in a batch or fed-batch reactor [54]. This is a useful basis for assessing suitability and requirements for conversion to a continuous reactor [45], Table 14.2, and formulate a short list of suitable reactors. Figure 14.3 [55]

Table 14.2 Thermal characteristics of reactions.

Reaction class	Batch or fed-batch reactor characteristics	Representative reaction time	Continuous reactor requirements/benefits
A	Near isothermal	Slow: >10 min	Long residence time. Plug flow or CSTR depending on kinetics and reaction network. Improved QC, lower inventory
B	Some accumulation of energy	1 s to 10 min kinetic influence	Some benefits (selectivity and process volume) from improved temperature control and better mixing/transport
C	>70% accumulation of energy	Very fast: <1 s mixing controlled	High quality mixing and mass transfer. High heat transfer area gives temperature control. Intensified and/or structured reactors.

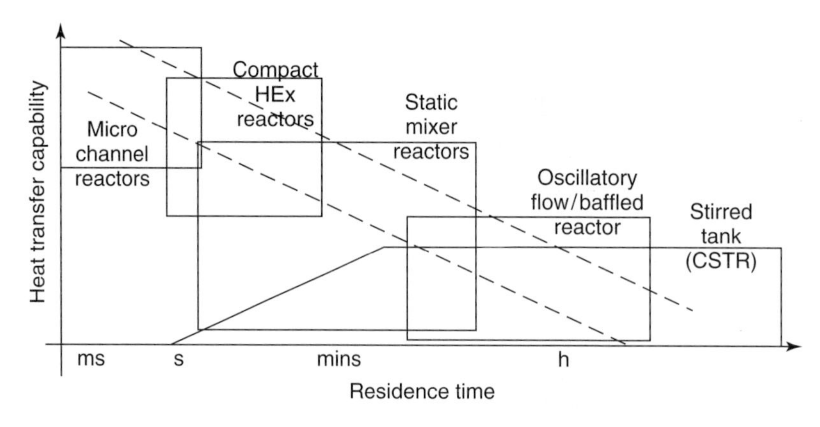

Figure 14.3 Reactor Characterization diagram. (Redrawn from Thomas [55].)

shows how this classification of reactions can be used to highlight the best reactor technology options based on the heat transfer requirements of the chemistry. This approach may equally be used for matching the reaction mixing or mass transfer requirements to the available reactor technologies.

14.7.3
Process Intensification – Reduction of Reaction Volumes

Theoretically, the volume of a batch reactor is equal to that of a plug flow reactor; however, batch reactors are never operated at 100% capacity and cycle times are always greater than reaction times. The combination of these leads to the conclusion that switching from B2C gives an inherent reduction in the reactor volume.

Switching away from the traditional batch autoclave also allows inherent intensification of reaction-important parameters. Consider heat transfer. The vast majority of reactions carried out in the fine and intermediate-scale chemicals industry are exothermic and in some cases highly exothermic. Managing this evolved heat is achieved by balancing the rate at which the reaction occurs, usually by slowly adding one of the reagents, with the heat removed through the available heat transfer area. Figure 14.4 shows the approximate ratio of available heat transfer area to the reaction volume for a dished bottom autoclave. The decline in this value is a well-known issue in scale-up of batch operations, and leads to the heat transfer limited modes of fed-batch operation referred to above. The values shown in Figure 14.4 should be contrasted with specific heat transfer areas available in continuous reactors: 100–500 $m^2\ m^{-3}$ for a typical shell and tube heat exchanger, up to 1000 $m^2\ m^{-3}$ for compact HEx reactors [56] and as high as 10 000 $m^2\ m^{-3}$ for microchannel HEx reactors. It is a mute point as to whether some of the Class A or B reactions (Table 14.2) are so due to solvent dilution in order to limit temperature excursion (or equally to limit the reaction rate to match the available rate of mixing or mass transfer) and so by use of intensified continuous equipment, with its

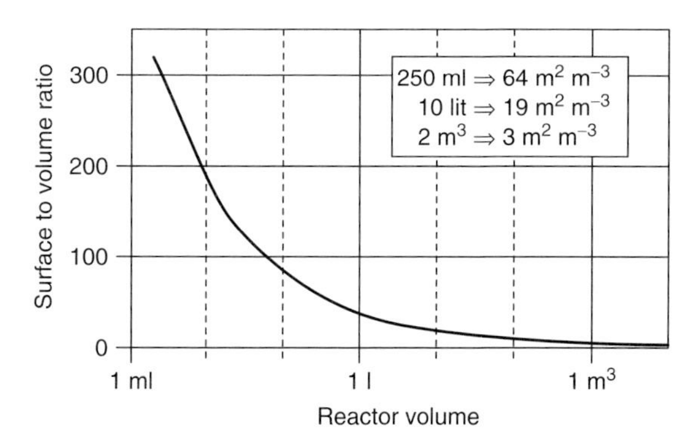

Figure 14.4 Heat transfer surface to reaction volume ratios for autoclaves.

associated increases in mass and heat transfer, the solvent level may be reduced or even eliminated [57, 58].

From the above discussion, switching from B2C leads inherently to a reduction in process vessel volumes, and thus a reduction also in the cost of the MPIs. Further gains may be possible by appropriate selection of the reactor. This should be based on consideration of the critical reactor design features required by the given reaction. If enhanced mixing, mass transfer, or heat transfer is a prerequisite for the design and economic gains, then an appropriate intensified reactor should be employed. If these are not required, then conventional continuous reactors may suffice to achieve most of the gains without the technical risk of novel technology (see Section 14.6 below). It should be remembered that all designs of reactor (irrespective of whether conventional or "novel") are a solution looking for a problem. It is the nature of the chemistry that determines the reactor type that should be used.

14.7.4 Reaction Selectivity

Significant gains in reaction selectivity may be obtained by switching from B2C. These not only result from improved heat/mass transfer or mixing but also through control of species concentrations. Consider the following two simple elementary reactions [59]:

- Parallel selectivity $A \rightarrow B$ and $2A \rightarrow C$
- Series selectivity $A \rightarrow B \rightarrow C$.

where A is the substrate, B is the desired product, and C is the by-product. High selectivity will be favored by low concentrations of A and B, respectively. This infers that for parallel reaction systems a fully back mixed reactor (i.e., CSTR) will favor good selectivity, whereas for series selectivity a plug flow reactor is preferred. A batch reactor follows the same concentration profile as a plug flow reactor indicating that they are inherently of the wrong design for parallel selectivity reactions (unless mitigated by a fed-batch strategy), and thus a continuous approach will have intrinsic benefits.

Other important aspects include the effect of temperature (via activation energy) and mixing, particularly for multiphase reactions. Both of these can impact on selectivity and thus can be improved in operating continuously using the inherent benefits of heat transfer area and mixing strategies discussed previously.

14.7.5 Operating Scale

It is a well-known fact (or urban myth?) that small-scale operations favor batch processes while large-scale manufacture requires continuous processing. Consider the typically assumed general trends (Figure 14.5) where the capital costs of batch

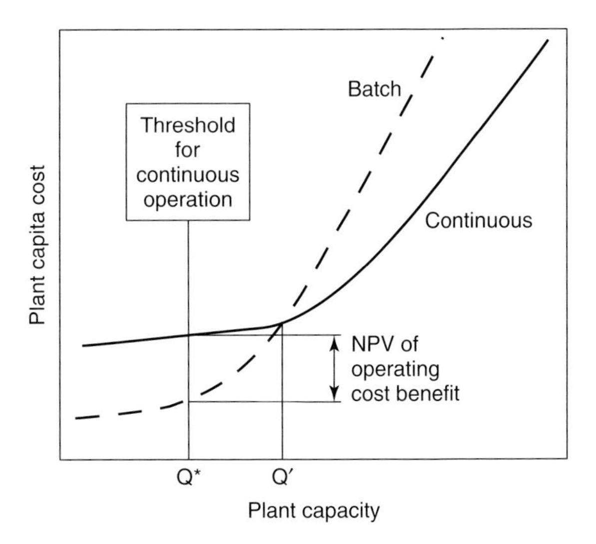

Figure 14.5 Cost benefits of continuous operation.

and continuous plant are shown as a function of production capacity. At a given throughput, Q′, the costs of the two plants are equal: above this the continuous plant has lower cost, and below it the converse. If the operating cost benefits of continuous plants are considered, for example, as a net present value (NPV), then the true plant capacity of economic equivalence, Q*, can be identified. This is always true: but quantification and prediction of Q* for any given product is exceedingly complex.

There are a number of attempts to provide quantitative answers in the literature, but these are generally flawed from either a cost engineering or chemical engineering perspective and their results should be treated with extreme caution. This caveat notwithstanding, it is probably reasonable to say that for most products the threshold lies in the range of 1000–10 000 tonne per year process stream (not product) rate. The threshold value will not, however, be the same for all products, processes, operating sites, and so on.

14.7.6
Process Analytics and Control

The selection of continuous over-batch processing is driven by two factors: economics and control [1]. Hitherto, this discussion had largely focused on the former. Continuous processes inherently require a greater degree of measurement and control (Section 14.4), a reflection of the shorter timescales over which process or product manipulation must be made in order to assure that intermediate and product streams remain to specification.

The control of a continuous process requires appropriate process measurement in a timescale consistent with the need to apply corrective action within the process space time. So, for a reaction with a residence time of several hours, analysis on a half-hourly basis may suffice. A reaction time of a few minutes, however, requires process measurement on a frequency in the order of a few seconds. This is in contrast to batch operation where an end of batch analysis may be done off-line as a precursor to that batch being forwarded to its next transformation or product isolation step. The option to extend the batch time to allow completion of the reaction, often a *de facto* control parameter in batch operations, is not relevant or practicable in continuous plant.

In designing a continuous plant, therefore, it is essential to establish the measurement and control strategy based on an understanding of which critical aspects are indicative of good or poor plant operation, and how deviation of these measurements can be exploited to perform a corrective process control action. Establishing suitable on-line real-time measurement techniques may be a blocker to the implementation of continuous processing. In contrast, implementation of a successful measurement and control strategy may be the enabler for improved product yields and product quality.

14.7.7 Process Economics

The discussions on product value and scale have already highlighted that there are different "wins" to be obtained from switching B2C. These depend on whether the product is commodity or proprietary in nature and the relative value of improved yields or reduced operating costs.

Three different economic drivers for B2C can be identified [34]:

1) **The chemical route**: a potential decrease in manufacturing costs is roughly proportional to the reduction in number of processing steps. B2C is therefore attractive if it is an enabler for a chemistry that cannot be carried out in classic batch equipment. This is the opportunity for novel reactor technologies. The economic impact is, however, hard to define quantitatively [34, 60].
2) **Overall yield**: This is not, however, a simple relationship. In a multistep process, the actual value for the increased yield of a given step depends on where it lies in the process chain [47]. The further down the manufacturing chain, the higher the (actual or transfer) price of the reactor feed and thus the higher the benefit of increased yield.
3) **Plant operating costs**: labor costs are particularly highlighted in this and are particularly important at the early part of the manufacturing chain, essentially when feed and product are closer to a cost-plus basis.

This breakdown of economic drivers is important – and parallels the discussion in Section 14.7.8. Thus, in assessing the benefits of B2C it is essential to consider whether they are most likely to be from cost reduction or yield increase. Specifically, it may be noted that [34] the adoption of "best reactor technology" will be favored where value is driven by yield increase. Where the emphasis is on cost minimization, then reactor selection will be based primarily on capital and operating cost considerations.

14.7.8 Innovation and Risk Management

Adopting a new process or technology inherently involves a technical risk (and by association an economic risk) that it will not achieve its design-specified productivity (in terms of either rate or efficiency). While many accuse industry of being unnecessarily conservative, it is frequently the technical risks, translated into a financial risk, that inhibit the adoption of new technology rather than a phobia toward new technology. At its simplest, the risk of new chemistry or technology when rolled into a single innovative project may be considered using basic probability-based mathematics. Say, for example, a new process has N innovations and the chance of each of these succeeding is C; then the probability of a successful project (P) is given by

$$P = C^N \tag{14.4}$$

So, for a project that has five innovative aspects each with a 90% degree of certainty, the overall probability of project success, $P = 0.9^5 = 59\%$. For avoidance of doubt, to put it another way, the project has a 41% chance of failure! This does not make for a good investment case.

This simple piece of arithmetic highlights the key strategies for risk management and its minimization in developing new technologies – to minimize the number of Ns and maximize their respective C-values.

The reality of risk assessment in investment for new processes is somewhat more complex than this. The specific innovations are often not discrete and the confidence of success of each item is a probability distribution rather than a single value. Techniques to handle the mathematical aspects have been available for many years [61] and computational tools are now readily available. A detailed coverage of managing uncertainty is beyond the scope of the current text and this simplistic approach suffices to address the key question of how to effectively manage the N- and C-values.

Increasing confidence of design and scale-up (viz., increasing the C-values) using a structured experimental and development program is essential; in order to minimize the total risk, it is essential to constrain parallel innovations to maintain some control over the N in Eq. (14.4). This may be easier said than done. It is always useful to ask whether a given innovation is necessary to achieve the overall project objectives. If, for example, a "novel" intensified reactor is a prerequisite to obtaining the process benefits, then so be it. If, however, the necessary rates and yields can be achieved using a proven reactor design then why take an additional risk? Essentially, all innovative aspects of the process need to be assessed using a risk–reward type analysis. Always ask – what does this perform that cannot be achieved using conventional, proven equipment?

14.8 Conclusions

A review indicates that there has been moderate progress in switching intermediate-scale and fine chemicals from B2C manufacture. Some of the reported activity is based on traditional reactors while some specifically exploits "novel" and intensified reactors. Consideration of the cost structure of batch and continuous plants shows that there may be significant economic and operating benefits in switching manufacture of fine and intermediate chemicals from B2C processing. For any given process, a threshold production (or process) rate can be identified, above which continuous processing is beneficial. Assessing this potential is, however, far from straightforward and is dependent on many factors. Selection of a reactor for continuous operation is primarily a function of the needs of the process chemistry, simultaneously meeting the mixing, heat and mass transfer needs of the reaction. While there are many alternative reactor designs available, for most cases only a very few will truly satisfy the needs of the process chemistry and allow the full potential of continuous operation to be achieved.

References

1. Pellek, A. and van Arnum, P. (2008) *Pharm. Technol.*, **9**, 52–58.
2. Stitt, E.H. (2002) *Chem. Eng. J.*, **90**, 47–60.
3. Stitt, E.H., Chisnall, R., Palmer, L., and Robbins, L. (2001) Services allied to catalyst supply. Presented at CatCon 2001, 5th Worldwide Catalyst Industry Conference, May 2001, Houston.
4. Kirschneck, D. (2008) *Spec. Chem. Mag.*, **28** (10), 52–54.
5. Ramakers, J. (2008) *Spec. Chem. Mag.*, **28** (2), 16–17.
6. Banat, M. *et al.* (2002) *Food Microbiol.*, **19**, 127–134.
7. Hornsey, I.S. (2003) *A History of Beer and Brewing*, Chapter 9, RSC, ISBN: 0854046305.
8. Campbell, S. (1998) *The Continuous Brewing of Beer*, in Food and Beverages, (eds. Packer, J.E. and Whiting, R.), New Zealand Institute of Chemistry, New Zealand, ISBN: 0908689586.
9. Godoy, A. *et al.* (2008) *Int. Sugar J.*, **119** (1311), 175.
10. Liker, J.K. (2004) *The Toyota Way*, McGraw-Hill.
11. Thomas, H. (2004) The Realities of Small Scale Continuous Processing, in *IChemE: Switching from Batch to Continuous Processing Conference*, November 2004, London.
12. Swichtenberg, W. (2008) *Pharma Manuf.*, **7** (1), 24–27.
13. Singh, I.P.. (2010) Continuous Processing, in *InformEx Speciality Chemical Conference*, May 11th 2010, New Jersey.
14. Flavell-While, C. (2009) *Chem. Eng.*, **812**, 15.
15. Cotterill, M. (2008) *Spec. Chem. Mag.*, **28** (7), 46–47.
16. Jähnisch, K. (2004) *Angew. Chem. Int. Ed.*, **43**, 406–446.
17. Giuliano, P. and De Panthou, F. (2007) *Spec. Chem. Mag.*, **42**, 44–45.
18. Warmington, A. (2008) *Spec. Chem. Mag.*, **28** (7), 18–24.
19. Enache, D.I., Hutchings, G.J., Taylor, S.H., Raymahasay, S., Winterbottom, J.M., Mantle, M.D., Sederman, A.J., Gladden, L.F., Chatwin, C., Symonds, K., and Stitt, E.H. (2007) *Catal. Today*, **128**, 26–35.

20. Anonymous (2008) *Spec. Chem. Mag.*, **28** (5), 10.
21. LaPorte, T., Hamedi, M., DePue, J., Shen, L., Watson, D., and Hsieh, D. (2008) *Org. Proc. Res. Dev.*, **12**, 956–966.
22. Licence, P., Ke, J., Sokolova, M., Ross, S.K., and Poliakov, M. (2003) *Green Chem.*, **5**, 99–104.
23. Houlton, S. (2003) Chemical Week, 24 September, 2003.
24. Hermitage, S.A. and Tilstam, U. (2003) *Org. Proc. Res. Dev.*, **8**, 9.
25. (2006) Elements. Degussa Science News Letter, p. 15.
26. Hessel, V., Löwe, H., and Löb, P. (2006) *Chem. Technol.*, **5**, 24.
27. Krummradt, H., Kopp, U., and Stoldt, J. (2000) Experiences with the use of microreactors in organic synthesis, in *Microreaction Technology – IMRET 3: Proceedings of the 3rd International Conference on Microreaction Technology* (ed. W. Ehrfeld), Springer, Berlin, p. 181.
28. Pieters, B. *et al.* (2007) *Chem. Eng. Technol.*, **30** (3), 407–409.
29. Wille, C. *et al.* (2004) *Chem. Eng. J.*, **101**, 179–185.
30. Kirschneck, D. and Tekautz, G. (2007) *Chem. Eng. Technol.*, **30** (3), 305–308.
31. Warmington, A. and Challener, C. (2008) *Spec. Chem. Mag.*, **28**, 40–46.
32. Anonymous (2009) *Chem. Eng. News*, **87**, 3.
33. Wearmington, A. and Challener, C. (2008) *Spec. Chem. Mag.*, **28** (3), 40–46.
34. Roberge, D.M., Zimmermann, B., Rainone, F., Gottsponer, M., Eyholzer, M., and Kockmann, N. (2008) *Org. Proc. Res. Dev.*, **12** (5), 905–910.
35. Sharrat, P.N. (1997) *Handbook of Batch Process Design*, Blackie Academic and Professional, Taylor & Francis.
36. Korovessi, E. and Linninger, A. (2005) *Batch Processes*, CRC Press.
37. Méndez, C. *et al.* (2006) *Comput. Chem. Eng.*, **30**, 913–946.
38. Seader, J.D. and Henley, E.J. (2006) *Separation Process Principles*, 2nd edn, Wiley International.
39. Abildgaard, O. (2003) *Comput. Control. Eng. J.*, **14** (3), 44–45.
40. Gerrard, M. (2000) *A Guide to Capital Cost Estimating*, 4th edn, Institute of Chemical Engineers, Rugby.
41. Peters, M.S., Timmerhaus, K.D., and West, R.E. (2003) *Plant Design and Economics for Chemical Engineers*, 5th edn, McGraw-Hill.
42. Sinnot, R.K. (1993) *Chemical Engineering*, Design, vol. 6, 2nd edn, Pergamon Press, Oxford.
43. Ulrich, G.D. (1988) *A Guide to Chemical Engineering Process Design and Economics*, John Wiley & Sons, Inc.
44. De la Mare, R.F. (1982) *Manufacturing Systems Economics*, Holt Reinhart and Winston.
45. Zhang, T.Y. (2006) *Chem. Rev.*, **106**, 2583–2595.
46. Roberge, D.M., Zimmermann, B., Rainone, F., Gottsponer, M., Eyholzer, M., and Kockmann, N. (2008) *Org. Proc. Res. Dev.*, **12**, 905–910.
47. Wessel, H.E. (1952) *Chem. Eng.*, **59**, 209.
48. Wessel, H.E. (1953) *Chem. Eng.*, **60**, 168.
49. Davidson, S. (2008) *Chem. Eng.*, **805**, 42–43.
50. Carey, J.S., Laffan, D., Thomson, C., and Williams, M.T. (2006) *Org. Biomol. Chem.*, **4**, 2337–2347.
51. Roberge, D.M., Ducry, L., Beiler, N., Cretton, P., and Zimmermann, B. (2005) *Chem. Eng. Technol.*, **28**, 318–323.
52. Atherton, J.H., Double, J.M., and Gourlay, B. (2005) Survey of PI requirements in the fine chemicals and pharmaceuticals sector. 7th World Congress Chemical Engineering, 2005, Glasgow.
53. Thomas, P.W. and Ramsay, A. (2004) Continuous manufacturing of a bulk active pharmaceutical ingredient. Presented at Switching from Batch to Continuous Processing Conference, November 22-23, 2004, London.
54. Roberge, D.M. (2004) *Org. Proc. Res. Dev.*, **8**, 1049–1053.
55. Thomas, H. (2008) *Chem. Eng.*, **805**, 38–40.
56. Anxionnaz, Z., Cassabaud, M., Gourdon, C., and Tochon, P. (2008) *Chem. Eng. Process.*, **47**, 2029–2050.
57. Enache, D.I., Hutchings, G.J., Taylor, S.H., Raymahasay, S., Winterbottom, J.M., Mantle, M.D., Sederman, A.J., Gladden, L.F.,

Chatwin, C., Symonds, K., and Stitt, E.H. (2007) *Catal. Today*, **128**, 26–35.

58. Enache, D.I., Thiam, W., Dumas, D., Ellwood, S., Hutchings, G.J., Taylor, S.H., Hawker, S., and Stitt, E.H. (2007) *Catal. Today*, **128**, 18–25.
59. Levenspiel, O. (1998) *Chemical Reaction Engineering*, 3rd edn, Wiley International.
60. Hessel, V., Lob, P., and Lowe, H. (2005) *Curr. Org. Chem.*, **9**, 765.
61. Allen, D.H. (1990) *A Guide to the Economic Evaluation of Projects*, 3rd edn, The Institutions Chemical Engineers, Rugby.

15
Progress in Methods for Identification of Micro- and Macroscale Physical Phenomena in Chemical Reactors: Improvements in Scale-up of Chemical Reactors

Bengt Andersson and Derek Creaser

15.1
Introduction

The traditional focus of chemical reaction engineers has been on developing design equations that can predict conversion and selectivity for different reactors. However, understanding the physical and chemical phenomena in chemical reactors is more important since that knowledge will help the engineer to develop completely new designs. The main objective of this chapter is to describe methods to measure or simulate critical steps in reactor scale-up. The focus is on understanding how fluid dynamics, mass and heat transfer, and kinetics interact in chemical reactors.

There are many nonintrusive experimental tools available that can help scientists to develop a good picture of fluid dynamics and transport in chemical reactors. Laser Doppler velocimetry (LDV), particle image velocimetry (PIV) and sonar Doppler for velocity measurement, planar laser induced fluorescence (PLIF) for mixing studies, and high-speed cameras and tomography are very useful for multiphase studies. These experimental methods combined with computational fluid dynamics (CFDs) provide very good tools to understand what is happening in chemical reactors.

However, even with the most advanced measuring and simulation tools, the most efficient methods are simple calculations that give an order-of-magnitude estimation of the influence of a phenomenon. Time constants for diffusion, heat conduction, and acceleration are very useful. For example, the time constant for diffusion $\tau_D = l^2/D$ is the time it takes to fill a cube of size l by diffusion, and the time for a particle to accelerate from zero velocity to approximately two-third of the velocity of the surrounding fluids is $\tau_u = \rho_d d^2/18\mu$, where ρ_d is the particle density, d is the diameter of the particle, and μ is the viscosity of the continuous phase.

Novel Concepts in Catalysis and Chemical Reactors: Improving the Efficiency for the Future.
Edited by Andrzej Cybulski, Jacob A. Moulijn, and Andrzej Stankiewicz

ISBN: 978-3-527-32469-9

15.2
Experimental Methods

15.2.1
Flow Characterization

PIV has become the most popular technique to measure velocity and turbulent properties (Figure 15.1). The movement of seed particles in a millimeter-thick laser sheet is measured by correlating two photos taken a few milliseconds apart. With two cameras, it is also possible to obtain a 3D vector of the velocity in that plane. The method gives, in general, very good resolution of the flow, but it requires optical access. Also, measurement close to walls can be problematic due to light reflections that disturb the measurements. One extension of PIV is the micro-PIV that uses fluorescent tracer particles, which allows all direct light, for example, reflections at the walls, to be filtered out [1].

LDV is the traditional method using tracer particles to measure velocity and one-point statistics of turbulent properties [2]. It is still a very useful technique and has the advantage that it can measure closer to walls compared to PIV. An inherent problem with LDV is that it does not measure at a specific point but rather at places

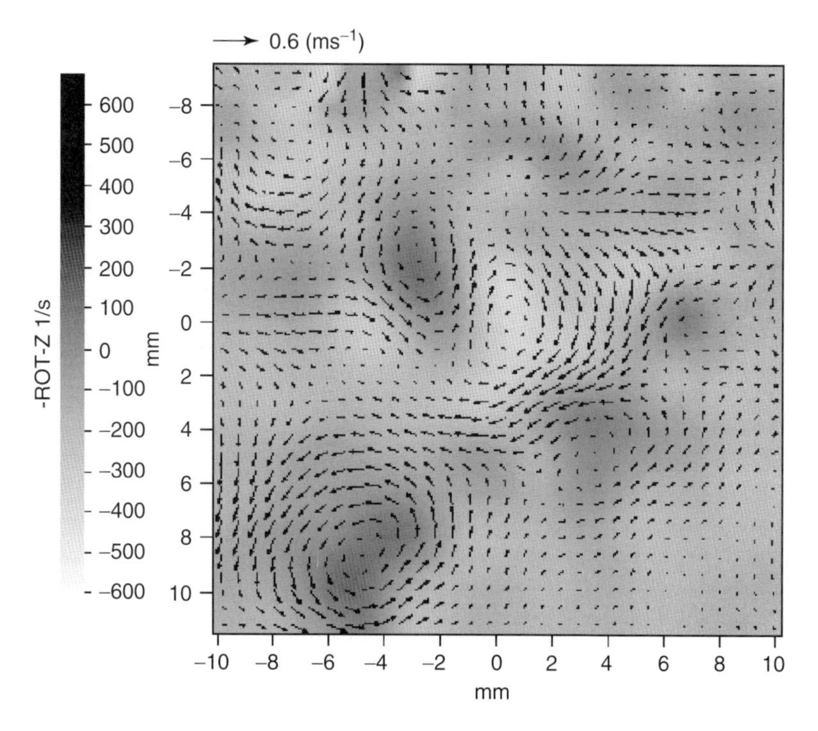

Figure 15.1 Instantaneous PIV measurement of velocity and vorticity in the impeller outflow from a Rushton turbine in a 1-l laboratory reactor.

where the particles are located. Usually, the offset is negligible and can, to some extent, be compensated for in postprocessing, but a sufficient number of tracer particles must be provided.

Optical systems can be used in multiphase flows at a very low volume fraction of the dispersed phase. Through a refractory index matching of liquid–liquid or liquid–solid systems, it is also possible to measure at high void fractions. However, it is not possible to obtain complete refractory index matching since the molecules at the phase boundary have different optical properties than the molecules in the bulk. Consequently, it is possible to measure at a higher fraction of the dispersed phase with larger drops and particles because of the lower surface area per volume fluid.

Ultrasonic Doppler velocimetry is a nonintrusive technique that has been developed into a very useful technique for opaque liquid flows [3]. This technique provides good measurement of velocity; new high-frequency techniques give a space resolution on the millimeter level, and even the large turbulent scales can be resolved.

The turbulent kinetic energy, k, is usually easily obtained in steady flow, but can be difficult to estimate in unsteady flows even with very good experimental techniques. In a stirred tank, the periodic variations from the impeller blades can be compensated for, for example, by making measurements only when the impeller blades are exactly in the same position. This technique should also give a statistically independent measurement that is required for good statistical evaluation. However, influence from the impeller blades is not the only nonrandom influence on velocities. The combination of periodic variations from the impeller, influence from the baffles, and the circulation time for the flow above and below the impeller give rise to higher order velocity correlations, which is not due to the turbulence [4]. These periodic variations must be removed from the velocity to obtain good estimations of the turbulent properties. Often LDV is a much better tool than PIV for this kind of measurement, since this technique allows a continuous measurement over a long time.

The rate of dissipation of turbulent kinetic energy, ε, is more difficult to measure. A very fine space resolution is required to measure the gradient of turbulent velocity fluctuations and calculate turbulent dissipation directly from the definition [5, 6]. An approximation of the flow field by a large eddy simulation (LES) formulation of the measured large eddies gives a reasonable estimation of the dissipation rate [7].

Velocity measurement of the dispersed phase in multiphase flow is possible using both PIV and LDV. In PIV, the particles can be masked according to size, and the velocity for each size fraction can be estimated [7]. The turbulent properties, for example, granular temperature, are more difficult to measure because of the low number of particles in the measured volume. With LDV it is also possible to obtain the velocity and size for the dispersed phase, but the turbulent properties for the dispersed phase are still difficult to measure accurately, owing to the low number of particles and also because the position of the particles is not exactly the same all the time.

15.2.2
Bubble and Drop Size Measurement

High-speed CCD cameras have become very useful and affordable. For most chemical reaction engineering purposes, a monochrome CCD camera with full 2000×2000 pixel resolution up to 2 kHz and with decreased resolution up to 64 kHz is sufficient. The limiting factor is most often optical access and the amount of light required at high speed. With light-intensified detectors almost every photon is counted, but still a few thousand photons per pixel in each frame are required to obtain good resolution. The selected measurement volume is determined either by using a laser sheet or using the camera focus. Both the focal point and the depth of the measured area can be adjusted with the optics.

The cameras are usually used in two different modes, front light and backlight. Standard images using the front light technique are very useful in tracking particle movements, collisions, breakup, and coalescence (Figure 15.2).

In backlight or shadowgraph technique, the light comes from behind and is directed toward the camera. This technique is very useful in measuring bubble or drop size distributions (Figure 15.3).

15.2.3
Concentration Measurements

PLIF has become a very useful technique to measure local concentrations. In PLIF, a laser sheet is formed and a high-speed CCD camera measures the excited light from a fluorescent dye that is mixed into the flow. Most commonly used is a simple laser at a fixed wavelength (532 nm for Nd:YAG). The excited light has a longer wavelength and all light reflections from the laser are filtered out. The camera is usually 10, 12,

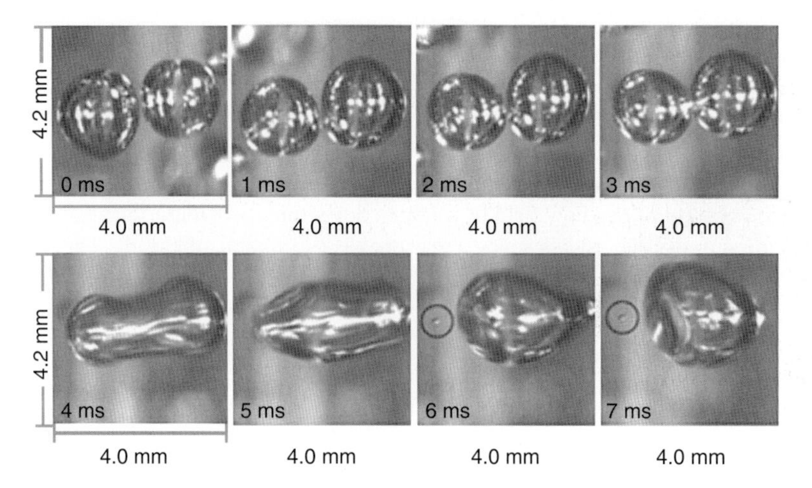

Figure 15.2 Bubble coalescence measured in a stirred tank reactor at 1000 Hz with a single 10-bit monochrome camera (From [8]).

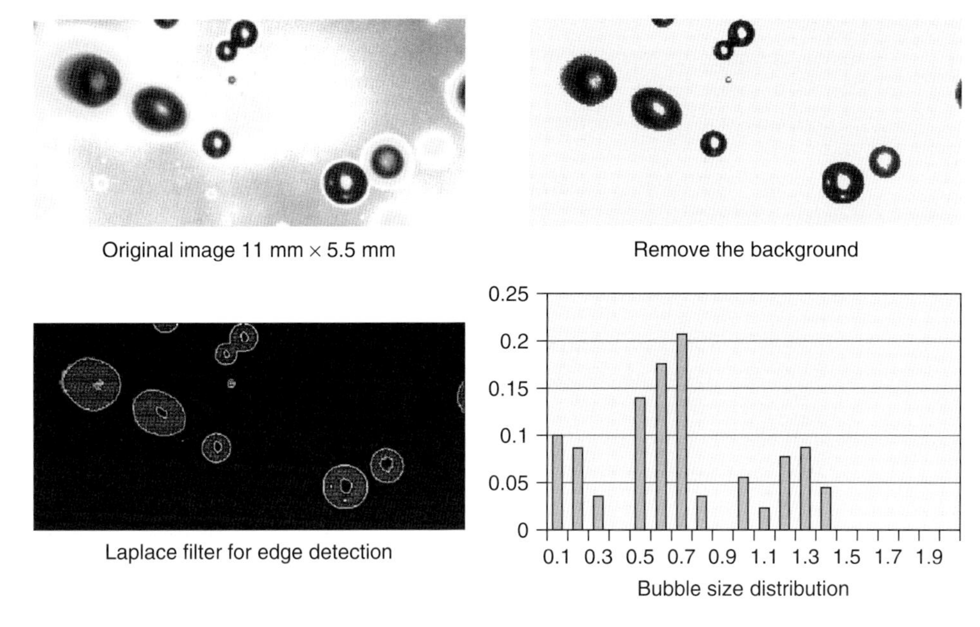

Figure 15.3 Bubble size measurements using shadowgraph technique (From [9]).

or 16 bit, giving high resolution, and, with good equipment, the background noise can be kept very low. The measured signal is a function of dye concentration and light intensity. A large fraction of the dye molecules can be excited, thus at high laser intensity, the signal will vary with the laser-sheet thickness, and, as a result, the light will not have a correspondingly higher signal. Consequently, a parallel laser sheet is important for good measurements, and the minimum thickness is limited to 20–30 μm due to light diffraction. More advanced instruments also exist that use the lifetime of the excited species and a combination of lasers with different wavelengths to identify different species, for example, in combustion.

Refractory index matching is important for PLIF measurements of multiphase flows, as seen in Figure 15.4.

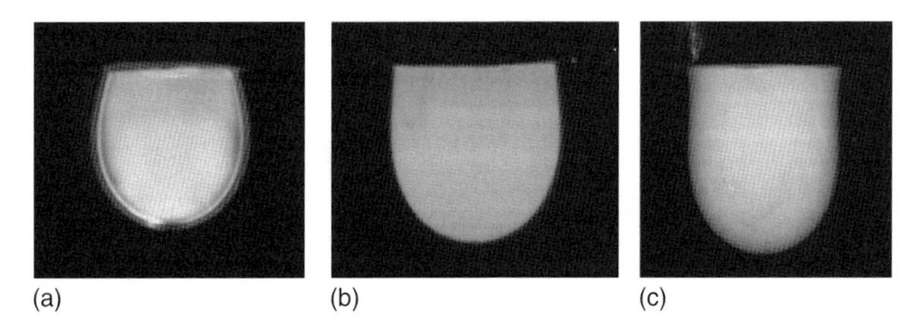

Figure 15.4 PLIF measurement of rhodamine in hanging water drops in octanol. (a) is unmatched and (b) and (c) are refractor-index-matched liquids using thin and thick laser sheets, respectively (From [10]).

15.2.4
Characterization of Opaque Multiphase Flows

The multiphase fluid systems of interest are often opaque, and thus noninvasive techniques based on optical methods or using laser beams are not effective. Various experimental techniques are available and continue to be developed to characterize opaque multiphase flows.

15.2.4.1 X-ray- and γ-Ray-Computed Tomography

These tomographic techniques that can obtain noninvasive three-dimensional images of multiphase fluid have evolved from the medical field. The attenuation of X-ray or γ-ray is measured, and because of the different attenuation of each phase, the distribution of phases in a scanned cross-section of a vessel is calculated. Typically, a single radiation source and a set of detectors are used to scan around a vessel, and thus only time-average results are obtained. The spatial resolution is usually on the order of a few millimeters. However, over the past decade, advances in this technique allowing high image acquisition rates and improved spatial resolution have made dynamic computed tomography a possibility [11]. γ-Ray is better suited when high radiation energy is needed, such as with large vessel diameters, dense packings, or thick walls due to pressurized operation.

15.2.4.2 Electrical Impedance Tomography (EIT)

These are tomographic and noninvasive techniques that are based on imaging the distribution of an electrical property within a medium. The electrical properties used in electrical impedance tomography (EIT) include capacitance, resistance, and inductance. Capacitance measurements are used for electrically insulating multiphase systems, whereas resistance and inductance are for electrically conducting materials. The technique can be used to measure concentration distributions of suspensions in pipe flow. It is implemented by placing an array of electrodes at regular positions around a pipe. A current is applied between each pair of electrodes, while all other electrodes measure the voltage difference. Each measurement is used to determine the electrical property of a single region of the pipe. The resulting measurements of the component distributions can be cross-correlated to obtain the velocity profile of the flow [12]. The high speed of the measurements through rapid switching of the pairs of electrodes allows also transient phenomena to be monitored [13]. The advantages with EIT technique include its relatively low cost, and the speed and safety of the measurements; however, it suffers from much poorer spatial resolution than that achievable with nuclear-based tomography techniques.

15.2.4.3 Magnetic Resonance Imaging (MRI)

Magnetic resonance imaging (MRI) is a versatile noninvasive tomographic technique offering possibilities to measure velocity fields as well as to determine concentration distributions in suspensions and emulsions, under both steady-state and dynamic conditions. A magnetic field gradient is applied to a sample and signals associated with proton spins are obtained at different frequencies depending on the location of the liquid. This technique can have high spatial resolution for concentration distributions down to some micrometers. Velocity measurements are carried out by applying a magnetic gradient pulse, which causes a change in the rotation of the proton spins and results in an accumulation which can be made proportional to the velocity vector or acceleration. The direction of the magnetic gradient determines the component of velocity measured. Instruments are usually equipped to apply gradients in three orthogonal directions, which can provide measurements of all three velocity components [11]. An additional unique feature of MRI is that is allows independent measurement of the acceleration field. In other techniques, the local acceleration can only be obtained by differentiating the instantaneous velocity with respect to time.

A disadvantage with MRI is that the equipment is costly and usually only available in medical research institutes. As a result, experiments must be designed around the existing equipment with competing applications. In addition, MRI is not applicable to pressurized vessels since transducers are in contact with the fluid and are thus mounted flush with the inside walls.

15.2.4.4 **Radioactive Particle Tracking (RPT)**

Radioactive particle tracking (RPT) can be used to map the velocity field by tracking the position of a single radioactive tracer particle in a reactor. The particle which may consist of a polypropylene shell contains a radionuclide that emits γ-rays. This technique is invasive; however, the particle can be designed to be neutrally buoyant so that it well represents the flow of the phase of interest. An array of detectors is positioned around the reactor vessel. Calibration must be performed by positioning the particle in the vessel at a number of known locations and recording each of the detector counts. During actual measurements, the γ-ray emissions from the particle are monitored over many hours as it moves freely in the system maintained at steady state. Least-squares regression methods can be applied to evaluate the temporal position of the particle and thus velocity field [13, 14]. This technique offers modest spatial resolutions of 2–5 mm and sampling frequencies up to 25 Hz.

15.2.4.5 **Ultrasonic Tomography**

Ultrasonic tomography uses differences in the acoustic impedance of different phases to construct information about the phase distribution in a vessel. The main limitation of the technique compared to the others discussed above is that measurement times are limited by the relatively slow speed of sound. An additional problem is the fact that in multiphase fluids, there is typically a large difference in the acoustic impedence of the phases. For example, in gas–liquid systems, the gas bubbles are nearly perfect reflectors if their diameters are larger than the acoustic wavelength [15]. This limits the application of ultrasonic tomography to imaging of low-density slurries and/or a small-size vessel. In order to mitigate the limitations posed by acoustic reflections and in order to obtain real-time images, an array of transducers is used around a pipe. The number of transducers has a large impact on the quality of the reconstructed images obtained [16]. A disadvantage that ultrasonic techniques share with MRI is that for high-quality signals, transducers must be mounted so that they are in contact with the fluid, which limits applications to low-pressure systems.

Ultrasonic methods can also be applied to velocity measurements based on measurement of the Doppler shift in the frequency of an ultrasonic wave scattered from a moving particle. The angle between the velocity vector and the direction of ultrasound propagation must be known, which practically limits the application of the technique to the measurement of unidirectional flows. However, this limitation may be overcome again by the use of an array of transducers [11].

15.3 Simulations

Simulations using CFDs are very useful both for simulation of equipment and for detailed simulations of different phenomena. Specific assumptions and models can be simulated and validated against measurements. The main limitation is that all scales are not resolved. All phenomena occurring on a scale below the grid resolution must be modeled. Using 10^6 computational cells in the simulation of a 10-m^3 stirred tank reactor gives an average of 10 ml for each cell, and models must be introduced for all phenomena occurring on a smaller scale, for example, momentum transfer due to particle collisions, bubble and drop breakup and coalescence, and mixing and reaction for fast reactions. Simulations using smaller volumes will give a better understanding of the details but then the overall picture is lost.

15.3.1 Single-Phase Simulations

For engineering purposes, there are two groups of CFD models for turbulent flows, Reynolds-averaged Navier–Stokes (RANSs) and LES'. The RANS models simulate time-averaged turbulent flow. The k–ε and k–ω models and a large number of similar models assume isotropic turbulence. This simplification is usually acceptable in bulk flow at high Reynolds numbers but fails at low *Re* and close to walls. The Reynolds stress model (RSM) simulates all Reynolds stresses and can predict a wider range of flows. However, the RSM model tends to be more unstable, and a good practice is usually to start with the k–ε model and change to RSM when convergence is reached. The RSM model can simulate nonisotropic turbulence and is also better than the k–ε type models in predicting transport of turbulence when production and dissipation are not in balance locally. All the RANS models have problems with transition from turbulent to laminar flow and vice versa.

The k–ω model can simulate flow close to walls while the other models require a low *Re* addition or wall functions to describe the flow below $y^+ < 30$. The models are optimized to give accurate flow predictions and the model parameters kinetic

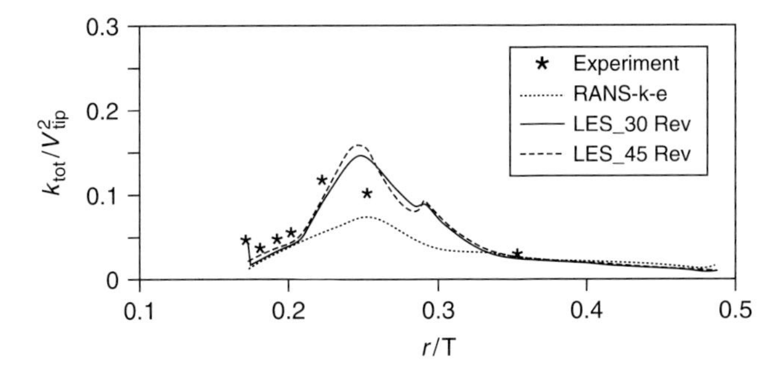

Figure 15.5 Measured and simulated turbulent kinetic energies (normalized with the impeller tip speed) at the impeller plane in a stirred tank reactor (From [17]).

energy, k, and rate of dissipation, ε, (and ω) are merely variables in the models and have only a weak coupling to the measurable physical entities (Figure 15.5). Models that use k and ε, for example, mixing or bubble breakup models, must be calibrated to the actual turbulence model.

LES has come within reach for engineering simulations even on PCs. In LES simulations, all large eddies are resolved and the smaller eddies that are assumed to be isotropic are modeled as a subgrid viscosity. This assumption is usually not valid [18] and a large fraction of the turbulence must be resolved. As a rule of thumb, at least 80% of the turbulent kinetic energy should be resolved. In the dynamic Smagorinsky model, the Smagorinsky constant, which is continuously recalculated, is used to calculate subgrid viscosity. It has been shown to describe turbulent flow with sufficient accuracy for engineering simulations. A reasonably good solution is usually obtained within a few hours, but since a nonlinear process is simulated, each simulation depends on the initial conditions, and days or even weeks of simulation can be required to obtain good statistics of the turbulent flow.

Turbulent inlet conditions for LES are difficult to obtain since a time-resolved flow description is required. The best solution is to use periodic boundary conditions when it is possible. For the remaining cases, there are algorithms for simulation of turbulent eddies that fit the theoretical turbulent energy distribution. These simulated eddies are not a solution of the Navier–Stokes equations, and the inlet boundary must be located outside the region of interest to allow the flow to adjust to the correct physical properties.

15.3.2
Multiphase Simulations

Simulations of multiphase flow are, in general, very poor, with a few exceptions. Basically, there are three different kinds of multiphase models: Euler–Lagrange, Euler–Euler, and volume of fluid (VOF) or level-set methods. The Euler–Lagrange and Euler–Euler models require that the particles (solid or fluid) are smaller than the computational grid and a finer resolution below that limit will not give a more accurate solution while the VOF model gives better predictions with a finer mesh.

In Euler–Lagrange modeling, the momentum equations are solved for both the continuous phase and for each particle. At a low volume fraction of the dispersed phase, this kind of model can be very accurate. Since an equation is solved for each particle, the number of particles is limited and a practical limit for simulation on a PC is a few hundred thousand particles. If the number of collisions between the particles is limited, it is possible to simulate a flow of packages of particles that have the same properties. These packages can contain a few hundred or thousand particles. At higher volume fractions, when the collisions between particles become important, it is necessary to limit the number of particles. Also very short time steps are required in order to accurately predict particle collisions. In general, all commercial CFD programs still have very poor models for Lagrangian simulations of flows where particle–particle collisions are important. The particles will also affect the continuous phase. The momentum transfer from particles to the continuous phase is straightforward, while the transfer of energy that will add to k in the k–ε model is more controversial. The turbulence eddies generated by the particles are usually much smaller than the turbulent eddies generated by the shear rate, and consequently, their effect on turbulent viscosity is also less. These smaller eddies will have less effect on momentum transfer and will dissipate faster than the larger eddies.

In the Euler–Euler model, a momentum balance is solved for each phase. The continuous phase is usually described as laminar or modeled by the k–ε model. The dispersed phase viscosity is typically calculated using kinetic theory for granular flow (KTGF). In this model, in analogy with the kinetic theory for molecules, granular temperature is introduced to quantify the random movement of the particles. This model works reasonably well for spherical particles that nearly ideally collide, that is, have a high restitution coefficient. For most other flows with a high volume fraction of the dispersed phase, the model needs to be calibrated and at least one simulation must be validated experimentally. This model usually has stability problems and requires very fine time resolution. A more stable simplification of the Euler–Euler model, the mixture or algebraic slip model, can be used for flows where the phases are in kinetic equilibrium and accelerate together. With this model, which can be used for, for example, bubble columns, it is sufficient to solve only the momentum equations for the mixture. The movement of the individual phases is then calculated using an algebraic equation. Flows at very high loading, where the particles are in constant contact, are usually simulated using a friction model. This model is very empirical and sensitive to how the material data are obtained.

Multiscale modeling is an approach to minimize system-dependent empirical correlations for drag, particle–particle, and particle–fluid interactions [19]. This approach is visualized in Figure 15.6. A detailed model is developed on the smallest scale. Direct numerical simulation (DNS) is done on a system containing a few hundred particles. This system is sufficient for developing models for particle–particle and particle–fluid interactions. Here, the grid is much smaller

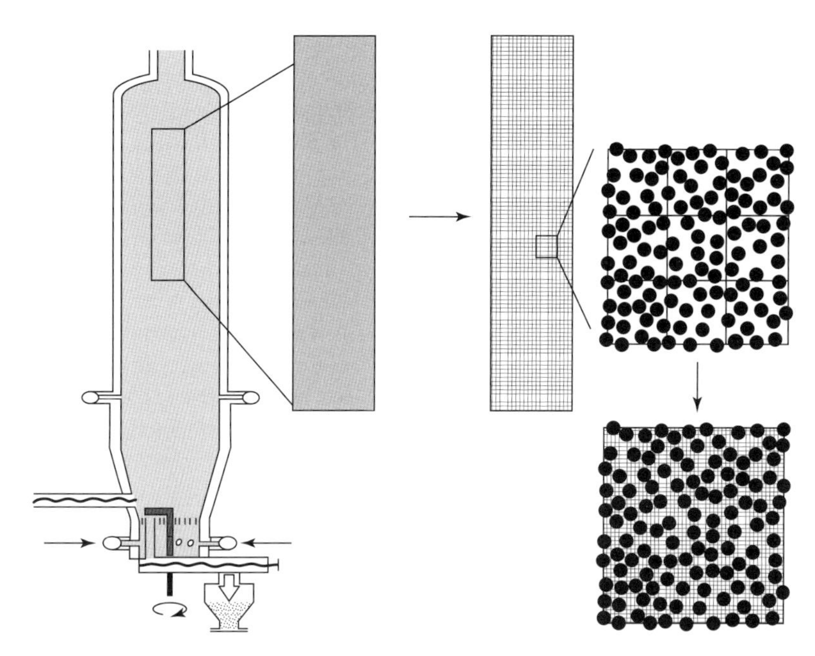

Figure 15.6 Multiscale CFD modeling (From [19]).

than the particles, but a sufficient number of particles is present to allow a model to describe a dense multiphase system. At the next level, the system is described using an Eulerian–Lagrangian model. This level contains a few hundred thousand particles, and with a few hundred particles in each computational cell the size of the system is limited to <0.1 m. This size allows Lagrangian simulation of particle collisions. Since the particles are much smaller than the grid, they are treated as point sources from the continuous phase point of view and particle–fluid momentum and energy transfer is modeled using a model developed on the finest scale. The individual particles cannot be resolved on the largest scale, and accordingly, a multifluid model is required, for example, the KTGF model, for which the parameters have been estimated using the models on the finer scales.

VOF or level-set models are used for stratified flows where the phases are separated and one objective is to calculate the location of the interface. In these models, the momentum equations are solved for the separated phases and only at the interface are additional models used. Additional variables, such as the volume fraction of each phase, are used to identify the phases. The simplest model uses a weight average of the viscosity and density in the computational cells that are shared between the phases. Very fine resolution is, however, required for systems when surface tension is important, since an accurate estimation of the curvature of the interface is required to calculate the normal force arising from the surface tension. Usually, VOF models simulate the surface position accurately, but the space resolution is not sufficient to simulate mass transfer in liquids.

For fluid particles that continuously coalesce and breakup and where the bubble size distributions have local variations, there is still no generally accepted model available and the existing models are contradictory [20]. A population density model is required to describe the changing bubble and drop size. Usually, it is sufficient to simulate a handful of sizes or use some quadrature model, for example, direct quadrature method of moments (DQMOM) to decrease the number of variables.

Owing to the high computational load, it is tempting to assume rotational symmetry to reduce to 2D simulations. However, the symmetrical axis is a wall in the simulations that allows slip but no transport across it. The flow in bubble columns or bubbling fluidized beds is never steady, but instead oscillates everywhere, including across the center of the reactor. Consequently, a 2D rotational symmetry representation is never accurate for these reactors. A second problem with axis symmetry is that the bubbles formed in a bubbling fluidized bed are simulated as toroids and the mass balance for the bubble will be problematic when the bubble moves in a radial direction. It is also problematic to calculate the void fraction with these models.

Ekambra *et al.* [21] compared the results from 1D, 2D, and 3D simulations of a bubble column with experimental results. They obtained similar results for holdup and axial velocity, while eddy viscosity, Reynolds stresses, and energy dissipation were very different in the three simulations as shown in Figure 15.7. This example also illustrates the importance of selecting the right variables for model validation. A 2D model will yield good results for velocity but will predict all variables based on turbulent characteristics poorly.

15.4 Microscale Measurement and Simulations

15.4.1 Flow Characterization

Almost all flows in chemical reactors are turbulent and traditionally turbulence is seen as random fluctuations in velocity. A better view is to recognize the structure of turbulence. The large turbulent eddies are about the size of the width of the impeller blades in a stirred tank reactor and about 1/10 of the pipe diameter in pipe flows. These large turbulent eddies have a lifetime of some tens of milliseconds. Use of averaged turbulent properties is only valid for linear processes while all nonlinear phenomena are sensitive to the details in the process. Mixing coupled with fast chemical reactions, coalescence and breakup of bubbles and drops, and nucleation in crystallization is a phenomenon that is affected by the turbulent structure. Either a resolution of the turbulent fluctuations or some measure of the distribution of the turbulent properties is required in order to obtain accurate predictions.

LES is suitable for the simulation of turbulence at moderate Reynolds numbers. Turbulence generation, transport, and dissipation are described very accurately

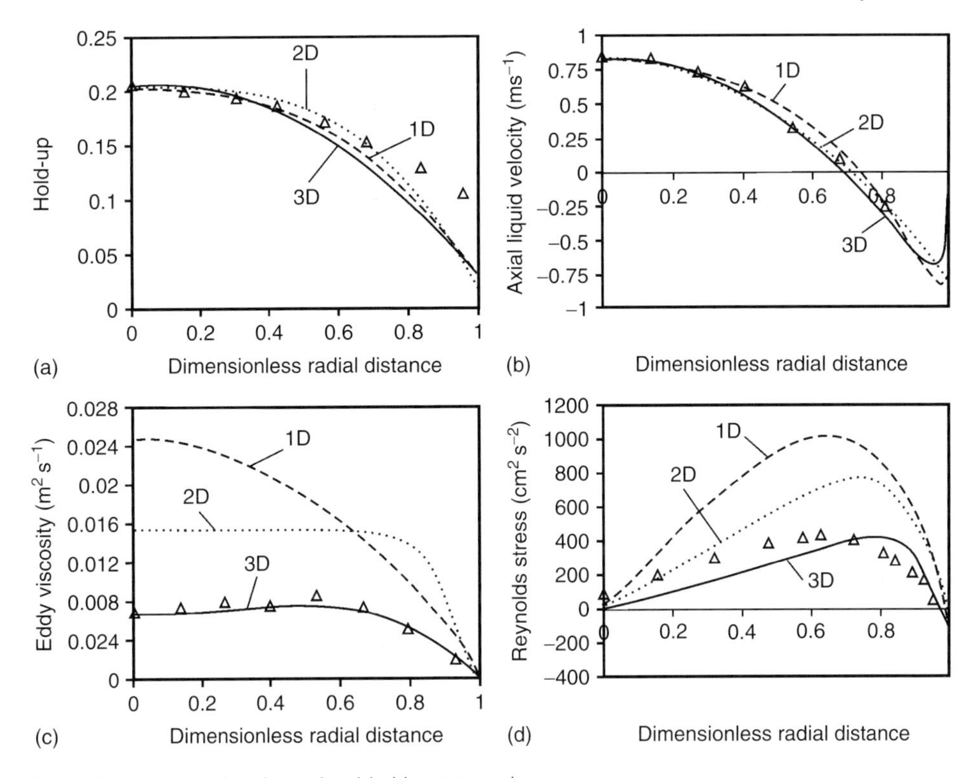

Figure 15.7 Measured and simulated holdup (a), axial velocity (b), eddy viscosity (c), and Reynolds stresses (d), using a 1D, 2D, and 3D simulations (From [21]).

when grid resolution is sufficient to resolve more than 80% of the turbulent energy and proper inlet conditions for turbulence is specified. RANS models give only the turbulent kinetic energy and rate of dissipation, and additional models must be provided for distribution of sizes and lifetime of the turbulent eddies.

15.4.2 Mixing and Mass Transfer

Many processes are affected by mixing, for example, fast reactions with selectivity problems and nucleation during crystallization. The rate of mixing is to a large extent determined by the turbulence. The mean flow distributes the reactants over the reactor. The large turbulent eddies transport the species transversely between the flow trajectories, and the smaller eddies are required for mixing within a reasonable time. In a gas phase, the rate of mixing is completely determined by the turbulence, whereas in a liquid phase with a 4-orders-of-magnitude lower molecular diffusivity, diffusion, and turbulence are important for the final mixing.

PLIF is very useful in studying mixing. It gives a very good insight into the process. Figure 15.8a shows the instantaneous concentration during mixing in liquids. It is

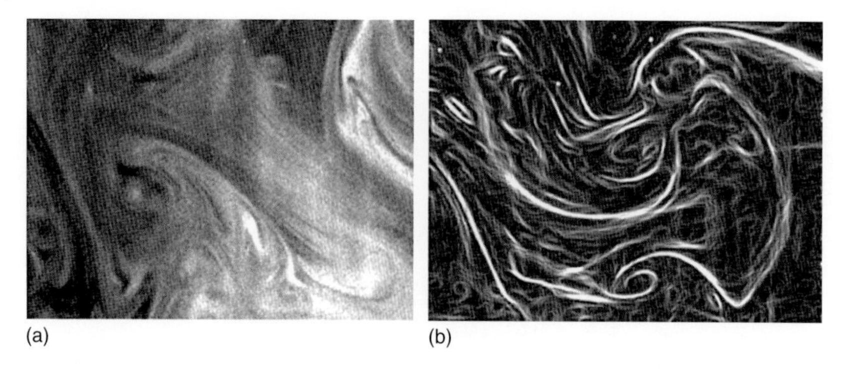

Figure 15.8 (a) Concentration of an inert tracer measured with PLIF. (b) Scalar dissipation rate (From [22]).

clearly seen that the average concentration is a very poor approximation for fast reactions. Figure 15.8b shows the scalar dissipation rate defined as

$$\langle N \rangle = \left\langle D \sum_{j=1}^{3} \frac{\partial \xi}{\partial x_j} \frac{\partial \xi}{\partial x_j} \right\rangle \tag{15.1}$$

where ξ is the mixture fraction or a normalized concentration C/C_{in} and D is the molecular diffusivity.

The scalar dissipation rate is a measure of mixing rate or reaction rate for fast, mixing-controlled reactions. As can be seen in Figure 15.8b, the mixing and reaction occur in the light streaks and the local concentration, where the reaction occurs, is very different from the average concentration. In RANS modeling, the average concentration is the average over both the computational cell volume and the timescale for the turbulent eddies. LES is time resolved and the average concentration is only the computational cell average. In gas phase, the time resolution from LES is sufficient to resolve most of the turbulent concentration fluctuations, but since the concentration is virtually zero at the position where the reactants meet for very fast reactions, the concentration will vary within the computational cell. For liquids, it is even more difficult since the relevant scale, the Bachelor scale $(\eta/Sc^{1/3})$, is an order of magnitude smaller than the Kolmogorov scale (η) in turbulence.

A model must be introduced to simulate fast chemical reactions, for example, flamelet, or turbulent mixer model (TMM), presumed mapping. Rodney Fox describes many proposed models in his book [23]. Many of these use a probability density function to describe the concentration variations. One model that gives reasonably good results for a wide range of non-premixed reactions is the TMM model by Baldyga and Bourne [24]. In this model, the variance of the concentration fluctuations is separated into three scales corresponding to large, intermediate, and small turbulent eddies.

An estimation of the importance of mixing on reaction rate can be done by calculating the Dahmkohler number

$$Da = \frac{\text{Time for mixing}}{\text{Time for reaction}} \tag{15.2}$$

According to the TMM model, the time constant for mixing on the large turbulent scale is on the order of $\tau = 0.5k/\varepsilon$, that is, $\approx$25 ms in a stirred tank reactor, and the mixing on intermediate scale is on the order of $\tau = 17.25\sqrt{\upsilon/\varepsilon}$, that is, $\approx$20 ms. The smallest scale will have an effect only for very large *Sc* numbers. If a noticeable conversion can be observed within 25 ms, the mixing will have an effect on reaction rate, that is, the reaction is mixing independent only if $Da \ll 1$.

15.4.3 Multiphase Systems

15.4.3.1 Fixed Bed

Traditionally, an average Sherwood number has been determined for different catalytic fixed-bed reactors assuming constant concentration or constant flux on the catalyst surface. In reality, the boundary condition on the surface has neither a constant concentration nor a constant flux. In addition, the Sh-number will vary locally around the catalyst particles and in time since mass transfer depends on both flow and concentration boundary layers. When external mass transfer becomes important at a high reaction rate, the concentration on the particle surface varies and affects both the reaction rate and selectivity, and consequently, the traditional models fail to predict this outcome.

Dixon and coworkers [25] have performed several CFD simulations of fixed beds with catalyst particles of different geometries (Figure 15.9). The vast number of surfaces and the problems with meshing the void fraction in a packed bed have made it necessary to limit the number of particles and use periodic boundary conditions to obtain a representative flow pattern. Hollow cylinders have a much higher contact area between the fluid and particles at the same pressure drop. However, with a random packing of the particles, there will be a large variation

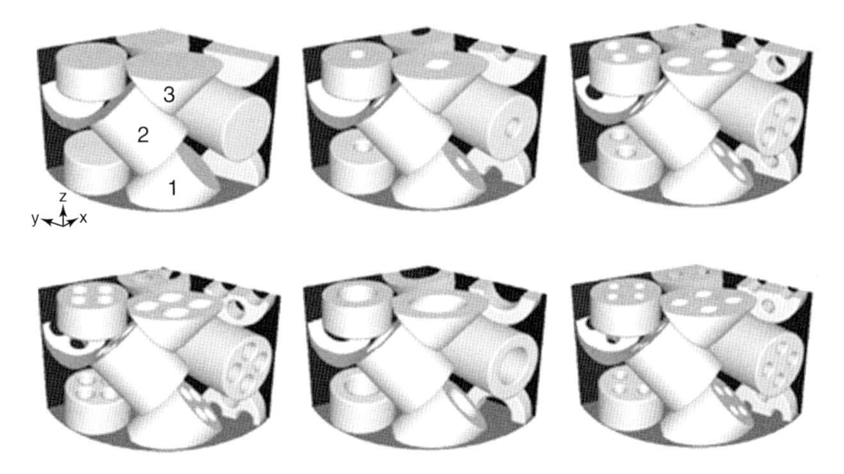

Figure 15.9 Particle designs used in the CFD study of mass and heat transfer in a fixed bed (From [25]).

of flow through the particles and the local *Sh*-number will have a correspondingly large variation.

Mass and heat transfer to the walls in turbulent flows is a complex mixture of molecular transport and transport by turbulent eddies. The generally assumed analogy between mass and heat transfer by assuming $Sh = Nu$, is not valid for turbulent flows [26]. Simulations and measurements have shown that there is a laminar film close to the surface where most of the mass transfer resistance for high *Sc* liquids is located. This film is located below $y^+ = 1$ and for low *Sc* fluids, and for heat transfer the whole boundary layer is important [27].

15.4.3.2 Solid Particles

Direct measurement of particle velocity and velocity fluctuations in fluidized beds or riser reactors is necessary for validating multiphase models. Dudukovic [14] and Roy and Dudukovic [28] have used computer-automated radioactive particle tracking (CARPT) to follow particles in a riser reactor. From their measurements, it was possible to calculate axial and radial solids diffusion as well as the granular temperature from a multiphase KTGF model. Figure 15.10 shows one such measurement.

15.4.3.3 Fluid Particles

Mass transfer in a gas–liquid or a liquid–liquid reactor is mainly determined by the size of the fluid particles and the interfacial area. The diffusivity in gas phase is high, and usually no concentration gradients are observed in a bubble, whereas large concentration gradients are observed in drops. An internal circulation enhances the mass transfer in a drop, but it is still the molecular diffusion in the drop that limits the mass transfer. An estimation, from the time constant, of the time it will take to empty a 5-mm drop is given by $\tau_D = d^2/4D = (10^{-3})^2/4 \times 10^{-9} = 6000\,\text{s}$. The diffusion timescale varies with the square of the diameter of the drop, so

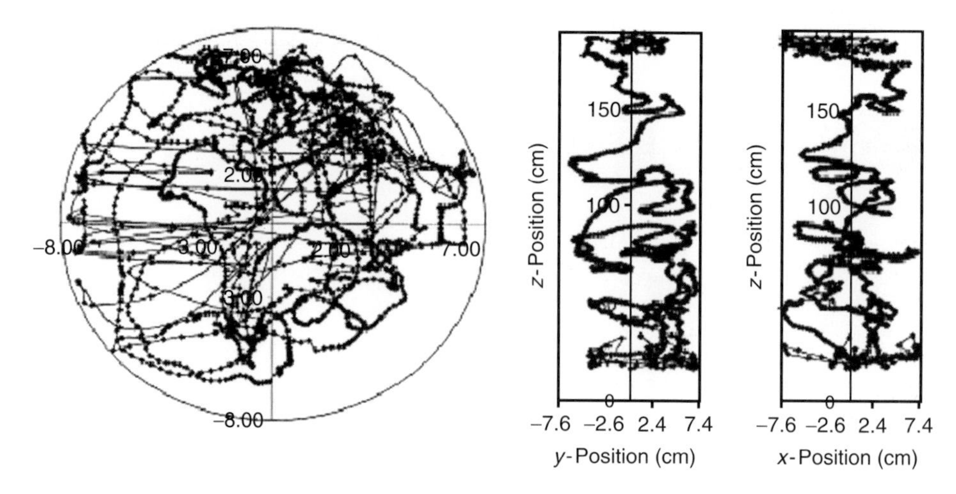

Figure 15.10 Lagrangian trace of the tracer particle during one stay in a riser (From [14]).

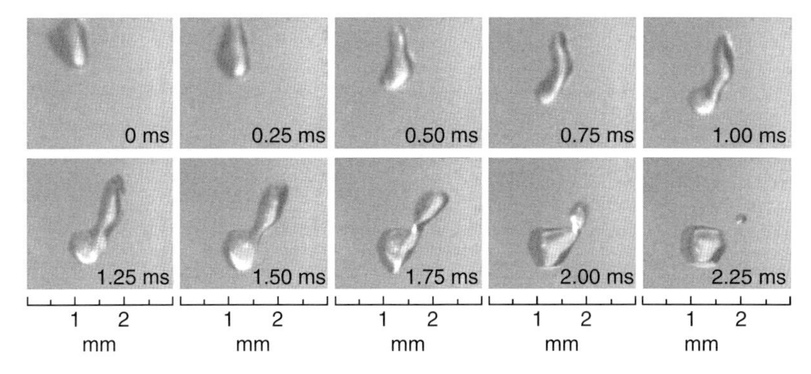

Figure 15.11 Breakup of an air bubble in turbulent water (From [8]).

that a 0.5-mm drop has a diffusion timescale of 60 s. For bubbles, the gas-phase diffusivities are about 4 orders of magnitude higher and the corresponding time constant for a 5-mm bubble is 0.5 s. For good mass transfer in liquid drops, it is necessary that the drops are very small or mix internally because of frequent breakup and coalescence to avoid large concentration gradients in the drop.

Mass transfer in the continuous phase is less of a problem for liquid–liquid systems unless the drops are very small or the velocity difference between the phases is small. In gas–liquid systems, the resistance is always on the liquid side, unless the reaction is very fast and occurs at the interface. The Sherwood number for mass transfer in a system with dispersed bubbles tends to be almost constant and mass transfer is mainly a function of diffusivity, bubble size, and local gas holdup.

Bubble and drop breakup is mainly due to shearing in turbulent eddies or in velocity gradients close to the walls. Figure 15.11 shows the breakup of a bubble, and Figure 15.12 shows the breakup of a drop in turbulent flow. The mechanism for breakup in these small surface-tension-dominated fluid particles is initially very similar. They are deformed until the aspect ratio is about 3. The turbulent fluctuations in the flow affect the particles, and at some point one end becomes

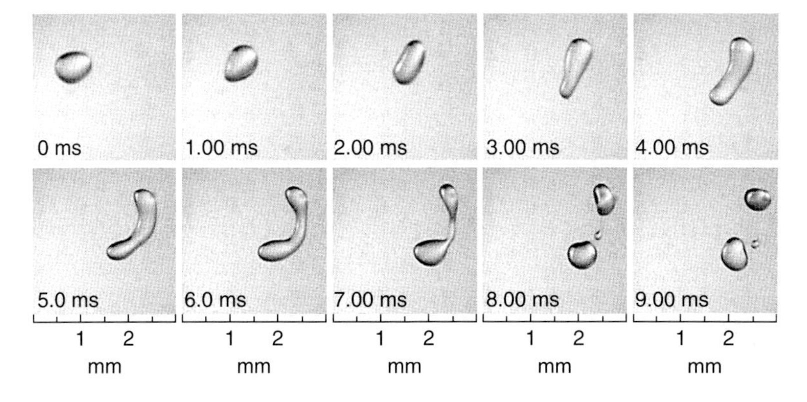

Figure 15.12 Breakup of an octanol drop in turbulent water (From [8]).

smaller than the other. The surface tension increases the pressure at that end and a pressure gradient arises across the particle, pushing fluid from the small end to the large end. When the bubble finally breaks, it will form one large and one small particle. For liquid drops, the density is much higher and the pressure gradient is not sufficient to move a significant part of the liquid from one side to the other, and the drop will break up into two almost equal-sized drops (Figure 15.12). The expected size distribution will be different for different systems since the breakup pattern depends on particle size, surface tension, and density of the dispersed phase.

Breakup can occur if the shear stress is large enough to deform the particle, that is, when $\tau > 2\sigma/d$. The surface area increases during breakup and sufficient energy must also be provided to compensate for the increase in surface energy. In turbulent flows, the available shear stress in the turbulent eddies of size λ can be estimated from [20]

$$\tau = \rho_c \frac{1}{2} u \frac{2}{\lambda} \approx \rho_c \varepsilon^{2/3} \lambda^{2/3} \tag{15.3}$$

The most efficient turbulent eddies for bubble breakup are eddies of the same size as the bubbles. Large eddies will merely move the bubbles and smaller eddies do not have sufficient energy to break up the bubbles. Assuming that the most efficient eddies to break up fluid particles are eddies of the same size as the bubble, that is, $\lambda \approx 2d$, gives the required turbulent energy dissipation

$$\varepsilon > \frac{2^{5/6} \sigma^{3/2}}{\rho_c d^{5/2}} \tag{15.4}$$

This equation gives only the energy required to break up a bubble. The rate of breakage will also involve the number density of eddies of size λ and a probability that the bubble will break up [20].

Coalescence can occur according to many different mechanisms. Figure 15.13 shows some of these mechanisms.

For systems with large density differences, the dominating mechanism for coalescence is accumulation of the low-density fluid in the low-pressure regions.

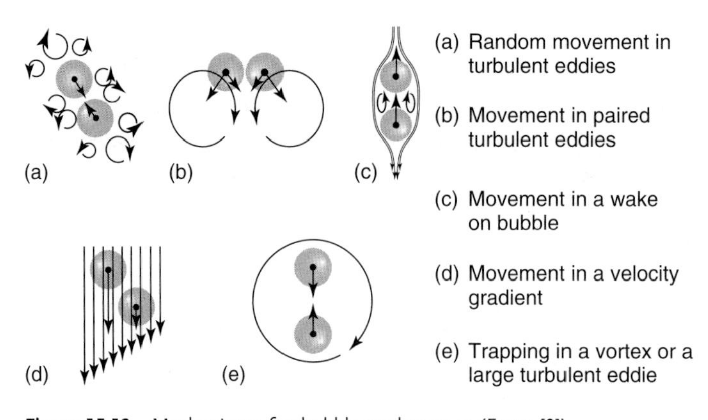

Figure 15.13 Mechanisms for bubble coalescence (From [9]).

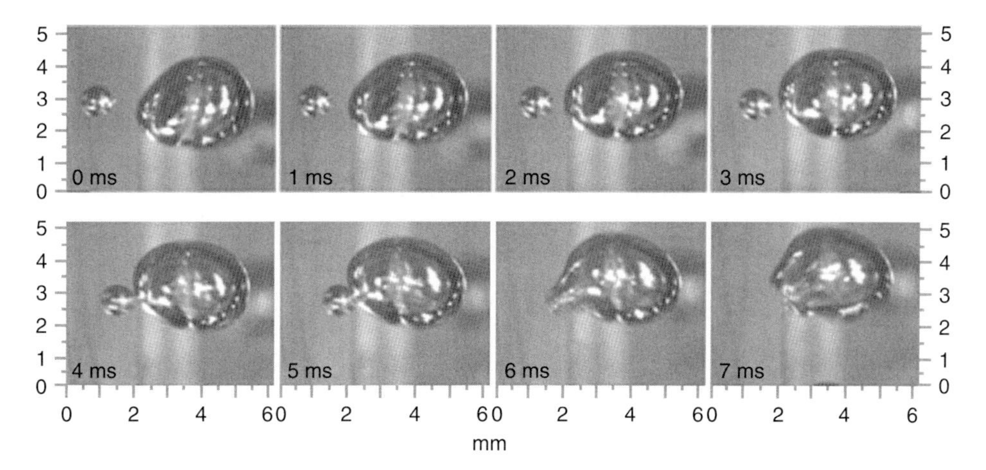

Figure 15.14 Coalescence of two air bubbles rising in a stirred tank reactor (From [29]).

In a stirred tank reactor, these low-pressure regions are behind the impeller blades, in the trailing vortices leaving the impeller blades, behind the baffles, and at the center of the large turbulent eddies.

In quiescent liquids and in bubble columns, buoyancy-driven coalescence is more important. Large fluid particles with a freely moving surface will also have a low-pressure region at the edge of the particle where the velocity is maximum. This low-pressure region will not only allow the bubble to stretch out and form a spherical cap but also allow other bubbles to move into that area and coalesce. Figure 15.14 shows an example of this phenomenon.

Fluid particles that are heavier than the continuous phase tend to move away from the low-pressure regions. They are thrown out of the turbulent eddies and the trailing vortices, and will thus collide more randomly.

15.5 Reactor Design

The optimal chemical reactor should be a small, continuous reactor that does not contain any moving parts. It should also be easy to change products, that is, easy to clean. The scale-up rules should be simple and scaling up from laboratory scale to industrial scale straightforward. Ideally, there should be no mass or heat transfer limitations on either scale. However, increasing production on the industrial scale will always push the process toward mass and heat transfer limitations, and accordingly reactors must be designed to provide high mass and heat transfer. In general, intense mixing for mixing-sensitive reactions and early upstream formation of stable fine dispersions in multiphase flows early upstream in the reactor will provide reactors that are easier to scale up.

15.5.1 Mixing-Controlled Reactions

In reactor design, it is very important to know how and where turbulence is generated and dissipated. In a liquid phase, it is also important that the smallest eddies are sufficiently small. The ratio between the reactor scale (L) and the smallest turbulent scale, the Kolmogorof scale (η), usually scales as $L/\eta \alpha Re^{3/4}$. The Kolmogorov scale can also be estimated from the viscosity and the power dissipation as $\eta = (\upsilon^3/\varepsilon)^{1/4} \approx 30\,\mu\text{m}$ in water with a power input of $1\,\text{W kg}^{-1}$ and from the Bachelor scale $\eta/Sc^{1/3} \approx 3\,\mu m$ in liquids. For a liquid, the estimation of the time constant for diffusion shows that it takes $\tau_D = l^2/D = (10^{-3})^2/10^{-9} = 1000\,\text{s} \approx 15$ min to completely mix a 1-mm layer by diffusion alone, and $\eta^2/D = 0.4\,\text{s}$ to mix on the Kolmogorov scale. If a substantial part of the reaction occurs within one second, the chemical reaction rate will be affected by mixing.

For effective mixing, high Reynolds numbers are required and the turbulence should be dissipated in the bulk of the flow and as little as possible on the walls. The source term for generating turbulent energy is in the $k{-}\varepsilon$ model written as

$$P_k = \upsilon_T \left[\left(\frac{\partial \langle U_i \rangle}{\partial x_j} + \frac{\partial \langle U_j \rangle}{\partial x_i} \right) \frac{\partial \langle U_i \rangle}{\partial x_j} \right] \tag{15.5}$$

The production of turbulence is maximum close to walls, where both shear rate and turbulent viscosity, υ_T, are high. In pipe flow, the maximum is close to $y^+ = 12$. A proper design of a chemical reactor for efficient mixing at low Re should allow the generated turbulence to be transported with the mean flow from the region where it is produced to the bulk of the fluid where it should dissipate.

The Alfa Laval ART© reactor that is designed to obtain good mixing at low Re has identical mixing cells along the reactor, and the cells are designed to generate turbulence effectively and dissipate it homogeneously in the cells. Figure 15.15 shows the turbulent structure simulated with LES.

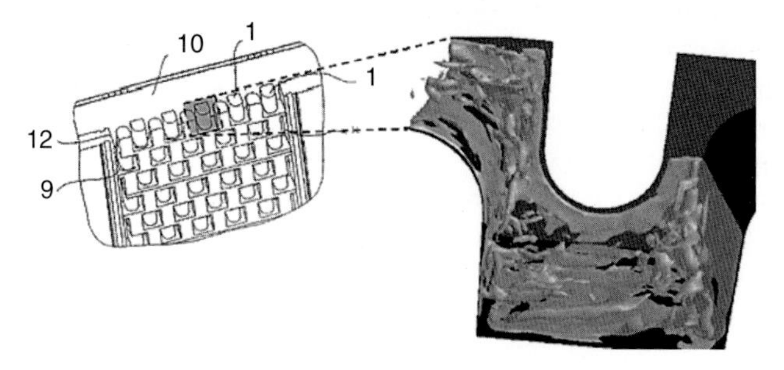

Figure 15.15 LES simulations of structures in turbulent flow in an Alfa Laval ART© reactor. Surface of iso-vorticity is shown in the figure.

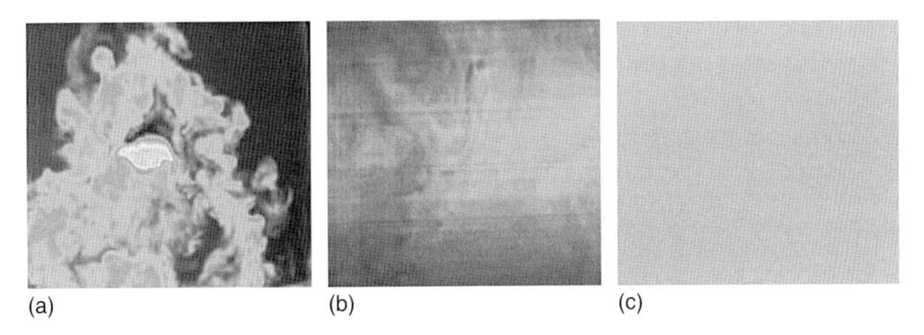

Figure 15.16 Mixing in an Alfa Laval ART© reactor at the inlet (a), and 3 (b) and 5 (c) cells downstream measured with PLIF (From [3]).

The mixing in the first five cells is illustrated in Figure 15.16. In the fifth cell, after 50 ms, the reactants are very well mixed. After the fifth cell, only the temperature must be controlled to keep the reactants well mixed [30]. The very good mixing properties were also verified with a mixing-sensitive reaction, that is, the mixing-sensitive diazo coupling between 1-naphthols, 2-Naphthols, and diazotized sulfanilic acid [30].

The stirred tank reactor with a Rushton turbine is not an optimal reactor from a mixing point of view. The turbulence is generated by the turbine, but unfortunately most of the energy is dissipated in the impeller region. Ng and Yianneskis [31] estimated that about 50% of the turbulent energy was dissipated in the impeller region, that is, more than 10 times higher dissipation rate per volume in the impeller region. This makes inlet positions for the reactant crucial for the final selectivity for mixing-sensitive reactions, and no simple scaling-up laws are possible.

15.5.2 Multiphase Reactors

15.5.2.1 Liquid–Liquid

An effective liquid–liquid reactor may be designed to obtain drops that continuously break up and coalesce, or it may be designed to obtain very small drops that have very efficient mass transfer and follow the continuous phase with a low rate of coalescence. The former will require a much larger reactor, but the separation of the phases after reaction is simpler.

Diffusion in liquids is very slow. Turbulent transport or very narrow channels are necessary for good contact between the phases. The droplets must also be very small to minimize transport limitations within the drops. An estimation of the time constant for diffusion in a 1-mm drop is

$$\tau_D = \frac{d^2}{4D} = \frac{(10^{-3})^2}{4 \cdot 10^{-9}} = 250\,\mathrm{s} = 4\,\mathrm{min}$$

whereas a 1-cm drop will need 400 min or 7 h.

Small drops will move with the flow and the response time for acceleration of a 1-mm liquid drop in a water-like liquid is on the order of

$$\tau_u = \frac{\rho d^2}{18\mu} = \frac{1000(10^{-3})^2}{18 \cdot 10^{-3}} = 0.05\,\mathrm{s}$$

This response time should be compared to the turbulent eddy lifetime to estimate whether the drops will follow the turbulent flow. The timescale for the large turbulent eddies can be estimated from the turbulent kinetic energy k and the rate of dissipation ε, $\tau_c = \frac{k}{\varepsilon} \approx 30\text{–}50\,\mathrm{ms}$, for most chemical reactors. The Stokes number is an estimation of the effect of external flow on the particle movement, $St = \tau_d/\tau_c$. If the Stokes number is above 1, the particles will have some random movement that increases the probability for coalescence. If $St \ll 1$, the drops move with the turbulent eddies, and the rates of collisions and coalescence are very small. Coalescence will mainly be seen in shear layers at a high volume fraction of the dispersed phase.

Smaller drops, preferably smaller than 1 mm, are required to obtain good mass transfer. Turbulent eddies may be ineffective in breaking up and forming drops with a size below 1 mm. According to Eq. (15.3), an energy input of $\varepsilon < 15\,\mathrm{W\,m^{-3}}$ is required to break a 1-mm drop with surface tension $0.04\,\mathrm{N\ m^{-1}}$. This high energy dissipation is usually not available in a chemical reactor, and an injector may be necessary. Efficient injectors are readily available, for example, injectors used for diesel sprays in diesel engines. These injectors give a pulsating flow at high frequency, but these pulses are quickly smoothened. Very small drops can be formed, and drops of 0.1 mm have efficient mass transfer and closely follow the flow. They have a time constant for diffusion of about 2.5 s and a Stokes number $\ll 1$.

15.5.2.2 **Gas–Liquid**

The diffusivity in gases is about 4 orders of magnitude higher than that in liquids, and in gas–liquid reactions the mass transfer resistance is almost exclusively on the liquid side. High solubility of the gas-phase component in the liquid or very fast chemical reaction at the interface can change that somewhat. The *Sh*-number does not change very much with reactor design, and the gas–liquid contact area determines the mass transfer rate, that is, bubble size and gas holdup will determine reactor efficiency.

The low density of gases makes it more difficult to keep the bubbles dispersed. The bubbles will move to the low-pressure areas, that is, behind the impellers, in the trailing vortices close to the impeller, behind the baffles, and at the inner side after a bend. The bubbles will coalesce in these areas with high gas holdup. It is very difficult to design reactors without low-pressure regions where the low-density fluid will accumulate. One such reactor is the monolith reactor for multiphase flow [32, 33].

Breakup will occur due to high turbulence and high shear rate. This will occur in the impeller region and close to the walls. In a stirred tank, almost all breakup of the bubbles occurs in the impeller region. According to Eq. (15.3), the energy required to break up a 5-mm bubble is on the order of $1\,\mathrm{W\ m^{-3}}$, while $35\,\mathrm{W\ m^{-3}}$ is required to break up a 1-mm air bubble in water. A high rate of bubble breakup and correspondingly high pressure drop are obtained in reactors with very narrow channels, for example, ceramic or metal foams. Unfortunately, these kinds of reactors tend to be unstable. A lower pressure drop is obtained when the phases are separated and spontaneous separation may occur.

Distributing a dispersed gas in a continuous liquid is very difficult, while spraying a liquid into a gas is more easily done. The inertia on the drops allows them to move over a wide area. The limitation is the reaction time in relation to the settling time. The drops should be on the order of 0.1 mm to obtain a diffusion time $\tau = 2.5$ s, which is less than the settling time.

15.5.2.3 **Fluid–Solid**

Catalytic beds are the traditional catalytic reactors. In academic education, spherical particles are popular since only a one-dimensional model is required to describe mass and heat transfer. However, a sphere is the geometrical body that has the lowest surface area per volume, and a better mass transfer can be obtained with most other geometries. The internal mass transfer can always be compensated for by making shell catalysts with the active material available only on the outer surface. However packed beds tend to be very difficult to model since the external mass transfer will have large local variation due to the accessibility of the surface. Owing to the random packing, the flow through and around the particles as well as the local Sherwood number will vary significantly. For hollow particles, this is even more problematic since some particles will orientate themselves with no pressure drop across the inner channels, resulting in no or very little flow. The local Sherwood number will be very small in these particles [25].

Open-channel monoliths are better defined. The Sherwood (and Nusselt) number varies mainly in the axial direction due to the formation of a hydrodynamic boundary layer and a concentration (temperature) boundary layer. Owing to the chemical reactions and heat formation on the surface, the local Sherwood (and Nusselt) numbers depend on the local reaction rate and the reaction rate upstream. A complicating factor is that the traditional Sherwood numbers are usually defined for constant concentration or constant flux on the surface, while, in reality, the catalytic reaction on the surface exhibits different behavior.

15.6 The Future

The objective of the chemical industry is to have robust and flexible chemical reactors. It should also be possible to reach large-scale production of new chemicals in a very short time. The available reactors should be able to produce a large variety of chemicals with high selectivity. To obtain this, we must not only understand how existing reactors work in detail but also develop new reactors.

Virtual prototyping will be the future method to develop new reactors and chemical processes. With a good description of the fluid dynamics, and mass and heat transfer in the reactor, the specific chemical reactions and physical properties of the fluid can be changed and a process optimization can be performed in virtual space. However, the available multiphase flow models are not sufficiently accurate at present, and the main challenge for the next decade is the further development of these models. Models for mixing and reaction of fast mixing-controlled reactions are also needed.

Scaling up will probably continue to be a problem since large reactors cannot be as efficient as small laboratory reactors. However, it may be possible to make laboratory or pilot-plant reactors that are more similar to large-scale reactors, allowing more reliable validation of the simulations and process optimization. The time from laboratory-scale to full-scale production should be shortened from years to months.

List of Symbols

D	Molecular diffusivity ($m^2\ s^{-1}$)
d	Particle diameter (m)
k	Turbulent kinetic energy ($m^2\ s^{-2}$)
l	Length (m)
N	Rate of scalar dissipation (s^{-1})
P	Rate of turbulence production ($m^2\ s^{-3}$)
u_λ	Velocity of a turbulent eddy of size λ
ε	Rate of turbulence dissipation ($m^2\ s^{-3}$)
η	Kolmogorov length scale (m)
λ	Size of turbulent eddy (m)
μ	Dynamic viscosity ($kg\ m^{-1}\ s^{-1}$)
ν	Kinematic viscosity ($m^2\ s^{-1}$)
ν_T	Turbulent viscosity ($m^2\ s^{-1}$)
ρ	Density ($kg\ m^{-3}$)
σ	Surface tension (Pa m)
τ_D	Diffusion time constant (s)
τ_u	Acceleration time constant (s)
τ	Shear stress ($kg\ m^{-1}\ s^{-2}$)

Subscript

c	Continuous phased
d	Dispersed phase

References

1. Hagsater, S.M. *et al.* (2008) A compact viewing configuration for stereoscopic micro-PIV utilizing mm-sized mirrors. *Exp. Fluids*, **45** (6), 1015–1021.
2. Bernard, P.S. and Wallace, J.M. (2002) *Turbulent Flow: Analysis, Measurement and Prediction*, John Wiley & Sons, Inc.
3. Hammoudi, M. *et al.* (2008) Flow analysis by pulsed ultrasonic velocimetry technique in SulzerSMX static mixer. *Chem. Eng. J.*, **139** (3), 562–574.

4. Kilander, J., Svensson, F.J.E., and Rasmuson, A. (2006) Flow instabilities, energy levels, and structure in stirred tanks. *AIChE J.*, **52** (12), 4039–4051.
5. Baldi, S. and Yianneskis, M. (2004) On the quantification of energy dissipation in the impeller stream of a stirred vessel from fluctuating velocity gradient measurements. *Chem. Eng. Sci.*, **59** (13), 2659–2671.
6. Virdung, T. and Rasmuson, A. (2008) Solid-liquid flow at dilute concentrations in an axially stirred vessel investigated using particle image velocimetry. *Chem. Eng. Commun.*, **195** (1), 18–34.
7. Tabib, M.V. and Joshi, J.B. (2008) Analysis of dominant flow structures and their flow dynamics in chemical process equipment using snapshot proper orthogonal decomposition technique. *Chem. Eng. Sci.*, **63** (14), 3695–3715.
8. Andersson, R. and Andersson, B. (2006) On the breakup of fluid particles in turbulent flows. *AIChE J.*, **52** (6), 2020–2030.
9. Sudiyo, R., Virdung, T., and Andersson, B. (2003) Important factors in bubble coalescence modeling in stirred tank reactors. 6th International Conference on Gas liquid and Gas-Liquid -Solid Reactor Engineering, 2003, Vancouver.
10. Andersson, R. (2005) Dynamics of fluid particles in turbulent flows: CFD simulations, model development and phenomenological studies, *Chemical and Biological Engineering*, Chalmers University of Technology, Gothenburg, p. 89.
11. Powell, R.L. (2008) Experimental techniques for multiphase flows. *Phys. Fluids*, **20** (4).
12. Ismail, I. *et al.* (2003) Tomography for multi-phase flow measurement in the oil industry. 4th International Symposium on Measurement Techniques for Multiphase Flows, 2003, Hangzhou.
13. Chaouki, J., Larachi, F., and Dudukovic, M.P. (1997) Noninvasive tomographic and velocimetric monitoring of multiphase flows. *Ind. Eng. Chem. Res.*, **36** (11), 4476–4503.
14. Dudukovic, M.P. (2001) Opaque multiphase flows: experiments and modeling. 4th International Congress on Multiphase Flow, 2001, New Orleans.
15. Schlaberg, H.I., Yang, M., and Hoyle, B.S. (1997) Ultrasound reflection tomography for industrial processes. 17th Ultrasonics International Conference (UI 97), 1997, Delft.
16. Yang, M. *et al.* (1999) Real-time ultrasound process tomography for two-phase flow imaging using a reduced number of transducers. *IEEE Trans. Ultrason. Ferroelectr. Freq. Control*, **46** (3), 492–501.
17. Yeoh, S.L., Papadakis, G., and Yianneskis, M. (2004) Numerical simulation of turbulent flow characteristics in a stirred vessel using the les and rans approaches with the sliding/deforming mesh methodology. *Chem. Eng. Res. Des.*, **82** (A7), 834–848.
18. Pope, S.B. (2000) *Turbulent Flows*, Cambridge University Press.
19. van der Hoef, M.A., Ye, M., van Sint Annaland, M., Andrews, A.T., IV, Sundaresan, S., and Kuipers, J.A.M. (2006) in *Advances in Chemical Engineering* (ed. C.B. Marin), Academic Press, Amsterdam, pp. 66–146.
20. Andersson, R. and Andersson, B. (2006) Modeling the breakup of fluid particles in turbulent flows. *AIChE J.*, **52** (6), 2031–2038.
21. Ekambara, K., Dhotre, M.T., and Joshi, J.B. (2005) CFD simulations of bubble column reactors: 1D, 2D and 3D approach. *Chem. Eng. Sci.*, **60** (23), 6733–6746.
22. Kornev, N., Zhdanov, V., and Hassel, E. (2007) Study of scalar macro- and microstructures in a confined jet. 5th International Symposium on Turbulence and Shear Flow Phenomena, 2007, Munich.
23. Fox, R.O. (2003) *Computational Models for Turbulent Reacting Flows*, Cambridge University Press.
24. Baldyga, J. and Bourne, J.R. (1999) *Turbulent Mixing and Chemical Reactions*, John Wiley & Sons, Inc.

25. Taskin, M.E. *et al.* (2008) CFD study of the influence of catalyst particle design on steam reforming reaction heat effects in narrow packed tubes. *Ind. Eng. Chem. Res.*, **47** (16), 5966–5975.
26. Churchill, S.W. (1999) The conceptual analysis of turbulent flow and convection. *Chem. Eng. Proces.*, **38** (4-6), 427–439.
27. Mathpati, C.S. and Joshi, J.B. (2007) Insight into theories of heat and mass transfer at the solid/fluid interface using direct numerical simulation and large eddy simulation. Joint 6th International Symposium on Catalysis in Multiphase Reactors/5th International Symposium on Multifunctional Reactors (CAMURE-6/ISMR-5-), 2007, Pune.
28. Roy, S. and Dudukovic, M.P. (2001) Flow mapping and modeling of liquid-solid risers, 8th Conference on Novel Reactor Engineering for the New Millennium, 2001, Barga.
29. Sudiyo, R. and Andersson, B. (2007) Bubble trapping and coalescence at the baffles in stirred tank reactors. *AIChE J.*, **53** (9), 2232–2239.
30. Bouaifi, M. *et al.* (2003) Experimental and numerical investigations of a jet mixing in a multifunctional channel reactor - Passive and reactive systems. 3rd International Symposium on Multifunctional Reactors (ISMR3)/18th Colloquia on Chemical reaction Engineering (CCRE18), 2003, Bath.
31. Ng, K. and Yianneskis, M. (1999) Observations on the distribution of energy dissipation in stirred vessels. Fluid Mixing 6 Symposium, 1999, Bradford.
32. Irandoust, S. and Andersson, B. (1988) Monolithic Catalysts for nonautomobile applications. *Catal. Rev. Sci. Eng.*, **30** (3), 341–392.
33. Kreutzer, M.T. *et al.* (2005) Multiphase monolith reactors: chemical reaction engineering of segmented flow in microchannels. 7th International Conference on Gas-Liquid and Gas-Liquid-Solid, 2005, Strasbourg.

Index

Novel Concepts in Catalysis and Chemical Reactors: Improving the Efficiency for the Future.
Edited by Andrzej Cybulski, Jacob A. Moulijn, and Andrzej Stankiewicz

ISBN: 978-3-527-32469-9

m

r

y